ACTION REFINEMENT IN PROCESS ALGEBRAS

Distinguished Dissertations in Computer Science

Edited by
C.J. van Rijsbergen University of Glasgow

The Conference of Professors of Computer Science (CPCS), in conjunction with the British Computer Society (BCS), selects annually for publication up to four of the best British PhD dissertations in computer science. The scheme began in 1990. Its aim is to make more visible the significant contribution made by Britain – in particular by students – to computer science, and to provide a model for future students. Dissertations are selected on behalf of CPCS by a panel whose members are:

M. Clint, Queen's University, Belfast
R.J.M. Hughes, University of Glasgow
R. Milner, University of Edinburgh (Chairman)
K. Moody, University of Cambridge
M.S. Paterson, University of Warwick
S. Shrivastava, University of Newcastle upon Tyne
A. Sloman, University of Birmingham
F. Sumner, University of Manchester

ACTION REFINEMENT IN PROCESS ALGEBRAS

LUCA ACETO
*Hewlett-Packard Research Laboratories,
Pisa Science Center*

CAMBRIDGE
UNIVERSITY PRESS

Published by the Press Syndicate of the University of Cambridge
The Pitt Building, Trumpington Street, Cambridge CB2 1RP
40 West 20th Street, New York, NY 10011, USA
10 Stamford Road, Oakleigh, Victoria 3166, Australia

First published 1992

Printed in Great Britain at the University Press, Cambridge

Library of Congress cataloguing in publication data available

British Library cataloguing in publication data available

ISBN 0 521 43111 5

Contents

Preface

This document constitutes the author's D.Phil. thesis, which was submitted to the University of Sussex in October, 1990. It is difficult to express how much I owe to my supervisor Matthew Hennessy. First of all, I should like to express my deep gratitude to him for giving me the opportunity to come and work with him at Sussex in October 1987. Since then, he has encouraged me in my research and given me moral and technical help in difficult moments.

My deepest gratitude goes to all the members of the "Theory Group" at Sussex, past and present. I have learned much from: S. Arun-Kumar, Simon Bainbridge, Marek Bednarczyk, Rance Cleaveland, Anna Ingólfsdóttir, Astrid Kiehn, Huimin Lin, Philip Merrick, Tim Regan, Edmund Robinson, Arne Skou and Allen Stoughton.

I thank Rocco De Nicola for introducing me to the world of semantics of concurrency, for showing continuous interest in my progress and for his encouragement in difficult moments.

Many thanks to Mark Millington and Robin Milner for their willingness to examine this thesis and their comments on a previous version of this document.

Conversations and correspondence with Gérard Boudol, Ilaria Castellani, Philippe Darondeau, Pierpaolo Degano, Uffe Engberg, Ursula Goltz, Roberto Gorrieri, Jeremy Gunawardena, Kim Larsen, Ugo Montanari and Frits Vaandrager were very helpful.

My work at Sussex has been sponsored by a grant from the United Kingdom Science and Engineering Research Council and the Esprit Basic Research Actions CONCUR and CEDISYS.

Parts of this book have previously appeared in [Ace89,90], [AH88,89,90].

Chapter 1

Introduction

The term *reactive systems*, [Pn85], has been coined to refer to systems which periodically interact with their environment. These systems may be purely software, such as processes written in the programming language OCCAM [OC84], purely hardware, such as VLSI systems, or a mixture of both. Indeed, most computational devices may be viewed in one way or the other as reactive systems and, consequently, a theory of these systems is also a general theory of concurrent, communicating systems. Many languages have been devised for describing and reasoning about reactive systems. In this thesis, we shall restrict our attention to the so-called *Process Algebras* such as **CCS** [Mil80,89], **CSP** [Hoare85], **ACP** [BK84,85] and **MEIJE** [AB84]. Process Algebras are prototype specification languages for concurrent, communicating systems; they consist of a set of combinators for constructing new systems from existing ones, a facility for the recursive definition of systems and are usually parameterized with respect to a set of uninterpreted action symbols, which stand for the basic tasks processes perform during their evolution. These languages evolved from Milner's original insight that concurrent processes have an algebraic structure and the combinational structure of process descriptions, which reflects the architectural structure of a machine performing a process, plays a fundamental rôle in their theories.

Following Milner [Mil80,89], the meaning of systems written in such process description languages is usually given in an operational style by means of Plotkin's *Structured Operational Semantics* [Pl81]. In this approach, processes are interpreted as *Labelled Transition Systems* (LTS's), [Kel76], and an LTS is associated to each process expression by means of structural rules, with the behaviour of a compound process being inferred from that of its components. Labelled Transition Systems are essentially edge-labelled directed graphs whose nodes represent the states of a system and whose edges stand for transitions between states. Transitions are labelled by actions which stand for interactions with the environment or internal computational tasks. Labelled Transition Systems are a simple, intuitive and flexible tool for defining the semantics of concurrent programs and, as shown by Plotkin [Pl81], they can be used to define an operational semantics for a wide range of programming languages by means of structural rules. The semantics for process expressions obtained by assigning an LTS to them is, however, too concrete. In order to abstract from unwanted details on the way processes compute, Milner proposed considering concurrent processes as "black boxes" which are experimented upon by their environment. Processes which cannot be told apart by means of experimentation are then deemed to be *observationally equivalent*. Notions of equivalence between process descriptions are an important component of the theories of Process Algebras. In fact, these languages are designed not only to describe actual systems, but also their specifications, and a notion of equivalence between process expressions can be used to give a formal proof of the correctness of an implementation with respect to a specification. In general, the process expression which denotes an implementation will be much more detailed than the one denoting the abstract specification; however, their equivalence with respect to the chosen notion of experimentation means that the two systems denote essentially the same behaviour, but at different levels of abstraction. Several different notions of equivalence for Process Algebras have been proposed in the literature; some of the most widely used amongst them are *trace equivalence*, [Hoare85], *observational equivalence*, [Mil80], *bisimulation equivalence*, [Pa81], [Mil83], *testing equivalence*, [DH84], [H88a], and *failure equivalence*, [BHR84]. Although these behavioural equivalences capture different ideas of "observable behaviour" of processes, the notions of experimentation underlying them have much in common. (See [Gl90a] for an extensive discussion of many of the behavioural equivalences proposed in the literature and of the notions of experimentation underlying them) In particular, all of these equivalences are based on the assumption that observations of the behaviour of processes are *sequential* and made up of *atomic* experiments. Moreover, as mentioned above, processes are considered as black boxes, whose internal structure cannot be observed. This means, for example, that, in this approach, the geographical distribution of systems is not observable. A major consequence of these assumptions on the way systems are experimented upon is that parallelism is semantically reduced to sequential nondeterminism, at least for finite processes. This is due to the fact that, in this discipline of observation, the concurrent occurrence of two observable actions cannot be distinguished from their occurrence in any order and, therefore, for each finite parallel process, there exists a purely sequential nondeterministic one which exhibits the same observable behaviour. Equivalences which perform the semantic reduction of parallelism to nondeterminism are usually called *interleaving* in the literature.

The simplifying assumptions underlying the interleaving semantic equivalences have led to elegant algebraic theories of nondeterminism in which verifications of large systems are possible. In order to fully exploit the algebraic nature of process descriptions, written in any of the standard Process Algebras, it is useful to work with notions of semantic equivalence which are *congruences* with respect to the operators in the language. In fact, congruence relations naturally induce *compositional proof techniques* for verifying equivalences between processes, thus exploiting the composition/decomposition techniques supported by Process Algebras, by guaranteeing the interchangeability of equivalent processes in all language contexts. This means that two equivalent processes can be substituted for one another in any context without affecting the overall behaviour of the composite system. Pragmatically, when working with notions of behavioural equivalence which are substitutive with respect to all the combinators in the algebra, proofs of equivalence between large systems can be usually reduced to proving the equivalence between their components: a generally simpler task.

An interesting by-product of the algebraic nature of the above mentioned approach is the fact that there exist complete axiomatic characterizations, usually in terms of equations or conditional equations, of the most common behavioural equivalences over different Process Algebras, [HM85], [BK85], [DeN85], [H88a,c]. The existence of such axiomatizations testifies the mathematical tractability of the interleaving equivalences and allows for algebraic verifications of the correctness of implementations with respect to specifications. (See [Va90a] for arguments in favour of an axiomatic approach to concurrency theory and examples of its applications)

As pointed out above, interleaving semantic equivalences offer an elegant semantic theory and usable proof techniques if one is interested in the description of the interactive behaviour of processes. However, the semantic reduction of parallelism to sequential nondeterminism makes them unsuitable for describing the nonsequential behaviour of distributed, concurrent systems. As argued by many authors in the literature, semantic theories for languages and models for the description of concurrent systems which treat parallelism as a primitive notion are best suited for this purpose. In recent years, many notions of equivalence for concurrent processes which distinguish concurrency from nondeterminism in the behaviour of processes have been proposed in the literature, e.g. *history preserving bisimulation* [GG88], *NMS bisimulation* [DDM86], *pomset bisimulation* [BC88], *distributed bisimulation* [CH89], [Ca88], *timed bisimulation* [H88c], *causal trees bisimulation* [DD89a] and *ST-bisimulation* [GV87], [Gl90]. These proposals can be roughly divided into three main classes, depending on the kind of experiments which are used to distinguish parallelism from nondeterminism in the behaviour of processes. The "causal equivalences", like history preserving bisimulation and pomset bisimulation, semantically distinguish concurrency from nondeterminism by admitting observations of the *causal structure* of the observed system. All the equivalences in this class distinguish the CCS processes $p = a|b$ and $q = a.b + b.a$, which are equated by all the above-mentioned interleaving equivalences, because q is capable of reacting positively to an experiment of the form "*do a causes b*", whilst p cannot as in its behaviour a and b are causally independent.

Equivalences like the distributed bisimulations of [CH89], [Ca88] are instead based on the assumption of observers that can detect the distributed nature of the observed systems. In particular, observations of systems will produce a *local* and a *global* residual, which together give the effect of the experimentation. For instance, a possible outcome of an experiment of the form "*do a*" on the process q given above will be to produce the process b as a local residual. As no such outcome can be obtained by performing the same experiment on p, the processes p and q are distinguished by these distributed equivalences.

Finally, equivalences like timed equivalence and ST-bisimulation equivalence distinguish concurrent from nondeterministic behaviour by taking a more refined view of the level of granularity of actions than the interleaving equivalences. These equivalences are based on the assumption that the actions processes perform during their evolution are not necessarily atomic in time and thus may have duration. In these equivalences, the processes p and q are distinguished because the actions a and b may overlap in the behaviour of p, but not in that of q.

The main aim of this thesis is to develop notions of semantic equivalence for Process Algebras which, like the ones mentioned above, do not equate parallelism to sequential nondeterminism and to study their use in giving suitable semantic theories for Process Algebras which incorporate an action refinement operator. We would like to develop semantic equivalences which are amenable to a mathematical analysis similar to that for the previously discussed interleaving equivalences and that support compositional proof techniques for establishing equivalences between processes. In particular, these semantic equivalences should be congruences with respect to the combinators in the languages we shall consider and should allow for behavioural and axiomatic verifications. We shall study several notions of equivalence, all variations on bisimulation equivalence, for Process Algebras based on **CCS** and **ACP** which differ in their view of the granularity of action occurrences. The main contributions of the thesis may be conveniently divided into three parts. In the first we consider a process algebra which contains explicit constants for termination, deadlock and divergence. We present both an equational and a behavioural semantics for it and show that they coincide. The behavioural semantics for this language is based on a variation of the notion of bisimulation preorder and takes the view that actions are atomic.

We then study behaviourally motivated semantic theories for two process algebras which incorporate operators which allow the refinement of actions by processes. In the presence of action refinement, semantic theories based on the atomicity of actions are, in general, no longer adequate. We show how suitable semantic theories for the languages we consider may be given in terms of variations of bisimulation equivalence which are based on the assumption that actions are not instantaneous. As an application of the theory, we show how action refinement can be used to give a behavioural characterization of a model for concurrent processes based on partial orders, the class of series-parallel pomsets.

Finally, we study and axiomatize a behavioural preorder for a simple subset of **CCS** based on actions which have a pomset structure. This preorder allows us to relate concurrency to nondeterminism without semantically reducing the former to the latter. In the next three sections we examine these three areas in more detail. We shall then give an outline of the thesis.

1.1 A Semantic Theory for Sequential Composition

The two main sequencing operators considered in the standard Process Algebras are *action prefixing* and *sequential composition*. Action prefixing, as favoured by **CCS** and **MEIJE**, is a restricted form of sequencing operator in which actions are used to prefix processes. For instance, following a **CCS**-like notation, $a.p$ denotes a process which is capable of performing action a and behaving like p thereafter. Sequential composition, familiar from sequential programming languages and favoured by **TCSP** and **ACP**, is a general form of sequencing operator in which processes are used to prefix other processes. For instance, $p;q$ denotes a process which behaves like q after the successful completion of the execution of p. Despite its intuitive simplicity, a wide array of semantic treatments of a general sequential composition in Process Algebras have been presented in the literature. (See [BV89] for a good survey of the proposals in the literature) In this thesis, we shall be concerned, amongst other things, with the development of Process Algebras which incorporate an operator for the refinement of actions by processes. This combinator may be seen as operating the substitution of actions by processes in process expressions and, consequently, cannot be supported by languages whose sequencing operator is action prefixing. For this reason, we shall pave the way to the semantic treatment of languages including action refinement combinators and a general sequential composition operator by proposing a simple behavioural semantics for a rich process algebra which includes such a sequencing combinator.

The correct semantic treatment of sequential composition in a language including a restriction operator requires the distinction between two forms of termination: successful termination and deadlock. Our semantic treatment of these two forms of termination will be based on the idea that both successfully terminated and deadlocked processes cannot perform any action. The only way of behaviourally distinguishing them is to place them in contexts built using the sequential composition operator.

1.2 Non-Atomic Actions and Action Refinement

The basic building block of process expressions given by means of Process Algebras is the set of uninterpreted action symbols these languages are usually parameterized upon. These actions correspond to the basic tasks processes perform during their evolution at a certain level of abstraction. However, concurrent systems may be viewed at different levels of abstraction, ranging from high level descriptions of their abstract properties to the level of firmware and circuitry. Several authors have recently argued that *action refinement* occurs naturally when developing specifications for processes or systems in a stepwise fashion, see, e.g., [GG88], [Gl90a]. At one level of abstaction, the specification might look like

$$SPEC \Longleftarrow \ldots;input;output;\ldots$$

where *input* and *output*, at this level of abstraction, may be considered as uninterpreted or unanalysed actions. At some stage during the refinement of the specification, it may be appropriate to describe in more detail "how" these actions are supposed to occur. These descriptions could be in terms of processes, say, P and Q, so that the new specification may look like

$$NSPEC \Longleftarrow \ldots;P;Q;\ldots$$

In this case, the more detailed specification *NSPEC* is obtained from the more abstract one *SPEC* by action refinement, refining the actions *input* and *output* to the processes P and Q, respectively. Refinement steps based on action refinement like the one outlined above are, however, not supported by any of the existing process algebras. In this thesis we shall aim at developing reasonable process algebras which incorporate an operator allowing us to refine actions by processes and to suggest suitable notions of semantic equivalence for specifications written in these process description languages. As argued above, suitable notions of semantic equivalence for process description languages should at least be congruences with respect to all the constructs in the language and in particular with respect to the refinement combinator. This requirement immediately rules out all of the standard interleaving equivalences. In fact, as it is well known, all of the interleaving equivalences satisfy the law

$$a|b \quad = \quad a.b + b.a. \tag{1.1}$$

On the other hand, if action a is refined by the process $c.d$ in both of the above processes, the resulting processes $(c.d)|b$ and $c.d.b + b.c.d$ are *not* equivalent with respect to any reasonable notion of semantic equivalence.

In this thesis, variations on the notion of bisimulation equivalence will be used as semantic equivalences upon which semantic theories for process description languages incorporating an action refinement combinator will be based. It will be shown how two bisimulation-like equivalences, based upon the assumption that actions are non-instantaneous, are suitable semantic equivalences for several process algebras with an action refinement combinator.

As pointed out above, suitable semantic equivalences for process algebras incorporating an action refinement combinator must treat parallelism as a primitive notion in their theories. However, as the work presented in [Gl90] and in this thesis shows, semantic equivalences based on the assumption that actions are non-atomic turn out to be sufficient for giving suitable semantics for languages which support action refinement. The full power of causal semantics, for instance of history preserving bisimulation [GG88], is not necessary for this purpose. As pointed out in [Gl90], this means that, in general, a notion of testing, [H88b], based on the refinement combinator does not suffice to reveal the full distinguishing power of such causal equivalences. We shall show, however, that the refinement combinator can indeed be used to give a behavioural characterization of a simple model based on partial orders: the series-parallel pomsets of Pratt and Gischer [Gi84].

1.3 Concurrency and Nondeterminism

As pointed out in the previous sections of this introduction, the wealth of notions of equivalence for concurrent systems proposed in the literature and those studied in this thesis may be roughly divided into two main classes: the one containing those equivalences which semantically reduce parallelism to sequential nondeterminism, the so-called *interleaving equivalences*, and that containing those equivalences which distinguish parallelism from nondeterminism in the behaviour of processes. The interleaving equivalences interpret the independence of concurrent events as their possibility to occur in any relative temporal order and in their respective semantic theories two processes are equated only if they exhibit the same degree of nondeterminism. On the other hand, the non-interleaving equivalences draw a sharp distinction between causal independence between events and their possibility to occur in any relative temporal order. Although their semantic theories differ in the way they capture the interplay between causality and the branching structure of processes, these equivalences identify two processes only if they exhibit the same degree of parallelism and do not relate nondeterministic specifications with concurrent implementations in any way. As it is frequently more natural to specify the behaviour of a system in terms of a sequential nondeterministic process and more efficient to implement it in a parallel fashion, it would be helpful to have a semantic theory of processes which allows us to relate these two notions without semantically reducing parallelism to sequential nondeterminism. For instance, one possible use of this feature of the theory could be in requiring that all the parallelism which is present in the specification be maintained in the implementation. This would not be possible by using interleaving semantic theories because of the reduction of parallelism to sequential nondeterminism they perform.

A semantic theory for a simple subset of **CCS** which allows us to relate concurrency to nondeterminism in the behaviour of processes without semantically reducing the former to the latter will be presented in this thesis. It will be shown how, following [BC88], such a theory can be obtained by means of a variation of strong bisimulation equivalence based on actions which have a pomset structure.

1.4 Outline of the Thesis

We now give a brief summary of the work presented in this thesis. In addition, each chapter will contain a section of introductory remarks which is meant to give a more detailed account of the work presented in it.

As pointed out above, we shall investigate several variations on the notion of bisimulation equivalence for process algebras based on **CCS** and **ACP** which differ in their view of the granularity of the actions processes perform during their evolution. We start by considering, in Chapter 2, a process algebra which contains explicit representations for successful termination, deadlock and divergence. The motivation for the work presented in this chapter is to establish a framework for giving an operational semantics for languages whose sequencing operator is general sequential composition rather than action-prefixing. Such an operator will be required for the syntactic treatment of action refinement presented in Chapters 3-5 and the semantic theory for it presented in Chapter 2 will serve as a touchstone for the ones to follow. We shall present both a behavioural and a denotational semantics for the language we consider in Chapter 2. The behavioural semantics is based on a variation on the notion of bisimulation preorder suitable for our language and takes the view that processes evolve by performing actions which are atomic. The denotational semantics is given in terms of the initial continuous algebra CI_E which satisfies a set of equations E. We shall prove that CI_E is fully abstract with respect to the behavioural semantics. As a corollary of this result we obtain a complete proof system for the behavioural preorder over the process algebra considered in this chapter and a c.p.o. model for a bisimulation preorder which abstracts from internal actions.

In Chapter 3, we shall begin the study of behaviourally motivated semantic theories for process algebras which include an operator for the refinement of actions by processes. We shall study a simple process algebra including combinators for sequential and nondeterministic composition and parallel composition without synchronization. In the presence of an operator for action refinement, semantic theories based on the atomicity of actions are, in general, no longer adequate. We shall show how a suitable semantic equivalence for the simple language with action refinement considered in this chapter may be given by means of a variation of bisimulation equivalence based on the assumption that actions have a beginning and an ending. The resulting semantic equivalence, denoted by $\sim_t$, is essentially strong bisimulation equivalence, but defined using subactions, $S(a)$ for the beginning of action a and $F(a)$ for its ending. This equivalence will be characterized as the largest congruence over the simple language which is contained in strong bisimulation equivalence. We shall present both an algebraic and a behavioural proof of this fact. The algebraic proof relies on a complete equational characterization of $\sim_t$ over the simple language. The behavioural proof makes an essential use of another notion of equivalence, called refine equivalence and denoted by $\sim_r$. Refine equivalence is a variation on $\sim_t$ in which we ensure that $p \sim_r q$ implies that whenever an occurrence of $S(a)$ in p is matched by an occurrence of $S(a)$ in q, their corresponding $F(a)$'s are also matched. For the simple language $\sim_t$ and $\sim_r$ coincide, but it is simpler to prove that $\sim_r$ is preserved by the action refinement combinator. The independent interest of $\sim_r$ will become clear in Chapter 4.

A semantic theory for a process algebra with action refinement, which includes hand-shake communication, an internal action and a restriction operator à la **CCS** is presented in Chapter 4. In this richer setting, care must be taken in restricting the allowed action refinements to those that afford the development of a reasonable semantic theory for the resulting language. For instance, action refinements will be required to reflect, in a formal sense, the synchronization algebra underlying the **CCS**-like parallel composition in the language. In addition, the application of action refinements to processes must take into account the rôle played by restriction in **CCS**-like languages. In these languages, the restriction combinator determines the scope of action symbols in process terms. We shall thus view it as a binding operator and define the application of action refinements to terms in our language by resorting to a standard theory of substitution in the presence of binders and to the theory of α-conversion. As it will be shown by means of an example, $\approx_t$, the version of $\sim_t$ which abstracts from the internal evolution of processes, is *not* a suitable notion of equivalence over the richer language. In fact, it is not preserved by the refinement operator. However, we shall show that the closure of $\approx_r$, the weak version of $\sim_r$, with respect to all $+$-contexts is preserved by action refinements over the richer process algebra considered in this chapter.

Chapter 5 will be devoted to showing how action refinement may be used to give a behavioural characterization of a simple model of concurrent computation based on partial orders, the series-parallel pomsets of Pratt and Gischer. More precisely, we shall show that series-parallel pomsets are fully abstract with respect to strong bisimulation equivalence over a simple process algebra with refinement. The proof of this result is algebraic and relies on showing that the behavioural equivalence and the one induced by the denotational semantics in terms of series-parallel pomsets have a common axiomatization over the language we consider. Moreover, using a novel technique developed by J.F. Groote, [Gro90], we prove that the axiomatization is also ω-complete. We end this chapter by giving a trace-theoretic characterization of the class of series-parallel pomsets. We shall prove that two series-parallel pomsets are equal iff they have the same set of ST-traces. As a corollary of this result we obtain a complete axiomatization of ST-trace equivalence over the simple language naturally associated with series-parallel pomsets.

In Chapter 6 we study and axiomatize a behavioural preorder over a simple subset of **CCS** based on actions which have a pomset structure. The pomset structure on the actions performed by processes will be used to give a measure of the degree of parallelism processes exhibit during their evolution. The measure of the degree of parallelism in computations will take the form of a preorder over pomsets, which will be given both an equational and a model-theoretic characterization. The preorder on computations will then be used to induce a natural bisimulation-like preorder on processes. This preorder will allow us to relate concurrency to nondeterminism in the behaviour of processes without reducing the former to the latter.

Finally, in Chapter 7, we shall give a brief overview of the results presented in the thesis and comment on some open problems and suggestions for further research.

In order to improve the readability of this document, we have tried to make each chapter readable independently and several technical definitions and results are explicitly recalled in different chapters. However, the readers interested in reading the thesis as a whole will find pointers between inter-related parts of the document.

1.5 Comparison with Related Work

In the remainder of this introduction, we briefly compare the work presented in this thesis with related work in the literature. The interested reader is referred to the sections of concluding remarks at the end of each chapter for more detailed comparisons.

Variations on the theme of bisimulation equivalence for languages containing an explicit representation of deadlock and successful termination have been considered by several authors in the literature, see, e.g. [BG87a,b], [Vra86]. In most of the proposals, processes are assumed to be capable of signalling their successful termination to the environment by performing a distinguished action symbol, $\surd$. Successfully terminated processes are then behaviourally distinguished from deadlocked ones because the former are capable of performing a "termination action" whilst the latter cannot. In this thesis, we take the view that both successfully terminated and deadlocked processes are "terminated" and thus are not capable of performing any action whatsoever. The only way of behaviourally distinguishing a successfully terminated process, say p, from a deadlocked one, say q, is to place the two processes in contexts built using the sequential composition operator. For example, letting a denote an action, we would expect that, as p is successfully terminated, the process $p; a$ is capable of performing action a. On the other hand, $q; a$ cannot proceed as q is deadlocked. Technically, this intuition about the behaviour of processes will be captured in the operational semantics by means of a "successful termination" predicate over process expressions. This way of describing the dichotomy successful termination/deadlock has proved to be rather flexible and permits the description of several notions of successful termination present in the literature.

* * *

The appearance of [CDP87] has given new impulse to the study of notions of equivalence for concurrent systems which support the refinement of actions by processes. A minimal requirement for these semantic equivalences is that they be *preserved by action refinement*, i.e. that equivalent processes should remain equivalent after replacing each occurrence of an action a by a process $\rho(a)$. As shown above by means of an example, semantic equivalences based on interleavings of atomic actions are, in general, not preserved by action refinement. This observation has led several researchers to investigate the behaviour of some of the semantic equivalences which distinguish concurrency from nondeterminism with respect to action refinement.

Refinement theorems for several of the equivalences mentioned in this introduction abound in the literature, see e.g. [NEL88], [GG88], [DD89b,90], [Gl90,90a], [GW89a], [Vo90]. Apart from a few exceptions (e.g. [NEL88]), most of this work has been carried out in the framework of *semantic models* for concurrent computation based on partial orders, like Event Structures (ES) [Win80,87], Petri Nets, [Rei85], and Causal Trees [DD89a]. Semantic equivalences have been defined over these mathematical structures and shown to be suitable for supporting *semantic action refinement* by proving that they are preserved by it. This line of research is justified by the fact that semantic equivalences based on some notion of causal observation are most naturally defined over semantic models based on partial orders and has led to a wealth of model-theoretic results, which we now briefly review without any pretence of completeness. The interested reader is invited to consult the above given references for more information and further pointers to the literature on action refinement.

History preserving bisimulation has been shown to be preserved under the refinement of actions by finite, conflict-free ES over the class of prime Event Structures, [Win87], in [GG88]. This refinement theorem by R. van Glabbeek and U. Goltz has been extended to unrestricted action-refinements over what the authors term *free Event Structures* in [DD90] by exploiting the close relationship between history preserving bisimulation and causal trees bisimulation. Prime ES are also the system model for which refinement theorems for ST-bisimulation and ST-trace equivalence are proven in [Gl90]. Again, due to the particular structure of the chosen system model, the theorem is shown for action refinements which allow the replacement of actions by finite, conflict-free ES. All of these results have been obtained in a setting without internal, unobservable computational steps. Refinement theorems for *branching bisimulation equivalence* [GW89a], a variation on the notion of weak bisimulation equivalence [Mil89], over process graphs, [BK85], and causal trees, [DD89a], have been proven in [GW89b] and [DD89b], respectively. A refinement theorem for a version of failures semantics based on interval semi-words has been proven in [Vo90] using safe Petri Nets as system model.

However, because of their essentially model-theoretic nature, the results briefly discussed above do not apply directly to existing process description languages like Process Algebras. In particular, theories of *semantic action refinement* do not immediately support compositional proof techniques at the level of syntactic descriptions of processes. Indeed, it is possible to lift equivalences and results obtained at the semantic level in a mathematical domain $\mathcal{D}$ to a process description language $\mathcal{L}$. However, semantic models for concurrent computations based on partial orders, like most of the above mentioned ones, give, in general, a very intensional description of the behaviour of processes and need to be factored out by means of some notion of equivalence. The lack of abstractness of these models makes the lifting of semantic equivalences to the syntactic level rather lengthy and complex. For example, any equivalence $\cong$ over a mathematical domain $\mathcal{D}$ can be lifted to a process description language $\mathcal{L}$ by defining a *meaning map* $[\![\cdot]\!] : \mathcal{L} \rightarrow \mathcal{D}$, which assigns a point in the mathematical domain to each process description in $\mathcal{L}$, and by defining $\cong$ over $\mathcal{L}$ as follows:

$$\text{for all programs } p \text{ and } q \text{ in } \mathcal{L},\ p \cong q \text{ iff } [\![p]\!] \cong [\![q]\!].$$

Proofs of the equivalence of two $\mathcal{L}$-programs p and q with respect to $\cong$ can now be given in two steps. First of all, it is necessary to determine their semantic counterparts $[\![p]\!]$ and $[\![q]\!]$, respectively. Secondly, the equivalence of $[\![p]\!]$ and $[\![q]\!]$ with respect to $\cong$ has to be proven by semantic reasoning. This proof strategy is rather involved and may be further complicated by the complexity of the objects in the semantic domain, be they Petri Nets, Event Structures and the like, and of the definition of the meaning map $[\![\cdot]\!]$ for process description languages, see e.g. [Win82]. The situation can be drastically improved if syntax based characterizations of $\cong$ are available over the language $\mathcal{L}$, e.g. in the form of equational axiomatizations or, more generally, proof systems. However, the derivation of such syntax based counterparts to $\cong$ from its semantic definition is, in general, still not a simple task and it may be argued that it would be better to develop equivalences that directly apply to process description languages in the first place. Moreover, the theories of semantic action refinement discussed above do not address syntactic issues like the treatment of action refinement in the presence of restriction or hiding operators and the interplay between action refinements and the synchronization structure of processes. As such issues are of fundamental importance in the description of processes by means of Process Algebras, we believe that it is advantageous to have theories of action refinement at the syntactic level.

A *syntactic* treatment of action refinement for a simple process algebra is presented in [NEL88], [Eng90]. There the authors provide fully abstract models for the languages they consider with respect to a trace based notion of equivalence. Similarly, one of the main aims of this thesis is the development of a theory of *syntactic action refinement* within the framework of process algebras. We shall study process algebras which incorporate a refinement combinator and define semantic equivalences, based on the notion of bisimulation, directly on these languages. This will allow us to import within our framework a wealth of compositional proof techniques from the theory of standard interleaving equivalences. The semantic equivalences for process algebras with refinement combinators which will be considered in this thesis will distinguish concurrency from nondeterminism by taking the view that actions are non-instantaneous and will not appeal to experimentations on the causal structure of the observed systems.

A related issue which has recently been the object of extensive investigation in the literature is that of endowing process description languages with operators which allow one to *define atomic actions*, see, e.g., [BC88], [Bou89] and [GMM88]. Such operators may be seen as playing a dual rôle to that of the action refinement combinator discussed above. In fact, they allow one to abstract the behaviour of a process to a single complex, but atomic, action. The usefulness of such a construct for the specification of concurrent systems has been demonstrated in [BC88], [Bou89], where, amongst other things, the authors show how to describe hand-shake communication à la **CCS** in terms of asynchronous communication. Although it may be argued that a full-grown process description language should support both refinement and abstraction operators, in this thesis we shall restrict ourselves to an investigation of the action refinement combinator.

* * *

Semantic theories for process description languages which relate concurrency to nondeterminism in the behaviour of processes, without semantically reducing the former to the latter, like the one considered in Chapter 6 of this thesis, have not received much attention in the literature. The only relevant reference known to the author is [Ab90], where a preorder, similar in spirit to the one presented in Chapter 6, has been independently defined over a rich finite Process Algebra. Both the work presented in Chapter 6 and the one in [Ab90] rely on a pomset operational semantics, [BC88], and exploit a preorder on pomsets to induce a notion of "concurrent refinement" on processes. The main difference between the two approaches stems from the way the operational semantics is defined. In [Ab90], whenever a process can perform a computation u, it can also perform any computation which is "at least as sequential as u". This property of the operational semantics does not hold for the pomset transition systems presented in [BC88] and in Chapter 6. An interesting property of Abramsky's preorder is that its kernel turns out to be the pomset bisimulation naturally associated with his pomset transition system semantics for the language considered in [Ab90]. As it will be shown by means of an example, this does not hold for the preorder considered in Chapter 6. Most of the work in [Ab90] is, however, devoted to a study of a notion of pomset bisimulation and the notion of "concurrent refinement" presented in that paper is not thoroughly investigated.

Chapter 2

A Semantic Theory Based on Atomic Actions

2.1 Introductory Remarks

In this chapter we wish to develop a theory for a process algebra which incorporates some explicit representation of termination, deadlock and divergence. The semantic theory will be based upon the interleaving assumption that processes evolve by performing actions which are atomic. We develop both an operational theory based on bisimulations, [Pa81], and an equational theory similar to those for **CCS**, **ACP**, [HM85], [H88c], [BK85].

The theory of **ACP**, [BK84], [BK85], deals with deadlock explicitly by introducing into the signature of the calculus a distinguished constant symbol δ. Deadlock can also occur directly in processes. If p can only perform actions from the set H then the process $\partial_H(p)$ is considered to be the same as the deadlocked process δ. But **ACP**, at least in its original formulation, does not have an explicit representation of successful termination.

On the other hand **CCS**, [Mil80], has a single "terminated" process, *nil*, which stands for both successful termination and deadlock. This choice is justified by the fact that in **CCS** these two kinds of termination are experimentally indistinguishable, due to the restricted form of sequential composition, *action-prefixing*, present in the calculus. As **ACP** allows sequential composition this is no longer the case. Consider, for example, the process $nil; p$, where *nil* is now used to denote a successfully terminated process. Then, since *nil* is successfully terminated, $nil; p$ can perform any action which p may perform. On the other hand, it is natural to assume that the process $\delta; p$ is deadlocked and will never perform any action. Thus, in the presence of sequential composition, there is an observable difference between the successfully terminated process *nil* and δ.

One may express desirable properties of processes by means of equations. For example

$$\delta; x = \delta$$

represents the fact that a deadlocked process can never proceed, and

$$nil; x = x$$

the fact that *nil* is a properly terminated process. Equational laws play a central rôle in the theory of **ACP**, where the approach is to isolate axioms expressing some *a priori* desirable properties that communicating systems should enjoy. A semantics for the resulting equational theory may then be obtained by constructing models for it, see e.g. [BK85], [BKO87], thus establishing its logical consistency. In this chapter, following previous work in the **CCS** literature [HM85], [H88c], we will derive the equations for our language from an operational view of processes. In this approach, the emphasis is on operational semantics as a framework within which different intuitions about the behaviour of processes may be discussed and compared. Sets of equations, for instance complete equational characterizations of some notion of behavioural equivalence over processes, may then be derived from and justified using the operational semantics. In the remainder of the chapter, we will examine the equational theory induced by one possible choice of operational semantics for our language; a more detailed comparison between our approach and the philosophy underlying **ACP** may be found in the concluding remarks (§ 2.5), where possible modifications to our operational semantics in order to obtain an equational theory closer to that of **ACP** are also discussed.

Many of the equations for our language are already well known either from **CCS** or **ACP**. However, the presence of the terminated process *nil* invalidates some of those from **ACP**. The equation

$$(x + y); z = x; z + y; z$$

is part of the theory of **ACP**, [BK85], but is not valid for our language, at least in its general form. In fact, if x is δ and y is nil then, assuming that $\delta + nil = \delta$, the left-hand side is equal to $\delta; z$, i.e. δ, whereas the righthand side is equal to $nil; z + \delta; z$, i.e. $z + \delta$. If z is a non-trivial process it is then reasonable to assume that δ and $z + \delta$ are different processes.

We will also have within our language processes which may diverge internally. We let Ω be a process which can only diverge internally. Using the usual notation for recursive terms this could also be represented by $rec\,x.\ \tau; x$, where τ is an internal unobservable move. The semantic identification of the totally undefined process Ω with the process that can only diverge internally $rec\,x.\ \tau;x$ is indeed open to debate. However, this choice may be supported both on behavioural and pragmatic grounds. In this chapter, we will follow Milner's experimental approach to the semantics of concurrent systems [Mil80]. This approach is based upon the idea that two processes that cannot be distinguished by means of experimentation based on observation should be deemed to be equivalent. With this in mind, it may be argued that the environment will never be able to elicit any information from a process which can only diverge internally by experimenting on it, i.e. such a process contains no observable information. Following Scott's approach to semantics [Sco82], a process that contains no information is considered less than any other process and thus identified with Ω. A similar choice is present in the theory of denotational semantics for imperative sequential programming languages, where Ω is usually given the same denotation as the program *while true do skip od*. Such a program can only embark on infinite internal idling and as such represents a natural counterpart of the process $rec\,x.\ \tau;x$. Pragmatically, the choice of semantically identifying Ω and $rec\,x.\ \tau;x$ allows us to rely on the standard body of techniques of *continuous algebraic semantics* [GTWW77], [Gue81], for instance to give a denotational semantics for our language and provide powerful proof techniques for it. Our choice has, however, some drawbacks in dealing with infinitary properties of processes such as *fairness*. However, a study of these properties is out of the scope of this thesis and, in general, can not be carried out within the framework of continuous semantics.

Obviously we would expect *nil* and Ω to be different processes and we will also demand that δ and Ω be different. The latter requirement is less defensible, but we are motivated by the *information-theoretic* view of computation as advocated by Scott. Here the process which can only diverge, Ω, contains no information and is therefore considered less than any other process. There is some information available about the process δ, namely that it is deadlocked; so Ω and δ should be considered different. In the presence of Ω, and in particular taking Scott's approach to semantics, it is natural to express our theory in terms of *inequations*. One inequation is

$$\Omega \leq x,$$

and more generally the equations given above could be viewed as shorthand for two inequations, $t = u$ representing $t \leq u$ and $u \leq t$.

The main purpose of this chapter is to show that an adequate semantic theory for a process algebra which contains divergence, termination and deadlock can be constructed using a suitable set of inequations, E. More specifically we propose as a denotational semantics the initial continuous algebra generated by E, CI_E, [GTWW77], [Gue81]. This is in contrast to [BK82] where metric spaces are used for this purpose in place of continuous partial orders. The advantage of the latter is that all of the usual operators found in process algebra may be interpreted, whereas using metric spaces we can only readily interpret operators which are contractive. For instance, unguarded recursive definitions give rise to operators which are not contractive; in addition to this drawback, silent actions and abstraction operators have never been dealt with satisfactorily in this framework. Moreover, we can apply the existing and well understood theory of algebraic cpo's, for example to show the existence of CI_E and to derive useful proof techniques such as Scott Induction, [LS87].

In order to show that CI_E is a reasonable model we develop a behavioural or observational view of processes and prove that this coincides with the interpretation given by CI_E. This is given in terms of a variation on bisimulation equivalence, [Pa81]. To take divergence into account we generalize bisimulation equivalence, $\approx$, to a preorder $\sqsubseteq$ which is often called pre-bisimulation preorder. Intuitively, $p \sqsubseteq q$ means that p and q are bisimilar except that at times p may diverge more frequently than q; in the absence of divergence $p \sqsubseteq q$ will imply $p \approx q$. This type of behavioural relation has been studied in [HP80], [Mil81], [Wal87], [Ab87a,b]. Here we modify it to take into consideration termination and deadlock and show that two processes are behaviourally related with respect to this new relation if and only if they are related in the equational model CI_E. In other words, CI_E is *fully abstract* with respect to this new behavioural preorder. There may be other fully abstract models, but CI_E is distinguished by being initial in the category of fully abstract models. In fact, it is initial in the category of models which are consistent with the behavioural preorder.

We now give a brief outline of the remainder of the chapter. In §2.2 we define the language whose semantic properties will be investigated in the chapter. The language is endowed with both an operational and a denotational semantics. The operational semantics is defined in §2.2.1 following standard lines by means of Plotkin's *Structural Operational Semantics* (**SOS**), [Mil80], [Pl81]. Section 2.2.1 also introduces several definitions and notational conventions which will be used throughout the chapter. The denotational semantics for the language is given in §2.2.2. The definition is based on the well-known techniques of *Initial Algebra Semantics*, [GTWW77], [Gue81]; as already mentioned, we propose as a denotational model for the language the initial continuous Σ-algebra which satisfies a set of equations E, CI_E. The following sections are entirely devoted to showing that CI_E is indeed a reasonable denotational model for our language. As argued by Milner, [Mil83], operational semantics should be the touchstone for assessing mathematical models for concurrent languages. The agreement between denotational models and operational ones is called *full abstraction* in [Mil77], [Pl77], [HP79]. In this chapter we follow Milner and Plotkin's paradigm and justify the choice of our denotational model by showing that CI_E is fully abstract with respect to a natural notion of an operational or behavioural preorder over our language. The behavioural preorder is introduced in §2.3, where several constraints which behavioural relations have to meet in order to be related to denotational ones are also discussed. In particular, it is argued in §2.3.1 that, in order to be related to $\leq_E$, a behavioural preorder should be *finitely approximable* ([H81],[Ab87b]) and *closed with respect to all contexts.*

All the remaining sections of the chapter are devoted to showing that our choice of a behavioural preorder, $\sqsubseteq^c_\omega$, possesses these two properties and coincides with $\leq_E$ over our language. Section 2.3.2 is devoted to an analysis of the preorder $\sqsubseteq_\omega$ and of its substitutive version $\sqsubseteq^c_\omega$. This analysis paves the way to the proof of our promised full abstraction result. The proof of full abstraction of CI_E with respect to $\sqsubseteq^c_\omega$ over our language is outlined in §2.3.3 and relies on two main results:

- finite approximability of $\sqsubseteq^c_\omega$, and
- partial completeness of $\leq_E$ with respect to $\sqsubseteq^c_\omega$.

The proof of the partial completeness result is given in full detail in §2.3.3. It relies on the usual machinery used in the proofs of equational completeness for bisimulation-like relations, [H81], [HM85], [Wal87], [H88c]. The proof of finite approximability of $\sqsubseteq^c_\omega$ occupies all of §2.4. It is given in two stages. The first, which is the topic of §2.4.1, consists of a modal characterization of the preorder $\sqsubseteq_\omega$ and is a simple adaptation of similar results present in the literature, [Mil81], [HM85], [St87], [Ab87b]. The second employs this modal characterization to prove that $\sqsubseteq_\omega$ is finitely approximable; this will allow us to conclude that $\sqsubseteq^c_\omega$ is finitely approximable as well.

We end with a section of concluding remarks in which we discuss the results of the chapter and relationships with related work.

2.2 The Language

Let *Act* be a countable set of atomic action symbols. It is assumed that *Act* has the structure $Act = \Lambda \cup \bar{\Lambda}$, where Λ $(\alpha, \beta \ldots \in \Lambda)$ denotes a countable set of *channel names* and $\bar{\Lambda} = \{\bar{\alpha} \mid \alpha \in \Lambda\}$ is the set of their complements. Complementation is extended to the whole of *Act* by assuming that $\bar{\bar{\alpha}} = \alpha$, for all $\alpha \in \Lambda$. The set *Act* will be called the set of *observable actions* and will be ranged over by $a, b, \ldots$. Let τ and δ be two distinguished symbols not occurring in *Act*. The symbol τ will stand for an internal, unobservable action; these actions will occur when processes communicate with each other. $Act_\tau =_{def} Act \cup \{\tau\}$ will be called the set of *actions* and will be ranged over by $\mu, \gamma \ldots$. The symbol δ will stand for a *deadlocked process*, a process that cannot perform any move but is not successfully terminated. Successful termination will be denoted by the constant symbol *nil*.

The set of constant symbols in the process algebra we will consider is completed by the symbol Ω; as discussed in the introductory section, Ω will stand for a process that can internally diverge. Alternatively, one may think of Ω as the totally undefined process, the process about which the environment has no information at all. Ω is not deadlocked and has not successfully terminated. The process combinators used to build new systems from existing ones will be the following:

- $+$ for *nondeterministic choice*,
- $;$ for *sequential composition*,
- $|$ for *parallel composition*,
- $\partial_H(\cdot)$ for the *encapsulation operator*. Intuitively, the process $\partial_H(p)$ behaves like p, but with the actions in H prohibited. A more detailed discussion of this operator may be found in, e.g., [BK85].

Formally:

Definition 2.2.1 *For each $n \in \omega$ let Σ_n, the set of operation symbols of arity n, be defined as follows:*

- $\Sigma_0 = \{nil, \delta, \Omega\} \cup Act_\tau$
- $\Sigma_1 = \{\partial_H(\cdot) \mid H \subseteq Act \wedge H = \overline{H}\}$
- $\Sigma_2 = \{+, ;, |\}$
- $\Sigma_n = \emptyset$, *for each* $n > 2$.

The signature Σ is defined as $\Sigma = \bigcup_{n \geq 0} \Sigma_n$.

Let `Var` *be a countable set of variables, ranged over by $x, y \ldots$. The syntax of* recursive terms *over Σ is then defined by*

$$t ::= f(t_1, \ldots, t_k)(f \in \Sigma_k) \mid x \mid rec\, x.\, t.$$

We assume the usual notions of free and bound variables in terms, with $rec\, x.\, \_$ as the binding constructor. The set of recursive terms over Σ will be denoted by $REC_\Sigma($`Var`*$)$ and will be ranged over by $t, u, \ldots$. The set of* closed recursive terms *over Σ will be denoted by REC_Σ and will be ranged over by $p, q, p' \ldots$. The set of* syntactically finite processes *(i. e. those not involving occurrences of $rec\, x.\, t$) will be denoted by $FREC_\Sigma$ and will be ranged over by $d, e, d' \ldots$.*

Notationally, all the binary operators will be used in infix form, with the assumption that ; binds stronger than |, which in turn binds stronger than +. The constructor $rec\, x.\ \_$ will have the lowest precedence among all the operators.

2.2.1 The Operational Semantics

The operational semantics for the language REC_Σ consists of three different components. The first is an interpretation of REC_Σ as a labelled transition system in Plotkin's SOS style, [Mil80], [Pl81]. This associates with each action symbol μ a binary infix relation. Intuitively, $p \xrightarrow{\mu} q$ means that p may perform the action μ and thereby be transformed into q. The second is a *successful termination predicate*, $\surd$, which will be written in a postfix manner. Intuitively, $p\surd$ if p has terminated successfully, which will mean, among other things, that p cannot perform any further actions. We would expect $nil\surd$ but not $\Omega\surd$, not $a;p\surd$ and not $\partial_{\{a\}}(a;p)\surd$. There is a choice of exactly how to define the termination predicate $\surd$. In what follows we will present one choice; another choice, which is more in keeping with the intuitions of [BG87b], will be discussed in the concluding remarks. The final component is a *convergence predicate*, $\downarrow$. Intuitively, $p\downarrow$ means that the set of actions which p can initially perform is fully specified. It will turn out that $nil\downarrow$ but not $\Omega\downarrow$.

Definition 2.2.2 *Let $\surd$ be the least subset of REC_Σ which satisfies:*

(i) $nil \in \surd$,

(ii) $p \in \surd$ *implies* $\partial_H(p) \in \surd$,

(iii) $p \in \surd$ *and* $q \in \surd$ *imply* $p + q, p; q, p|q \in \surd$,

(iv) $t[rec\, x.\ t/x] \in \surd$ *implies* $rec\, x.\ t \in \surd$.

In what follows we will write $p\surd$ iff $p \in \surd$. Note that the process $nil + \delta$ is not considered successfully terminated. Intuitively, the process is "stagnating" on a branch of its computation and the environment has no way of discarding this branch. In the semantic theory that we shall present in what follows, the process $nil + \delta$ will be equated to the deadlocked process δ.

Definition 2.2.3 *Let $\downarrow$ be the least subset of REC_Σ which satisfies*

(i) $nil\downarrow, \delta\downarrow, \mu\downarrow$

(ii) $p\downarrow$ *implies* $\partial_H(p)\downarrow$

(iii) $p \downarrow, q \downarrow$ *imply* $(p + q) \downarrow, (p|q) \downarrow$

(iv) $t[rec\, x.\ t/x] \downarrow$ *implies* $rec\, x.\ t \downarrow$

(v) $p\surd, q \downarrow$ *imply* $(p; q) \downarrow$

(vi) $\neg(p\surd), p \downarrow$ *imply* $(p; q) \downarrow$.

Intuitively, $p \downarrow$ iff p is a completely specified process, i.e. if we can expand the recursive definition of p a finite number of times to obtain at the top level all the possible moves of p. Clause vi) of the definition of the predicate $\downarrow$ deserves some comment. It expresses the intuition that, if p is not successfully terminated, $p; q$ is a completely specified process if p is; in this case, in fact, the set of initial moves of $p; q$ is determined by that of p. $\uparrow$, the *divergence* predicate, will denote the complement of $\downarrow$, i.e. $p \uparrow$ iff $\neg(p \downarrow)$.

Example 2.2.1 *The following processes are divergent:*

- $rec\, x.\ a + x$
- $rec\, x.\ a; x + \Omega$
- $rec\, x.\ a; x + rec\, x.\ a + a|x$.

The predicate $\downarrow$ is used to detect a form of "syntactic divergence". Roughly, $p \uparrow$ if p contains unguarded recursive definitions, [Mil80], or unguarded occurrences of the divergent process Ω. One can show that $p\surd$ implies $p \downarrow$ using induction on the proof of $p\surd$. Of course the converse is not true; for instance, $a; q \downarrow$ but $a; q \notin \surd$.

Definition 2.2.4 *For each* $\mu \in Act_\tau$, *let* $\xrightarrow{\mu}$ *be the least binary relation on* REC_Σ *which satisfies the following axiom and rules:*

(1) $\mu \xrightarrow{\mu} nil$

(2) $p \xrightarrow{\mu} p'$ *implies* $p + q \xrightarrow{\mu} p'$, $q + p \xrightarrow{\mu} p'$

(3) $p \xrightarrow{\mu} p'$ *implies* $p; q \xrightarrow{\mu} p'; q$

(4) $p\surd$, $q \xrightarrow{\mu} q'$ *imply* $p; q \xrightarrow{\mu} q'$

(5) $p \xrightarrow{\mu} p'$ *implies* $p|q \xrightarrow{\mu} p'|q$, $q|p \xrightarrow{\mu} q|p'$

(6) $p \xrightarrow{a} p'$, $q \xrightarrow{\bar{a}} q'$ *imply* $p|q \xrightarrow{\tau} p'|q'$

(7) $p \xrightarrow{\mu} p'$, $\mu \notin H$ *imply* $\partial_H(p) \xrightarrow{\mu} \partial_H(p')$

(8) $t[rec\,x.\,t/x] \xrightarrow{\mu} p'$ *implies* $rec\,x.\,t \xrightarrow{\mu} p'$.

For any p, let $Sort(p) = \{\mu \in Act_\tau \mid \exists \sigma \in Act_\tau^\star, q \in REC_\Sigma : p \xrightarrow{\sigma\mu} q\}$, where, for $\sigma \in Act_\tau^\star$, $\xrightarrow{\sigma}$ is defined in the natural way. One can check that, for each p, $Sort(p)$ is finite. That is, according to the terminology of [Ab87a,b], the transition system $\langle REC_\Sigma, Act_\tau, \longrightarrow\rangle$ is *sort finite*. Some of our results will depend on this fact.

The three concepts defined above take no account of the special nature of τ. Following Milner [Mil80], τ is meant to be an internal invisible action. We now define three weaker versions of $\xrightarrow{\mu}$, $\surd$ and $\downarrow$ which use this assumption.

Let $\xRightarrow{\mu}$ denote $(\xrightarrow{\tau})^\star \circ \xrightarrow{\mu} \circ (\xrightarrow{\tau})^\star$. So $p \xRightarrow{\mu} q$ means that p may evolve to q performing the action μ and possibly silent moves. We will also use the relation $\xRightarrow{\varepsilon}$, defined as $(\xrightarrow{\tau})^\star$. In what follows, we will write $p \xrightarrow{\tau^\omega}$ iff there exists a sequence $\langle p_i \mid i \geq 0\rangle$ such that $p_0 = p$ and $p_i \xrightarrow{\tau} p_{i+1}$, for each $i \geq 0$.

Let $Stable(p) = \{q \mid p \xRightarrow{\varepsilon} q \text{ and } q \not\xrightarrow{\tau}\}$. Then the weak counterpart to $\surd$ is defined by

$$p\surd\!\!\surd \text{ if, for each } q \in Stable(p),\ q\surd.$$

For example $nil\surd\!\!\surd$, $\tau + \delta\surd\!\!\surd$, but not $\delta\surd\!\!\surd$. This relation is characterized by:

$$p\surd\!\!\surd \iff \begin{cases} i) & p \not\xrightarrow{\tau} \text{ and } p\surd, \text{ or} \\ ii) & p \xrightarrow{\tau} \text{ and, for each } q,\ p \xrightarrow{\tau} q \text{ implies } q\surd\!\!\surd. \end{cases}$$

Note that $rec\,x.\ \tau; x + a\surd\!\!\surd$ which is somewhat anomalous. However, we will only apply the "weak tick" predicate $\surd\!\!\surd$ to processes which cannot perform an infinite sequence of τ-actions and, for such processes, no such counterintuitive cases arise. Processes that can perform an infinite sequence of τ-actions are semantically divergent, which brings us to our final weak predicate. Let $\Downarrow$ be the least predicate over REC_Σ which satisfies

$$p \downarrow \text{ and (for each } q,\ p \xrightarrow{\tau} q \text{ implies } q \Downarrow) \text{ imply } p \Downarrow.$$

Intuitively, $p \Downarrow$ means that p cannot perform τ-actions indefinitely and a syntactically divergent process cannot be reached by performing these actions. Formally, one can prove

$$p \Downarrow \Leftrightarrow (p \not\xrightarrow{\tau^\omega} \text{ and, for each } q,\ p \xRightarrow{\varepsilon} q \text{ implies } q \downarrow).$$

Note also that $p\surd$ implies $p \Downarrow$. This follows because we already know that $p\surd$ implies $p \downarrow$ and one can also show that it implies $p \xrightarrow{\mu}$ for no μ, including τ.

In the semantic preorder to be defined in §2.3 we will use versions of $\Downarrow$ which are parameterized by actions:

- $p \Downarrow \tau$ if $p \Downarrow$
- $p \Downarrow a$ if $p \Downarrow$ and, for each q, $p \stackrel{a}{\Longrightarrow} q$ implies $q \Downarrow$.

This concludes our operational description of a semantics for the language REC_Σ. It defines a Labelled Transition System with divergence and termination predicates $\langle REC_\Sigma, Act_\tau \cup \{\varepsilon\}, \Longrightarrow, \surd\!\!/, \Downarrow \rangle$. In §2.3 this LTS will be used to define an operational preorder on processes.

2.2.2 Denotational Semantics

As pointed out in the introductory remarks, the main purpose of this chapter is to show that an adequate semantic theory for the process algebra described in the previous section can be constructed using a suitable set of inequations, E. Following [CN76], [GTWW77], [Gue81], [H88a], we propose as a denotational semantics for REC_Σ the initial continuous algebra generated by a set of equations E, CI_E.

In order to show that CI_E is a reasonable model for REC_Σ, in subsequent sections we will develop a behavioural theory of processes and prove that this corresponds to the interpretation given by CI_E. In other words, CI_E is fully-abstract with respect to the behavioural preorder we will introduce in the next section. We assume the reader is familiar with the basic notions of continuous algebras (see, e.g., the above quoted references); however, in what follows we give a quick overview of the way a denotational semantics can be given to $REC_\Sigma(\mathtt{Var})$ following the standard lines of algebraic semantics, [Gue81]. The interested reader is invited to consult [H88a] for an explanation of the theory.

Let Σ be the signature introduced in definition 2.2.1 and A be any Σ-cpo. A denotational semantics for the language $REC_\Sigma(\mathtt{Var})$ is given by the mapping

$$A[\![\cdot]\!] : REC_\Sigma(\mathtt{Var}) \to [ENV_A \to A],$$

where $ENV_A = [\mathtt{Var} \to A]$ is the set of A-environments, ranged over by the metavariables $\rho, \rho' \ldots$. As usual, $\rho[x \to a]$ will denote the environment which is defined as follows

$$\rho[x \to a](y) = \begin{cases} a & \text{if } x = y \\ \rho(y) & \text{otherwise.} \end{cases}$$

For completeness' sake we define $A[\![\cdot]\!]$ by structural induction on recursive terms as follows:

(i) $A[\![x]\!]\rho = \rho(x)$

(ii) $A[\![f(t_1, \ldots, t_k)]\!]\rho = f_A(A[\![t_1]\!]\rho, \ldots, A[\![t_k]\!]\rho)$ $(f \in \Sigma_k)$

(iii) $A[\![rec\,x.\ t]\!]\rho = Y\lambda a.\ A[\![t]\!]\rho[x \to a]$,

where Y denotes the least fixed-point operator.

Note that for each $p \in REC_\Sigma$, $A[\![p]\!]\rho$ does not depend on the environment ρ. The denotation of a closed term p will be denoted by $A[\![p]\!]$ and we write $p \leq_A q$ iff $A[\![p]\!] \leq_A A[\![q]\!]$ and $p =_A q$ iff $p \leq_A q$ and $q \leq_A p$.

As already pointed out, a natural choice of A would be the initial Σ-cpo CI_E in the class of Σ-cpo's which satisfy some set of equations, or inequations, E defined over the signature Σ. The equations that we will consider will express desirable properties of processes; many of them are already well known from **CCS** or **ACP**. Some of the equations which are part of the theory of **ACP** have had to be modified due to our different treatment of successful termination. For example, note that by considering equation (A4) in Figure 2.1 for $x = \delta$ we obtain that $\delta + nil = \delta$. This identity captures the main difference between our *nil* and the empty process ε recently investigated in the literature on **ACP**, [BG87a,b]; the intuition underlying it has been discussed after definition 2.2.2. A more detailed comparison between our equations and the ones used in the theory of **ACP** with the empty process may be found in the concluding remarks. Let $\mathcal{C}(E)$ denote the category of Σ-cpo's which satisfy the equations in Figure 2.1 and continuous Σ-homomorphisms. The following result is then standard, [CN76], [GTWW77], [Gue81], [H88a].

Proposition 2.2.1 *$\mathcal{C}(E)$ has an initial object CI_E.*

2.3 The Behavioural Semantics

2.3.1 The Behavioural Preorder

This section is devoted to an operational preorder which will be the behavioural counterpart of the denotational relation $\leq_{CI_E}$ (the ordering relation in the initial model CI_E) over REC_Σ. The existence of such a behavioural preorder, defined using a by now well-established mathematical tool, will reinforce CI_E as a reasonable model for the language REC_Σ.

The behavioural preorder will be defined using a variation of bisimulation equivalence, [Pa81], [Mil83], suitable for our language REC_Σ. Let *Rel* denote the set of binary relations over REC_Σ. We define a functional $\mathcal{F} : Rel \to Rel$ as follows:

given $\mathcal{R} \in Rel$, $p\mathcal{F}(\mathcal{R})q$ iff, for each $\mu \in Act_\tau$,

(i) if $p \stackrel{\mu}{\Longrightarrow} p'$ then, for some q', $q \stackrel{\mu}{\Longrightarrow} q'$ and $p'\mathcal{R}q'$

(**A1**) $x + y = y + x$

(**A2**) $x + (y + z) = (x + y) + z$

(**A3**) $x + x = x$

(**A4**) $x + nil = x$

(**A5**) $x + \delta = x$ if $x \notin \surd$

(**B1**) $x; nil = x = nil; x$

(**B2**) $\delta; x = \delta$

(**B3**) $x; (y; z) = (x; y); z$

(**B4**) $(x + y); z = x; z + y; z$ if $x, y \notin \surd$

(**C1**) $\delta | \delta = \delta$

(**C2**) Let $x \equiv \sum_{i \in I} \mu_i; x_i \{+\Omega\}$,

$$\delta | x = x | \delta = \begin{cases} \delta\{+\Omega\} & \text{if } I = \emptyset \\ \sum_{i \in I} \mu_i; (\delta | x_i)\{+\Omega\} & \text{otherwise} \end{cases}$$

(**Exp**) Let $x \equiv \sum_{i \in I} \mu_i; x_i \{+\Omega\}$
and $y \equiv \sum_{j \in J} \gamma_j; y_j \{+\Omega\}$,

$$x | y = \sum_{i \in I} \mu_i; (x_i | y) + \sum_{j \in J} \gamma_j; (x | y_j) + \sum_{(i,j): \mu_i = \overline{\gamma}_j} \tau; (x_i | y_j)\{+\Omega\}$$

(**E1**) $\partial_H(nil) = nil$

(**E2**) $\partial_H(\delta) = \delta$

(**E3**) $\partial_H(\mu) = \begin{cases} \delta & \text{if } \mu \in H \\ \mu & \text{otherwise} \end{cases}$

(**E4**) $\partial_H(x; y) = \partial_H(x); \partial_H(y)$

(**E5**) $\partial_H(x + y) = \partial_H(x) + \partial_H(y)$

(**Ω1**) $\Omega \leq x$

(**Ω2**) $\tau; (x + \Omega) \leq x + \Omega$

(**Ω3**) $\partial_H(\Omega) \leq \Omega$

(**Ω4**) $\Omega; x \leq \Omega$

(**T1**) $\mu; \tau = \mu$

(**T2**) $\tau; x + x = \tau; x$

(**T3**) $\mu; (x + \tau; y) = \mu; (x + \tau; y) + \mu$

Note: The summation notation in axiom (**Exp**) is justified by axioms (**A1**)-(**A2**). In axiom (**Exp**) an empty sum is understood as nil, $\{+\Omega\}$ indicates that Ω is an optional summand of a term and Ω is a summand of the right hand side iff it is either a summand of x or of y.

Figure 2.1. The set of inequations E

(ii) if $p \Downarrow \mu$ then

a) $q \Downarrow \mu$

b) if $q \stackrel{\mu}{\Longrightarrow} q'$ then, for some p', $p \stackrel{\hat{\mu}}{\Longrightarrow} p'$ and $p'\mathcal{R}q'$

iii) if $p \Downarrow$ then $p\checkmark\!\!\!/ \Leftrightarrow q\checkmark\!\!\!/$.

The notation $\hat{}$ is used to simplify the definition: $\hat{\tau}$ stands for ε and $\hat{a}$ stands for a.

The functional $\mathcal{F}$ is one of the methods for adapting the usual defining functional of bisimulation equivalence. A number of variations are discussed in [Wal87], [Ab87a,b]. There are also a number of ways of defining a behavioural preorder using $\mathcal{F}$. An established method is to take $\sqsubseteq$ to be the largest relation $\mathcal{R} \in Rel$ such that $\mathcal{R} \subseteq \mathcal{F}(\mathcal{R})$, [Mil83], [Mil88]. This relation is easily seen to be a preorder, i.e. a reflexive and transitive relation, and is, in fact, the maximum fixed-point of the equation $\mathcal{R} = \mathcal{F}(\mathcal{R})$. The preorder $\sqsubseteq$ also satisfies many of the properties that we have already discussed in the introduction. For example, for every $p \in REC_\Sigma$, $\Omega \sqsubseteq p$; also $\delta; p \simeq \delta$ and $nil; p \simeq p$, where $\simeq$ is the kernel of $\sqsubseteq$, i.e. $\simeq = \sqsubseteq \cap \sqsubseteq^{-1}$. The processes nil and δ are incomparable with respect to $\sqsubseteq$. In fact, it is easy to see that $nil \Downarrow$ and $\delta \Downarrow$, but $nil\checkmark\!\!\!/$ whereas $\delta \notin \checkmark\!\!\!/$. Note also that $\delta + \Omega \simeq \Omega$. This follows from the definition but it is also perfectly reasonable. Intuitively, we would expect that, for every action a, $\delta + \Omega \sqsubseteq \delta + a$. But $\delta + a \simeq a$, so that $\delta + \Omega$ should be less than a for each a; the only such process is Ω.

Clause (iii) in the definition of $\mathcal{F}(\mathcal{R})$ takes care of deadlock considerations and there are a number of equivalent ways of stating it. Suppose we say that

p *must terminate* if $p \Downarrow$ and $p \stackrel{\varepsilon}{\Longrightarrow} p' \stackrel{\tau}{\not\longrightarrow}$ implies $p'\surd$.

Then (iii) could be replaced by:

(iiia)′ p *must terminate* implies q *must terminate*

(iiib)′ $p \Downarrow$, q *must terminate* imply p *must terminate*.

Alternatively, suppose we say that p is *deadlocked* if $p \Downarrow$, $p \stackrel{\mu}{\longrightarrow}$ for no μ, but $p \notin \surd$ and

p *may deadlock* if $p \stackrel{\varepsilon}{\Longrightarrow} p'$ for some p' such that p' is deadlocked.

Then clause (iii) could also be replaced by:

(iiia)′ p *may deadlock* implies q *may deadlock*

(iiib)′ $p \Downarrow$, q *may deadlock* imply p *may deadlock*.

However, replacing clause (iii) with clauses such as

$$\text{if } p \Downarrow \text{ then } p\surd \text{ iff } q\surd$$

or

$$\text{if } p \Downarrow \text{ then } p \text{ is deadlocked iff } q \text{ is deadlocked}$$

would lead to a different semantic preorder. The terms $\tau;\delta$ and δ would be distinguished as would $a;\tau;\delta$ and $a;\delta$. Since $a;\tau$ and a are identified this would mean that the revised semantic preorder would not be preserved by ;.

An alternative method for using $\mathcal{F}$ to obtain a behavioural preorder is to apply it inductively as follows:

- $\sqsubseteq_0 = REC_\Sigma \times REC_\Sigma$ (the top element in the lattice $(Rel, \subseteq)$)
- $\sqsubseteq_{n+1} = \mathcal{F}(\sqsubseteq_n)$

and finally $\sqsubseteq_\omega = \bigcap_{n\geq 0} \sqsubseteq_n$.

The two relations $\sqsubseteq$ and $\sqsubseteq_\omega$ are in general different. For example, [Ab87b], take the synchronization trees p and q defined as follows:

$$p \equiv a^\omega + \Omega,\ q \equiv \sum_{k\in\omega} a^k + \Omega.$$

Then it is easy to see that $p \sqsubseteq_\omega q$, but $p \not\sqsubseteq q$. Two equivalent terms in our language are $rec\,x.\ a;x+\Omega$ and $rec\,x.\ x;a+a$, respectively. All the properties of $\sqsubseteq$ discussed above are also true of $\sqsubseteq_\omega$. In deciding which preorder to use, we will take into account the type of semantic model we discussed in the previous section. We wish to define a behavioural preorder $\lesssim$ which satisfies

$$p \lesssim q \iff p \leq_{CI_E} q, \tag{2.1}$$

where $p \leq_{CI_E} q$ means $CI_E[\![p]\!] \leq CI_E[\![q]\!]$, for the set of inequations E in Figure 2.1. This requirement induces certain constraints on $\lesssim$, the most important of which is called *finite approximability*. For any binary relation $\mathcal{R}$ over REC_Σ, let $\mathcal{R}^F$ be defined by:

$$p\mathcal{R}^F q \text{ if, for every finite term } d,\ d\mathcal{R}p \text{ implies } d\mathcal{R}q.$$

We say that $\mathcal{R}$ is *finitely approximable* (*fa*) if $\mathcal{R} = \mathcal{R}^F$. Note that, for every transitive relation $\mathcal{R}$, $\mathcal{R} \subseteq \mathcal{R}^F$; thus, in order to show that such relations are *fa*, it is sufficient to prove that $\mathcal{R}^F \subseteq \mathcal{R}$. Intuitively, the finite approximability of a relation $\mathcal{R}$ means that $\mathcal{R}$ is essentially determined by how it behaves on finite terms. By the general construction of CI_E, [H88a], it follows that $\leq_{CI_E}$ is *fa* and therefore, to meet (2.1), we must also choose a behavioural preorder which is also *fa*. The above example shows that $\sqsubseteq$ is not *fa*, as $p \sqsubseteq^F q$ but $p \not\sqsubseteq q$.

There is one further complication caused by requirement (2.1). The relation $\leq_{CI_E}$ is, by definition, *closed with respect to all contexts.* To explain this we need some notation. For any binary relation $\mathcal{R}$ over REC_Σ, let $\mathcal{R}$ be extended to $REC_\Sigma(\mathtt{Var})$ by:

$t\mathcal{R}u$ if, for every closed substitution ρ, $t\rho\mathcal{R}u\rho$.

For any $\mathcal{R}$ over $REC_\Sigma(\mathtt{Var})$ define the new relation $\mathcal{R}^c$ by:

$t\mathcal{R}^c u$ if, for every context $C[\cdot]$ such that $C[t]$ and $C[u]$ are closed, $C[t]\mathcal{R}C[u]$.

Then $\mathcal{R}$ is said to be closed with respect to contexts if $\mathcal{R} = \mathcal{R}^c$. By construction, it follows that $\leq_{CI_E}$ is closed with respect to contexts. However, this is not true of $\sqsubseteq$ or $\sqsubseteq_\omega$. The usual counter-example associated with the **CCS** + operator, [Mil80], works:

$$a \sqsubseteq_\omega \tau; a \text{ but } a + b \not\sqsubseteq_\omega \tau; a + b.$$

We may sum up this discussion by saying that in order to reflect the semantic ordering $\leq_{CI_E}$ behaviourally, it is necessary to choose a behavioural preorder which is both finitely approximable and preserved by contexts. We shall show that $\sqsubseteq_\omega$ is *fa* and therefore it is appropriate to take as our behavioural preorder $\sqsubseteq_\omega^c$, its closure with respect to all contexts. The proof that $\sqsubseteq_\omega$ is *fa* depends on the fact that our operational semantics is sort finite. In a transition system which is not sort finite $\sqsubseteq_\omega$ may not be *fa*. For instance, consider the following synchronization trees from [Ab87b]:

$$\begin{aligned} p &\equiv a(\textstyle\sum_{n\in\omega} b_n nil + \Omega) + \Omega \\ q &\equiv \textstyle\sum_{n\in\omega} a(\sum_{m\in\omega-\{n\}} b_m nil + \Omega) + \Omega, \end{aligned}$$

where, for each $n \neq m$, $b_n \neq b_m$. Then $p \sqsubseteq^F q$, but $p \not\sqsubseteq_2 q$.

The remainder of the chapter is devoted to proving that our behavioural and denotational view of processes do agree on REC_Σ, i.e. that, for $p, q \in REC_\Sigma$,

$$p \leq_{CI_E} q \iff p \sqsubseteq_\omega^c q.$$

In the next section we analyze the preorder $\sqsubseteq_\omega^c$, giving an equivalent but more manageable definition. Using this equivalent formulation we show that $\sqsubseteq_\omega^c$ satisfies all of the equations in E.

2.3.2 Analysis of the Preorder

In this section we give a reformulation of $\sqsubseteq$ and use the more manageable definition to prove some of its properties. We are mainly interested in $\sqsubseteq_\omega$ but, as it turns out, most of the technical development concerns $\sqsubseteq$ rather than $\sqsubseteq_\omega$.

Let $\mathcal{G} : Rel \rightarrow Rel$ be the functional defined as follows:

for each $\mathcal{R} \in Rel$, $p\mathcal{G}(\mathcal{R})q$ iff, for each $\mu \in Act_\tau$,

(i) if $p \xrightarrow{\mu} p'$ then, for some q', $q \xRightarrow{\hat{\mu}} q'$ and $p'\mathcal{R}q'$

(ii) if $p \Downarrow \mu$ then

a) $q \Downarrow \mu$

b) if $q \xrightarrow{\mu} q'$ then, for some p', $p \xRightarrow{\hat{\mu}} p'$ and $p'\mathcal{R}q'$

(iii) if $p \Downarrow$ then $p\surd\!\!/$ iff $q\surd\!\!/$.

Let $\lesssim$ denote the maximum fixed-point of the functional $\mathcal{G}$, whose existence can be easily shown following standard lines, [Mil88].

Proposition 2.3.1 *For* $p, q \in REC_\Sigma$, $p \sqsubseteq q \iff p \lesssim q$.

Proof: Standard and thus omitted. □

This proposition allows us to investigate the properties of $\sqsubseteq$ using the technically simpler relation $\lesssim$. As a first application we show that $\lesssim$, and consequently $\sqsubseteq$, is preserved by many of the operators of the calculus.

Lemma 2.3.1 *If* $p \lesssim q$ *then*

(a) $p; r \lesssim q; r$

(b) $p|r \lesssim q|r$

(c) $\partial_H(p) \lesssim \partial_H(q)$.

Proof: We examine only two of the operators leaving the remaining case to the reader.

(a) To show that $p \lesssim q$ implies $p; r \lesssim q; r$ it is sufficient to prove that the relation $\mathcal{R}$ defined as follows:

$$\mathcal{R} =_{def} \{(p;r,q;r) \mid p \lesssim q \text{ and } r \in REC_\Sigma\} \cup Id_{REC_\Sigma}$$

is a pre-bisimulation with respect to the functional $\mathcal{G}$, i.e. $\mathcal{R} \subseteq \mathcal{G}(\mathcal{R})$. We only check that the clauses of the definition of the functional $\mathcal{G}$ are met for $(p;r,q;r)$ such that $p \lesssim q$.

(i) Assume $p; r \xrightarrow{\mu} x$. There are two cases to examine:

- $p \xrightarrow{\mu} p'$ and $x \equiv p'; r$. Then, as $p \lesssim q$, there exists q' such that $q \overset{\hat{\mu}}{\Longrightarrow} q'$ and $p' \lesssim q'$. Thus $q; r \overset{\hat{\mu}}{\Longrightarrow} q'; r$ and $(p'; r, q'; r) \in \mathcal{R}$, by the definition of $\mathcal{R}$.
- $p\surd$ and $r \xrightarrow{\mu} x$. As $p\surd$, we may assume that $p \Downarrow$, $q \surd\!\!\surd$ and $q \Downarrow$. By the definition of $\surd\!\!\surd$, $\forall q' \in Stable(q)\; q'\surd$. Now, since $q \Downarrow$, $Stable(q) \neq \emptyset$. Let $q' \in Stable(q)$; then $q; r \overset{\varepsilon}{\Longrightarrow} q'; r \xrightarrow{\mu} x$ and $(x, x) \in \mathcal{R}$.

(ii) Assume $(p;r) \Downarrow \mu$. First of all we show that this implies $(q;r) \Downarrow \mu$. Note that $(p;r) \Downarrow \mu$ implies $p \Downarrow \mu$. As $p \lesssim q$, we have that $q \Downarrow \mu$. Suppose $(q;r) \Uparrow \mu$. We distinguish two cases:

$(\mu = \tau)$ This is equivalent to

$$q; r \xrightarrow{\tau^\omega} \text{ or } \exists y : q; r \overset{\varepsilon}{\Longrightarrow} y \text{ and } y \uparrow .$$

Assume that $q; r \xrightarrow{\tau^\omega}$. As $q \Downarrow$, it must be the case that there exists q' such that $q \overset{\varepsilon}{\Longrightarrow} q'$, $q'\surd$ and $r \xrightarrow{\tau^\omega}$. By induction on the derivation $q \overset{\varepsilon}{\Longrightarrow} q'$ we can show that there exists p' such that $p \overset{\varepsilon}{\Longrightarrow} p'$ and $p' \lesssim q'$. As $q'\surd$ and $p \Downarrow$, $p' \surd\!\!\surd$. This would mean $(p;r) \Uparrow$ which contradicts the hypothesis.

Checking that the other possibility leads to a contradiction as well is omitted.

$(\mu = a)$ By definition, $(q;r) \Uparrow a$ iff $(q;r) \Uparrow$ or $\exists y : q; r \overset{a}{\Longrightarrow} y$ and $y \Uparrow$. The case $(q;r) \Uparrow$ is dealt with as above. We assume that $\exists y : q; r \overset{a}{\Longrightarrow} y$ and $y \Uparrow$. By the definition of $\Uparrow$, $y \Uparrow$ iff $y \xrightarrow{\tau^\omega}$ or $\exists \bar{y} : y \overset{\varepsilon}{\Longrightarrow} \bar{y}$ and $\bar{y} \uparrow$.

Thus either $q; r \overset{a}{\Longrightarrow} y \xrightarrow{\tau^\omega}$ or $q; r \overset{a}{\Longrightarrow} y \uparrow$, for some y.

If $q; r \overset{a}{\Longrightarrow} y \xrightarrow{\tau^\omega}$ then, as $q \Downarrow a$, $\forall q' : q \overset{a}{\Longrightarrow} q'$, $q' \not\xrightarrow{\tau^\omega}$. Thus there must exist q' such that

$$(q \overset{a}{\Longrightarrow} q'\surd \text{ and } r \xrightarrow{\tau^\omega}) \text{ or } (q \overset{\varepsilon}{\Longrightarrow} q'\surd \text{ and } r \overset{a}{\Longrightarrow} y \xrightarrow{\tau^\omega}).$$

In both cases, from $p \Downarrow a$ and $p \lesssim q$ we may deduce that $(p;r) \Uparrow a$. Against the hypothesis.

If $q; r \overset{a}{\Longrightarrow} y \uparrow$ we proceed by analyzing the move $q; r \overset{a}{\Longrightarrow} y$. There are three possibilities:

(a) $q \stackrel{a}{\Longrightarrow} q'$ and $y \equiv (q';r) \uparrow$. As $q \Downarrow a$ it must be the case that $q'\surd$ and $r \uparrow$. By induction on the derivation $q \stackrel{a}{\Longrightarrow} q'$ we get p' such that $p \stackrel{a}{\Longrightarrow} p'$ and $p' \lesssim q'$. As $p \Downarrow a$ it must be the case that $p' \surd\!\!\!/$ and $p' \Downarrow$. Hence there exists p'' such that $p' \stackrel{\varepsilon}{\Longrightarrow} p''\surd$. Thus $p;r \stackrel{a}{\Longrightarrow} (p'';r) \uparrow$. This contradicts the hypothesis that $(p;r) \Downarrow a$.

(b) $q \stackrel{a}{\Longrightarrow} q'\surd$ and $r \stackrel{\varepsilon}{\Longrightarrow} y$.

(c) $q \stackrel{\varepsilon}{\Longrightarrow} q'\surd$ and $r \stackrel{a}{\Longrightarrow} y$.

Both (b) and (c) follow the pattern of case (a). The μ-moves of $q;r$ can be matched by those of $p;r$ as in case (i) above.

(iii) Assume $(p;r) \Downarrow$. We have to show that $(p;r)\surd\!\!\!/$ iff $(q;r)\surd\!\!\!/$. By clause (ii) above, $(q;r) \Downarrow$ and this, together with $p \lesssim q$, implies $p \Downarrow$ and $q \Downarrow$. It is easy to see that $(p;r)\surd\!\!\!/$ implies $p\surd\!\!\!/$ and $p\surd$ implies $r\surd\!\!\!/$.

Assume now $(p;r)\surd\!\!\!/$. By the above observation we get that $p\surd\!\!\!/$ and $p\surd$ implies $r\surd\!\!\!/$. As $p \Downarrow$, $p\surd\!\!\!/$ and $p \lesssim q$ imply $q\surd\!\!\!/$. Hence, if $q;r \notin \surd\!\!\!/$ it must be the case that, for some $x \in \mathit{Stable}(r)$, $q;r \stackrel{\varepsilon}{\Longrightarrow} q';r \stackrel{\varepsilon}{\Longrightarrow} x$, where $q'\surd$ and $x \notin \surd$.

It is now easy to see that this would imply $p;r \notin \surd\!\!\!/$, contradicting the hypothesis. The proof of the converse implication is similar.

(b) It is sufficient to prove that $\mathcal{R} \subseteq \mathcal{G}(\mathcal{R})$, where

$$\mathcal{R} =_{def} \{(p|r, q|r) \mid p \lesssim q \text{ and } r \in REC_\Sigma\}.$$

Most of the proof is identical to that of the corresponding case of Lemma 1 in [Wal87] (pp. 4-5). The only new property to check is that $(p|r) \Downarrow$ implies $p|r\surd\!\!\!/$ iff $q|r\surd\!\!\!/$. Assume $p|r \Downarrow$ and $q|r \notin \surd\!\!\!/$. Then there exists $q'|r'$ such that $q|r \stackrel{\varepsilon}{\Longrightarrow} q'|r'$, $q'|r'$ is stable and $q'|r' \notin \surd$. This implies that either $q' \notin \surd$ or $r' \notin \surd$. As $q|r \stackrel{\varepsilon}{\Longrightarrow} q'|r'$, there exist sequences $\langle q_i \mid 0 \leq i \leq n\rangle$, $\langle r_i \mid 0 \leq i \leq n\rangle$ and $\langle a_i \mid 0 \leq i \leq n\rangle$ such that

$$q_0 \equiv q, r_0 \equiv r, \forall i < n\ q_i \stackrel{a_i}{\Longrightarrow} q_{i+1} \text{ and } r_i \stackrel{\bar{a_i}}{\Longrightarrow} r_{i+1}, q_n \stackrel{\varepsilon}{\Longrightarrow} q' \text{ and } r_n \stackrel{\varepsilon}{\Longrightarrow} r'.$$

As $p|r \Downarrow$ implies $q|r \Downarrow$, we have that $q_i \Downarrow$, for each i. Hence we may inductively construct a sequence $\langle p_i \mid 0 \leq i \leq n\rangle$ such that $p_i \stackrel{a_i}{\Longrightarrow} p_{i+1}$ and $p_i \lesssim q_i$. Thus there exists p' such that $p_n \stackrel{\varepsilon}{\Longrightarrow} p'$ and $p' \lesssim q'$. As $p|r \Downarrow$, $p' \Downarrow$. If $q' \notin \surd$ then $p' \notin \surd\!\!\!/$ and $p|r \notin \surd\!\!\!/$– a contradiction. If $r' \notin \surd\!\!\!/$ then, as $q'|r'$ is stable, $p' \Downarrow$ and $p' \lesssim q'$ there exists a stable state of the form $p''|r'$ such that $p' \stackrel{\varepsilon}{\Longrightarrow} p''$. Now, $p''|r' \notin \surd$ implies $p|r \notin \surd\!\!\!/$, again a contradiction. The converse implication is similar.

Checking that $\lesssim$ is preserved by $\partial_H(\cdot)$ is left to the reader. □

As it is well known from the theory of bisimulation equivalence for **CCS**, [Mil88], and **ACP**, [BK85], $\lesssim$ is not preserved by $+$. For example, $nil \lesssim \tau$ but it is not the case that $a+nil \lesssim a+\tau$. In fact, $(a + nil) \Downarrow$ but $a + \tau \xrightarrow{\tau} nil$ and $a + nil \not\lesssim nil$. However, following Milner, [Mil80], we have a standard way of associating a precongruence with $\lesssim$. It is sufficient to close $\lesssim$ with respect to all the operators in Σ. The resulting precongruence, which we denote by $\lesssim^{fc}$, is known to be the largest Σ-precongruence contained in $\lesssim$. Note that $\lesssim^{c}$ and $\lesssim^{fc}$ are a priori different. In the latter we only close with respect to contexts built from the operators in Σ, but in the former we also close with respect to contexts involving $rec\, x.\, \_$. We will eventually prove that they coincide, but, for the moment, we concentrate on $\lesssim^{fc}$.

Let us now define the following preorder over REC_Σ:

$$p \lesssim^{+} q \iff \forall r \in REC_\Sigma\ p + r \lesssim q + r.$$

By analogy with one of the characterizations of the congruence associated with bisimulation equivalence, [Mil88], we might expect that $\lesssim^{+}$ and $\lesssim^{fc}$ coincide over REC_Σ. In order to prove that this is indeed the case, it will be useful to introduce an alternative characterization of $\lesssim^{+}$. The following definition is adapted from [Wal87].

Definition 2.3.1 *For each* $p, q \in REC_\Sigma$, $p \lesssim^{*} q$ *iff*

(i) $\forall a \in Act$, *if* $p \xrightarrow{a} p'$ *then, for some* q', $q \overset{a}{\Longrightarrow} q'$ *and* $p' \lesssim q'$

(ii) *if* $p \overset{\tau}{\Longrightarrow} p'$ *then*

  (a) $p' \Downarrow$ *implies, for some* q', $q \overset{\tau}{\Longrightarrow} q'$ *and* $p' \lesssim q'$

  (b) $p' \Uparrow$ *implies, for some* q', $q \overset{\varepsilon}{\Longrightarrow} q'$ *and* $p' \lesssim q'$

(iii) $p \Downarrow \mu$ *implies*

  (a) $q \Downarrow \mu$

  (b) *if* $q \xrightarrow{\mu} q'$ *then, for some* q', $p \overset{\mu}{\Longrightarrow} p'$ *and* $p' \lesssim q'$

(iv) *if* $p \Downarrow$ *then* $p\surd$ *iff* $q\surd$.

The relation $\lesssim^{*}$ is easily seen to be a preorder. The following theorem states that $\lesssim^{*}$ and $\lesssim^{+}$ coincide over REC_Σ. The proof of the theorem uses the following technical lemma.

Lemma 2.3.2 *If* $p \Uparrow$, $p \lesssim q$ *and* $\neg \exists p' : p \overset{\tau}{\Longrightarrow} p' \wedge p' \Downarrow$ *then* $p \lesssim q + r$, *for each* $r \in REC_\Sigma$.

Proof: It is sufficient to check that the relation

$$\mathcal{R} =_{def} \lesssim \cup \{(p', q+r) \mid p \stackrel{\varepsilon}{\Longrightarrow} p' \text{ and } p' \lesssim q\}$$

is such that $\mathcal{R} \subseteq \mathcal{F}(\mathcal{R})$. Checking the clauses of the definition of $\lesssim$ is routine. □

Theorem 2.3.1 *For each* $p, q \in REC_\Sigma$, $p \lesssim^\star q$ *iff* $p \lesssim^+ q$.

Proof: The proof is identical to the one of the corresponding result in [Wal87] (Lemma 3). Thus we just check the cases not covered by that lemma.

($\Leftarrow$) Assume that $p \not\lesssim^\star q$ because

$$p \Downarrow \text{ and } [(p\surd\!\!\!\surd \wedge q \notin \surd\!\!\!\surd) \vee (p \notin \surd\!\!\!\surd \wedge q\surd\!\!\!\surd)].$$

We have to find a process $r \in REC_\Sigma$ such that $p + r \not\lesssim q + r$. Assume, wlog, that $p \Downarrow$, $p\surd\!\!\!\surd$ and $q \notin \surd\!\!\!\surd$. Take $r \equiv nil$. Then $(p + nil) \Downarrow$ and $(p + nil)\surd\!\!\!\surd$ whilst, as it may be easily checked, $(q + nil) \notin \surd\!\!\!\surd$. Hence $p + nil \not\lesssim q + nil$ and $p \not\lesssim^+ q$.

($\Rightarrow$) Suppose that $p \lesssim^\star q$. We show that, for each $r \in REC_\Sigma$, $(p + r) \Downarrow$ implies $(p + r)\surd\!\!\!\surd$ iff $(q + r)\surd\!\!\!\surd$.

Now, $(p + r) \Downarrow$ implies $p \Downarrow$. As $p \Downarrow$ and $p \lesssim^\star q$ we have that $p\surd\!\!\!\surd$ iff $q\surd\!\!\!\surd$. The claim then follows from the fact that, for each $p, r \in REC_\Sigma$,

$$Stable(p+r) = \begin{cases} \{p + r\} & \text{if } p \stackrel{\tau}{\not\longrightarrow} \text{ and } r \stackrel{\tau}{\not\longrightarrow} \\ \{x \in Stable(p) \cup Stable(r) \mid p \stackrel{\tau}{\Longrightarrow} x \text{ or } r \stackrel{\tau}{\Longrightarrow} x\} & \text{otherwise.} \end{cases}$$ □

As an easy corollary of the above theorem we get that $\lesssim^\star \subseteq \lesssim$. In fact,

$$p \lesssim^\star q \Longleftrightarrow p \lesssim^+ q \Longrightarrow p + nil \lesssim q + nil \Longleftrightarrow p \lesssim q.$$

The next lemma establishes the fact that $\lesssim^\star$ is a Σ-precongruence.

Lemma 2.3.3 $\lesssim^\star$ *is a* Σ*-precongruence.*

Proof: We examine each operator separately.

(;) Assume $p \lesssim q$ and $r \in REC_\Sigma$. We check that the clauses of the definition of $\lesssim^\star$ are met by $p; r$ and $q; r$.

(i) Suppose that $p; r \stackrel{a}{\longrightarrow} x$. By the operational semantics there are two cases to examine:

- $p \xrightarrow{a} p'$ and $x \equiv p'; r$. Then, as $p \lesssim^\star q$, $q \overset{a}{\Longrightarrow} q'$ and $p' \lesssim q'$, for some q'. By the operational semantics, $q; r \overset{a}{\Longrightarrow} q'; r$ and, as by Lemma 2.3.1 $\lesssim$ is preserved by ;, $p'; r \lesssim q'; r$.
- $p\surd$ and $r \xrightarrow{a} x$. Now $p\surd$ implies $p \Downarrow$ and since $p \lesssim^\star q$ we may deduce that

$$q \Downarrow,\ q \not\surd \text{ and, } \forall \mu \in Act_\tau,\ q \not\xrightarrow{\mu}.$$

As $q \not\xrightarrow{\tau}$ we get that $Stable(q) = \{q\}$. Thus $q\surd$. Hence $q; r \xrightarrow{a} x$ and $x \lesssim x$.

(ii) Suppose that $p; r \overset{\tau}{\Longrightarrow} x$. Following the definition of $\lesssim^\star$ we distinguish two cases:

- $x \Downarrow$. We proceed by analyzing the move $p; r \overset{\tau}{\Longrightarrow} x$.

  (1) $p \overset{\tau}{\Longrightarrow} p'$ and $x \equiv p'; r$. Then $x \Downarrow$ implies $p' \Downarrow$. As $p \lesssim^\star q$, there exists q' such that $q \overset{\tau}{\Longrightarrow} q'$ and $p' \lesssim q'$. By the operational semantics, $q; r \overset{\tau}{\Longrightarrow} q'; r$ and, by Lemma 2.3.1, $p'; r \lesssim q'; r$.

  (2) $p \overset{\varepsilon}{\Longrightarrow} p'\surd$ and $r \overset{\tau}{\Longrightarrow} x$. Left to the reader.

- $x \Uparrow$. This case can be checked using the pattern used in the above case.

(iii) Assume $(p; r) \Downarrow \mu$. Reasoning as in the corresponding case of Lemma 2.3.1 we get that $(q; r) \Downarrow \mu$. Suppose that $q; r \xrightarrow{\mu} x$. By the operational semantics, there are two cases to examine:

- $q \xrightarrow{\mu} q'$ and $x \equiv q'; r$. As $(p; r) \Downarrow \mu$ and $(q; r) \Downarrow \mu$, we get that $p \Downarrow \mu$ and $q \Downarrow \mu$. As $p \lesssim^\star q$, there exists p' such that $p \overset{\mu}{\Longrightarrow} p'$ and $p' \lesssim q'$. By the operational semantics, $p; r \overset{\mu}{\Longrightarrow} p'; r$ and, by Lemma 2.3.1, $p'; r \lesssim q'; r$.
- $q\surd$ and $r \xrightarrow{\mu} x$. As $(p; r) \Downarrow \mu$ we get that $p \Downarrow$. Thus, $p \lesssim^\star q$ and $q\surd$ imply $p\surd$. By the operational semantics, $p; r \xrightarrow{\mu} x$ and $x \lesssim x$.

(iv) As in the corresponding case of Lemma 2.3.1.

(+) To prove that $\lesssim^\star$ is preserved by +, it is sufficient to notice that + is associative with respect to $\lesssim$ and use the alternative characterization of $\lesssim^\star$ given by the previous theorem.

(|) The proof is identical to that of the corresponding case of Lemma 4 in [Wal 87]. Checking clause (iv) of the definition of $\lesssim^\star$ can be done as in Lemma 2.3.1.

Checking that $\lesssim^\star$ is preserved by $\partial_H(\cdot)$ is left to the reader. □

As a corollary of the above result we get that $\lesssim^\star$, and consequently $\lesssim^+$, coincides with $\lesssim^{fc}$.

Corollary 2.3.1 *The relations $\lesssim^{fc}$, $\lesssim^+$ and $\lesssim^\star$ all coincide over REC_Σ.*

Proof: By Theorem 2.3.1 we know that $\lesssim^\star = \lesssim^+$. By the definition of $\lesssim^{fc}$ we derive that $\lesssim^{fc} \subseteq \lesssim^+$. Moreover, by the above lemma, $\lesssim^\star$ is a Σ-precongruence and this implies $\lesssim^\star \subseteq \lesssim^{fc}$, as $\lesssim^{fc}$ is the largest Σ-precongruence contained in $\lesssim$ and $\lesssim^\star \subseteq \lesssim$. □

The results that we have presented so far seem to imply that $\lesssim^\star$ is a suitable notion of semantic preorder for the language REC_Σ. However, difficulties arise when we try to relate $\lesssim^\star$ with the denotational model CI_E outlined in §2.2. This is discussed more fully in the concluding remarks.

As we have already seen, $\lesssim$ (and consequently $\sqsubseteq$) is not finitely approximable [Ab87b]. Fortunately, however, we are able to relate $\sqsubseteq$, $\sqsubseteq_\omega$ and $\sqsubseteq_\omega^F$ over an important subset of REC_Σ^2; in fact, our next aim is to show that the three preorders coincide over $FREC_\Sigma \times REC_\Sigma$. The following technical result is standard.

Fact 2.3.1 *For each $n \in \omega$, $\sqsubseteq_{n+1} \subseteq \sqsubseteq_n$.*

Theorem 2.3.2 *For each $d \in FREC_\Sigma$, $p \in REC_\Sigma$, $d \sqsubseteq p$ iff $d \sqsubseteq_\omega p$.*

Proof: The "only if" implication is easily seen to hold by induction on n. To prove that $d \sqsubseteq_\omega p$ implies $d \sqsubseteq p$ we define the depth of a finite process as follows:

$$\mathtt{dt}(d) =_{def} \begin{cases} 2 & \text{if } \nexists \mu \in Act_\tau : d \stackrel{\mu}{\Longrightarrow} \\ 1 + max\{\mathtt{dt}(d') \mid \exists \mu : d \stackrel{\mu}{\Longrightarrow} d'\} & \text{otherwise.} \end{cases}$$

This is well defined because all the $d's$ are finite (hence $\{d' \mid d \stackrel{\mu}{\Longrightarrow} d'\}$ is finite for each $\mu \in Act_\tau$) and the transition system that we are considering is sort-finite. Note that, for each $d \in FREC_\Sigma$, $\mu \in Act_\tau$, $d \stackrel{\mu}{\Longrightarrow} d'$ implies $\mathtt{dt}(d') \leq \mathtt{dt}(d) - 1$. By induction on $\mathtt{dt}(t)$ we will prove that $d \sqsubseteq_{\mathtt{dt}(d)} p$ implies $d \sqsubseteq p$.

Base case, $\mathtt{dt}(d) = 2$. Assume $\mathtt{dt}(d) = 2$ and $d \sqsubseteq_2 p$. We will show that the relation

$$\mathcal{R} =_{def} \{(d, p') \mid p \stackrel{\varepsilon}{\Longrightarrow} p'\}$$

is a prebisimulation. Consider $(d, p') \in \mathcal{R}$; we proceed to check the defining clauses of the functional $\mathcal{F}$. Clause (i) is trivially met as $\mathtt{dt}(d) = 2$ implies $d \stackrel{\mu}{\not\Longrightarrow}$ for each $\mu \in Act_\tau$. Assume $d \Downarrow \mu$. As $d \sqsubseteq_2 p$ we have that $p \Downarrow \mu$. It is easy to see that this implies $p' \Downarrow \mu$.

Assume $d \Downarrow \mu$, $p' \Downarrow \mu$ and $p' \stackrel{\mu}{\Longrightarrow} p''$. As $d \sqsubseteq_2 p$ we have that $p \Downarrow \mu$. Hence, it must be the case that $\mu = \tau$ (otherwise $p \stackrel{\varepsilon}{\Longrightarrow} p' \stackrel{\mu}{\Longrightarrow} p''$ whilst, as $\mathtt{dt}(d) = 2$, $d \stackrel{\mu}{\not\Longrightarrow}$. This would contradict the hypothesis that $d \sqsubseteq_2 p$). Thus $(d, p'') \in \mathcal{R}$ by the definition of $\mathcal{R}$. To check clause (iii) assume that $d \Downarrow$. Then, as $d \sqsubseteq_2 p$, $p \Downarrow$ and $d\surd$ iff $p\surd$. Suppose

$$(d\surd \text{ and } p' \notin \surd) \text{ or } (d \notin \surd \text{ and } p'\surd).$$

If $d\surd$ and $p' \notin \surd$ then $p \notin \surd$, contradicting the hypothesis. If $d \notin \surd$ and $p'\surd$ then $p \stackrel{\tau}{\Longrightarrow} p'$ and $d \not\sqsubseteq_1 p'$, again contradicting the hypothesis that $d \sqsubseteq_2 p$. Hence $d\surd$ iff $p'\surd$. Thus $\mathcal{R} \subseteq \mathcal{F}(\mathcal{R})$ and, as $(d,p) \in \mathcal{R}$, $d \sqsubseteq p$.

Inductive step. Assume $\mathtt{dt}(d) > 2$ and $d \sqsubseteq_{\mathtt{dt}(d)} p$. Then:

(i) if $d \stackrel{\mu}{\Longrightarrow} d'$ then there exists p' such that $p \stackrel{\hat{\mu}}{\Longrightarrow} p'$ and $d' \sqsubseteq_{\mathtt{dt}(d)-1} p$. As $\mathtt{dt}(d') \leq \mathtt{dt}(d) - 1$ we have that $d' \sqsubseteq_{\mathtt{dt}(d)-1} p'$ implies $d' \sqsubseteq_{\mathtt{dt}(d')} p'$. By induction this implies that $d' \sqsubseteq p'$.

(ii) Assume $d \Downarrow \mu$. Then $p \Downarrow \mu$ as $d \sqsubseteq_{\mathtt{dt}(d)} p$. Assume now that $p \stackrel{\mu}{\Longrightarrow} p'$. Then reasoning as in case (i) we can find d' such that $d \stackrel{\mu}{\Longrightarrow} d'$ and $d' \sqsubseteq p'$.

(iii) Follows from the assumption that $d \sqsubseteq_{\mathtt{dt}(d)} p$.

Thus $d \sqsubseteq_\omega p \Longrightarrow d \sqsubseteq_{\mathtt{dt}(d)} p \Longrightarrow d \sqsubseteq p$. □

Corollary 2.3.2 *For each $d \in FREC_\Sigma$, $p \in REC_\Sigma$, $d \sqsubseteq p \Longleftrightarrow d \sqsubseteq_\omega p \Longleftrightarrow d \sqsubseteq^F_\omega p$.*

Proof: An easy consequence of the above theorem and of the definition of $\sqsubseteq^F_\omega$. □

As a consequence of the above result we have that $\sqsubseteq^{fc}_\omega$ and $\lesssim^\star$ coincide over $FREC_\Sigma \times REC_\Sigma$. This explains our interest in $\lesssim^\star$; it is technically more manageable than $\sqsubseteq^F_\omega$ and it will be used in the next section to show that for $d \in FREC_\Sigma$, $p \in REC_\Sigma$

$$d \sqsubseteq^{fc}_\omega p \Longleftrightarrow CI_E[\![d]\!] \leq CI_E[\![p]\!].$$

This result can be lifted to the entire language in a fairly standard way, once we know that $\sqsubseteq_\omega$ is finitely approximable.

As a final result in this section we examine the equations in E and show that they are satisfied by $\sqsubseteq^{fc}_\omega$. Let $\leq_E$ denote the least Σ-precongruence over REC_Σ which satisfies the equations E.

Proposition 2.3.2 *For $p, q \in REC_\Sigma$, $p \leq_E q$ implies $p \sqsubseteq^{fc}_\omega q$.*

Proof: Since $\sqsubseteq^{fc}_\omega$ contains $\lesssim^\star$ and $\lesssim^\star$ is a Σ-precongruence, it is sufficient to establish that $\lesssim^\star$ satisfies all of the equations E. This we leave to the reader. □

2.3.3 Full-Abstraction

In this section we outline the proof of full-abstraction, namely

$$p \sqsubseteq_\omega^c q \iff CI_E[\![p]\!] \leq CI_E[\![q]\!]. \qquad (2.2)$$

For convenience we will abbreviate $CI_E[\![p]\!]$ to $[\![p]\!]_E$. The first point to note is that it is sufficient to prove (2.2) for $\sqsubseteq_\omega^{fc}$ rather than $\sqsubseteq_\omega^c$. For, in this case, we can show that $\sqsubseteq_\omega^{fc}$ and $\sqsubseteq_\omega^c$ coincide: we already know that $\sqsubseteq_\omega^c \subseteq \sqsubseteq_\omega^{fc}$ and the fact that $\sqsubseteq_\omega^{fc}$ coincides with the preorder generated by the model means that it is preserved by contexts, i.e. $\sqsubseteq_\omega^{fc} = \sqsubseteq_\omega^{fc,c}$. We know $\sqsubseteq_\omega^{fc} \subseteq \sqsubseteq_\omega$, from which it now follows that $\sqsubseteq_\omega^{fc} \subseteq \sqsubseteq_\omega^c$.

One crucial property of $\sqsubseteq_\omega^{fc}$ which we require is *finite approximability*.

Theorem 2.3.3 $\sqsubseteq_\omega^{fc}$ *is finitely approximable.*

The proof of this theorem is quite involved and uses a characterization of $\sqsubseteq_\omega$ in terms of a modal property language similar to that in [HM85], [Ab87b]. The next section is entirely devoted to the exposition of the proof. Since it is independent of the rest of the chapter, we assume the theorem in the remainder of this section. Another property which we will establish is a partial completeness result, namely:

Theorem 2.3.4 *For* $d \in FREC_\Sigma$, $p \in REC_\Sigma$, $d \lesssim^* p$ *implies* $[\![d]\!]_E \leq [\![p]\!]_E$.

Using these two theorems we now show how to establish full-abstraction (2.2). The proof actually requires some general results about the semantic mappings defined in §2.2.2, which may be found in [CN76], [Gue81], [H88a]. The first states that for any $p \in REC_\Sigma$ there exists an infinite sequence of *finite approximations* $p^n \in FREC_\Sigma$ such that, for any interpretation A,

$$A[\![p]\!] = \bigsqcup_{n \geq 0} A[\![p^n]\!].$$

The second states that for interpretations of the form CI_E, for each $d, e \in FREC_\Sigma$,

$$[\![d]\!]_E \leq [\![e]\!]_E \text{ if, and only if, } d \leq_E e.$$

Finally, every finite approximation p^n may be generated syntactically from p. Let $\prec$ be the least Σ-precongruence over REC_Σ which satisfies

$$\begin{array}{llll} (\Omega) & \Omega & \leq & x \\ (\textbf{Rec}) & rec\, x.\, t & = & t[rec\, x.\, t/x]. \end{array}$$

Then, for every $n \geq 0$, $p^n \prec p$. These general results may now be applied to prove:

Theorem 2.3.5 (Full-Abstraction) *For* $p, q \in REC_\Sigma$, $[\![p]\!]_E \leq [\![q]\!]_E$ *iff* $p \sqsubseteq_\omega^{fc} q$.

Proof: Suppose $[\![p]\!]_E \leq [\![q]\!]_E$. Since we may assume that $\sqsubseteq_\omega^{fc}$ is *fa*, it is sufficient to show that, for finite d, if $d \sqsubseteq_\omega^{fc} p$ then $d \sqsubseteq_\omega^{fc} q$. From the partial completeness result and the coincidence of $\lesssim^\star$ and $\sqsubseteq_\omega^{fc}$ over $FREC_\Sigma \times REC_\Sigma$, we may assume that, for such a d, $[\![d]\!]_E \leq [\![p]\!]_E$ and therefore $[\![d]\!]_E \leq [\![q]\!]_E$. Now, since $d \in FREC_\Sigma$, $[\![d]\!]_E$ is a finite element in the algebraic cpo CI_E. This means that, for some $n \geq 0$, $[\![d]\!]_E \leq [\![q^n]\!]_E$, i. e. $d \leq_E q^n$. From Proposition 2.3.2 it follows that $d \sqsubseteq_\omega^{fc} q^n$. Now, it is trivial to check that both of the laws (Ω) and (Rec) are satisfied by $\lesssim^\star$, i. e. $\prec \subseteq \lesssim^\star$, from which it follows that $q^n \sqsubseteq_\omega^{fc} q$. We may therefore conclude that $d \sqsubseteq_\omega^{fc} q$.

The converse is even more straightforward as it is an immediate consequence of the partial completeness theorem. Let us recall that $\leq_{CI_E}$ denotes the relation over $REC_\Sigma \times REC_\Sigma$ defined by $p \leq_{CI_E} q$ iff $[\![p]\!]_E \leq [\![q]\!]_E$. By the construction of CI_E, it is a finitely approximable relation. So, it is sufficient to show that if $d \sqsubseteq_\omega^{fc} p$, for $d \in FREC_\Sigma$, then $d \leq_{CI_E} p$. But, as $\sqsubseteq_\omega^{fc}$ and $\lesssim^\star$ coincide over $FREC_\Sigma \times REC_\Sigma$, this is precisely the statement of Theorem 2.3.4. □

There may be other models which are fully abstract with respect to $\sqsubseteq_\omega^{fc}$ over REC_Σ. However, CI_E is characterized by being initial in the category of fully abstract models and continuous Σ-homomorphisms. A Σ-cpo A is called *consistent* if, for each p, $q \in REC_\Sigma$,

$$p \sqsubseteq_\omega^{fc} q \text{ implies } A[\![p]\!] \leq A[\![q]\!].$$

The next theorem states that CI_E is initial in the category of consistent models and continuous Σ-homomorphisms. As every fully abstract model is obviously consistent, this implies that CI_E is, up to isomorphism, the initial fully abstract model.

Theorem 2.3.6 *Let A be a consistent Σ-cpo. Then there exists a unique continuous Σ-homomorphism* $h_A : CI_E \longrightarrow A$.

Proof: Assume A is a consistent Σ-cpo. By the full abstraction theorem, $[\![p]\!]_E \leq [\![q]\!]_E$ iff $p \sqsubseteq_\omega^{fc} q$, for each p, $q \in REC_\Sigma$. As A is consistent, for each p, $q \in REC_\Sigma$, $[\![p]\!]_E \leq [\![q]\!]_E$ implies $A[\![p]\!] \leq A[\![q]\!]$. Thus $A \in C(\leq_{CI_E})$, the category of Σ-cpo's which satisfy the relation $\leq_{CI_E}$ and continuous Σ-homomorphisms. As CI_E is initial in $C(\leq_{CI_E})$ and $A \in C(\leq_{CI_E})$, we thus have that there exists a unique continuous Σ-homomorphism $h_A : CI_E \longrightarrow A$. □

So we have reduced full-abstraction to two theorems, Theorem 2.3.3 and Theorem 2.3.4. As already stated, the former is the subject of the next section, so in the remainder of this subsection we prove the latter. The proof of the partial completeness theorem follows the lines of the completeness theorems for finite terms in [HM85], [Wal87], [H88c], except that some care must be taken in the form of induction used— one of our terms may be infinite.

We first show that all finite terms may be reduced to a suitable normal form. The following facts will be useful in the syntactic manipulations to follow:

Lemma 2.3.4 *The following are derived laws of the set of inequations E:*

$$\begin{array}{lll} \textbf{(D1)} \ \tau;\Omega & = & \Omega \\ \textbf{(D2)} \ \tau;(x+\Omega) & = & x+\Omega \end{array}$$

Proof: Left to the reader. The proof of (D2) uses axioms (Ω1), (Ω2) and (T2). □

Definition 2.3.2 *The set of* normal forms *(nf's) is the least subset of REC_Σ which satisfies:*

(i) *δ is a nf,*

(ii) *$\sum \mu_i; p_i\{+\Omega\}$ is a nf if*

- **(a)** *each p_i is a nf, and*
- **(b)** *if, for some i, μ_i is τ then $p_i \Downarrow$.*

The notation $\{+\Omega\}$ is used, as usual, to indicate that Ω is an optional summand.

Note that, according to the definition, if n is a nf then $n \Uparrow$ iff Ω is a summand of n.

Proposition 2.3.3 (Normalization) *For each $d \in FREC_\Sigma$ there exists a nf,* $\mathtt{nf}(d)$, *such that* $\mathtt{nf}(d) =_E d$ *and* $\mathtt{dt}(d) = \mathtt{dt}(\mathtt{nf}(d))$.

Proof: By induction on the *depth* of d. We assume, as inductive hypothesis that the statement is true for all d' such that $\mathtt{dt}(d') < \mathtt{dt}(d)$. The proof proceeds by a further induction on the structure of d. We examine only two cases leaving the others to the reader.

$d \equiv e; f$ By the inner inductive hypothesis, $e =_E \mathtt{nf}(e)$. If $\mathtt{nf}(e) \equiv \delta$ then $d \equiv e; f =_E \delta; f =_E \delta$ which is a nf. If $\mathtt{nf}(e) \equiv nil$ then, as $\mathtt{dt}(e; f) = \mathtt{dt}(f)$, we may apply the inner inductive hypothesis to obtain that $f =_E \mathtt{nf}(f)$. Thus $d \equiv e; f =_E \mathtt{nf}(f)$. Otherwise

$$\begin{array}{llll} e; f & =_E & (\sum \mu_i; e_i\{+\Omega\}); f & \\ & =_E & \sum(\mu_i; e_i); f\{+\Omega; f\} & \text{by repeated application of (B4)} \\ & =_E & \sum \mu_i; (e_i; f)\{+\Omega\} & \text{by } (\Omega 1), (\Omega 4) \text{ and repeated application of (B3).} \end{array}$$

As the depth of $e_i; f$ is less than that of $e; f$, for each i, we may apply the inner inductive hypothesis to obtain a *nf* π_i such that $\pi_i =_E e_i; f$, for each i. Thus $e; f =_E \sum \mu_i; \pi_i\{+\Omega\}$. Assume now that $\mu_j = \tau$ and $\pi_j \Uparrow$ for some j. Then Ω is a summand of π_j. Let π_j be $\bar{\pi} + \Omega$. Then

$$\begin{aligned} e; f &=_E \textstyle\sum_{i \neq j} \mu_i; \pi_i\{+\Omega\} + \tau; (\bar{\pi} + \Omega) \\ &=_E \textstyle\sum_{i \neq j} \mu_i; \pi_i\{+\Omega\} + \bar{\pi} + \Omega \qquad \text{by induction and (D2).} \end{aligned}$$

This procedure can be iterated to obtain a *nf*.

$d \equiv e|f$ By the outer inductive hypothesis, $e =_E \mathtt{nf}(e)$ and $f =_E \mathtt{nf}(f)$. If $\mathtt{nf}(e) \equiv \delta$ then there are two cases to examine:

- $\mathtt{nf}(f) \equiv \delta$. Then $e|f =_E \delta|\delta =_E \delta$, by axiom (C1).
- $\mathtt{nf}(f) \equiv \sum_{i \in I} \mu_i; f_i\{+\Omega\}$. We distinguish two cases:

  If $I = \emptyset$ then, by axiom (C2), $e|f =_E \delta\{+\Omega\}$. A normal form may now be obtained by possibly applying axioms (A1) and (A5).

  If $I \neq \emptyset$ then, by axiom (C2),

$$e|f =_E \delta|(\sum_{i \in I} \mu_i; f_i\{+\Omega\}) =_E \sum_{i \in I} \mu_i; (\delta|f_i)\{+\Omega\}.$$

  By the inductive hypothesis, for each $i \in I$, there exists a normal form π_i such that $\pi_i =_E \delta|f_i$. Thus $e|f =_E \sum_{i \in I} \mu_i; \pi_i\{+\Omega\}$. Assume now that there exists $j \in I$ such that $\mu_j = \tau$ and $\pi_j \Uparrow$. Then, as π_j is a *nf*, $\pi_j =_E \pi + \Omega$, where $\pi \Downarrow$. Thus

$$\tau; \pi_j =_E \tau; (\pi + \Omega) =_E \pi + \Omega \text{ by (D2).}$$

  Iterating this procedure we may generate a normal form.

So by symmetry we may assume that $\mathtt{nf}(e)$ and $\mathtt{nf}(f)$ are both different from δ and have the form $\mathtt{nf}(e) \equiv \sum_{i \in I} \mu_i; e_i\{+\Omega\}$ and $\mathtt{nf}(f) \equiv \sum_{j \in J} \gamma_j; f_j\{+\Omega\}$. Then

$$\begin{aligned} e|f &=_E (\sum_{i \in I} \mu_i; e_i\{+\Omega\})|(\sum_{j \in J} \gamma_j; f_j\{+\Omega\}) \\ &=_E \sum_{i \in I} \mu_i; (e_i|\mathtt{nf}(f)) + \sum_{j \in J} \gamma_j; (\mathtt{nf}(e)|f_j) + \sum_{(i,j):\mu_i = \bar{\gamma}_j} \tau; (e_i|f_j)\{+\Omega\}, \end{aligned}$$

by applying axiom (Exp). By the inductive hypothesis, for each $i \in I$, $j \in J$ and (i, j) such that $\mu_i = \bar{\gamma}_j$ there exist normal forms π_i, π_j and π_{ij} such that

$$\pi_i =_E e_i|\mathtt{nf}(f),\ \pi_j =_E \mathtt{nf}(e)|f_j \text{ and } \pi_{ij} =_E e_i|f_j.$$

Thus $e|f =_E \sum_{i\in I}\mu_i;\pi_i + \sum_{j\in J}\gamma_j;\pi_j + \sum(i,j) : \mu_i = \bar{\gamma}_j\tau;\pi_{ij}\{+\Omega\}$. Assume now that there exists $i_1 \in I$ such that $\mu_{i_1} = \tau$ and $\pi_{i_1} \Uparrow$. Then, as π_{i_1} is a *nf*, $\pi_{1_1} =_E \bar{\pi} + \Omega$, where $\bar{\pi} \Downarrow$. Thus

$$\mu_{i_1};\pi_{i_1} =_E \tau;(\bar{\pi}+\Omega) =_E \bar{\pi}+\Omega, \text{ by (D2)}.$$

The same applies to each $j \in J$ such that $\gamma_j = \tau$ and $\pi_j \Uparrow$ and to each π_{ij} such that $\pi_{ij} \Uparrow$. Iterating this procedure we may generate a *nf*. □

The above proposition tells us that, wlog, we can deal with the set of normal forms instead of $FREC_\Sigma$. This is first applied in proving the following technical result.

Fact 2.3.2 *For each $d \in FREC_\Sigma$, $d \Downarrow$ and $d \Uparrow a$ imply $d =_E d + a;\Omega$.*

Proof: By the above proposition we may assume, wlog, that d is a *nf*. The proof is by induction on the depth of the *nf* $d \equiv \sum_{i\in I}\mu_i;d_i$. By the definitions of the divergence predicates, $d \Uparrow a$ and $d \Downarrow$ imply $d \stackrel{a}{\Longrightarrow} e$ for some $e \Uparrow$. There are two cases to examine:

(1) $d \stackrel{a}{\longrightarrow} e$ and $e \Uparrow$. Then there exists $i \in I$ such that $\mu_i = a$ and $d_i \equiv e$. As e is a *nf*, $e \Uparrow$ implies $e =_E e + \Omega$. Hence

$$\begin{aligned} d &=_E d + a;(e+\Omega) \\ &=_E d + a;(e+\tau;\Omega) && \text{by (D1) and substitutivity} \\ &=_E d + a;(e+\tau;\Omega) + a;\Omega && \text{by (T3)} \\ &=_E d + a;\Omega. \end{aligned}$$

(2) $d \stackrel{\tau}{\longrightarrow} d' \stackrel{a}{\Longrightarrow} e$ and $e \Uparrow$. Then $d' \Uparrow a$. Moreover, as d is a *nf* and $d \Downarrow$, d' is a *nf* and $d' \Downarrow$. Thus, by the inductive hypothesis, $d' =_E d' + a;\Omega$. We can then calculate

$$\begin{aligned} d &=_E d + \tau;d' && \text{as } \mu_i = \tau \text{ and } d_i \equiv d', \text{ for some } i \in I, \\ &=_E d + \tau;(d' + a;\Omega) \\ &=_E d + \tau;(d' + a;\Omega) + d' + a;\Omega && \text{by (T2)} \\ &=_E d + a;\Omega && \text{by (T2) and (A1)-(A3)}. \end{aligned}$$

This completes the proof. □

We would not expect arbitrary terms from REC_Σ to have normal forms since the process of normalization may not terminate. However, for weakly convergent terms we have a weaker form of normal forms, called *head normal forms*; these look like normal forms at the topmost level.

Definition 2.3.3 *The set of* head normal forms *(hnf's) is the least subset of* REC_Σ *which satisfies:*

- δ *is a hnf,*
- $\sum \mu_i; p_i$ *is a hnf if, for each* i, $\mu_i = \tau$ *implies* p_i *is a hnf.*

It is easy to check that if h is a hnf then $h \Downarrow$. It would not be reasonable to expect all terms to be reducible to hnf's using the equations E. For example, $p \equiv rec\, x.\ a; x$ is not a hnf and applying equations to it will not help. However, if we are allowed to expand recursive definitions, i.e. use the axiom (Rec), then p can be rewritten to $a; rec\, x.\ a; x$ which is a hnf. Let us use $\leq_{Er}$ to denote the precongruence obtained by allowing the use of axiom (Rec). In this extended rewrite system all weakly convergent terms may be reduced to hnf's.

Proposition 2.3.4 *For each* $p \in REC_\Sigma$ *such that* $p \Downarrow$, *there exists a hnf* $\mathtt{h}(p)$ *such that* $p =_{Er} \mathtt{h}(p)$.

Proof: We assume that, for each q such that $p \stackrel{\tau}{\Longrightarrow} q$, q has a hnf. The proof then proceeds by induction on why $p \downarrow$. We examine only two cases of the inductive definition of $\downarrow$.

$p \equiv q; r$ Then $(q; r) \downarrow$ iff **(i)** $q\surd$ and $r \downarrow$ or **(ii)** $q \not\in \surd$ and $q \downarrow$.

If **(i)** holds then, by induction, $r =_{Er} \mathtt{h}(r)$. It is easy to show that $q\surd$ implies $q =_{Er} nil$. Hence $q; r =_{Er} nil; \mathtt{h}(r) =_{Er} \mathtt{h}(r)$.

If **(ii)** holds then, by induction, $q =_E \delta$ or $q =_{Er} \mathtt{h}(q) \equiv \sum \mu_i; q_i$. If $q =_E \delta$ then apply (D2) to obtain a hnf. Otherwise $q; r =_{Er} \mathtt{h}(q); r$. Recall that all hnf's are weakly convergent. Then

$$\begin{aligned} \mathtt{h}(q); r \equiv (\textstyle\sum \mu_i; q_i); r &=_{Er} \textstyle\sum(\mu_i; q_i); r && \text{by repeated use of (B4)} \\ &=_{Er} \textstyle\sum \mu_i; (q_i; r) && \text{by repeated use of (B3).} \end{aligned}$$

Assume that there exists j such that $\mu_j = \tau$. Then $\mathtt{h}(q); r \stackrel{\tau}{\longrightarrow} (nil; q_j); r$. By hypothesis, there exists a hnf $\bar{h}$ such that $\bar{h} =_{Er} (nil; q_j); r =_{Er} q_j; r$, by (B3) and (B1). Thus each $q_j; r$ such that $\mu_j = \tau$ has a hnf $\bar{h}_j$. Let $J = \{j \mid \mu_j = \tau\}$. Then

$$q; r =_{Er} \sum_{j \in J} \tau; \bar{h}_j + \sum_{j \not\in J} \mu_j; (q_j; r),$$

which is a hnf.

$p \equiv rec\, x.\ t$ Then $(rec\, x.\ t) \downarrow$ iff $t[rec\, x.\ t/x] \downarrow$. By the inductive hypothesis, there exists a hnf h such that

$$t[rec\, x.\ t/x] =_{Er} h.$$

Thus, by axiom (Rec), $rec\, x.\ t =_{Er} t[rec\, x.\ t/x] =_{Er} h$.

The other cases can be checked as in Proposition 2.3.3. □

A standard ingredient in the proofs of the equational characterization of bisimulation type relations is the so called "derivation lemma", [HM85], [H88b], [Wal87]. We also need such a lemma and, although it is not strictly necessary, it will be convenient to extend the set of equations with

$$\begin{array}{llcl} \textbf{(X1)} & (x+\mu;y)|z & = & (x+\mu;y)|z+\mu;(y|z) \\ \textbf{(X2)} & (x+a;y)|(z+\bar{a};w) & = & (x+a;y)|(z+\bar{a};w)+\tau;(y|w) \\ \textbf{(X3)} & z|(x+\mu;y) & = & z|(x+\mu;y)+\mu;(z|y). \end{array}$$

These equations are satisfied by the interpretation CI_E and so including them is harmless. Let F denote the extended set of equations obtained by augmenting E with axioms (X1), (X2), (X3) and (Rec).

Proposition 2.3.5 *For $p \in REC_\Sigma$, $p \stackrel{\mu}{\Longrightarrow} q$ implies $p =_F p + \mu; q$.*

Proof: By induction on the derivation $p \stackrel{\mu}{\Longrightarrow} q$.

Basis: $p \stackrel{\mu}{\longrightarrow} q$. We proceed by a sub-induction on why $p \stackrel{\mu}{\longrightarrow} q$. Most cases are straightforward, so we just give the proof for a few selected cases in which the auxiliary axioms are used.

- $p \equiv p_1|p_2$, $p_1 \stackrel{\mu}{\longrightarrow} q_1$ and $q \equiv q_1|p_2$. By the sub-inductive hypothesis, $p_1 =_F p_1 + \mu; q_1$. Hence

$$\begin{array}{lll} p_1|p_2 & =_F & (p_1+\mu;q_1)|p_2 \\ & =_F & (p_1+\mu;q_1)|p_2+\mu;(q_1|p_2) \quad \text{by (X1)} \\ & =_F & p_1|p_2+\mu;(q_1|p_2). \end{array}$$

- $p \equiv p_1|p_2$, $p_1 \stackrel{a}{\longrightarrow} q_1$ and $p_2 \stackrel{\bar{a}}{\longrightarrow} q_2$. Then, by the sub-inductive hypothesis, $p_1 =_F p_1 + a; q_1$ and $p_2 =_F p_2 + \bar{a}; q_2$. Thus

$$\begin{array}{lll} p_1|p_2 & =_F & (p_1+a;q_1)|(p_2+\bar{a};q_2) \\ & =_F & (p_1+a;q_1)|(p_2+\bar{a};q_2)+\tau;(q_1|q_2) \quad \text{by (X2)} \\ & =_F & p_1|p_2+\tau;(q_1|q_2). \end{array}$$

- $p \equiv rec\, x.\, t$ and $t[rec\, x.\, t/x] \stackrel{\mu}{\longrightarrow} q$. By the sub-inductive hypothesis, $t[rec\, x.\, t/x] =_F t[rec\, x.\, t/x] + \mu; q$. The claim now follows by axiom (Rec).

Inductive step. We distinguish two cases:

(i) $p \stackrel{\mu}{\longrightarrow} p' \stackrel{\tau}{\Longrightarrow} q$. By induction, $p =_F p + \mu;p'$ and $p' =_F p' + \tau;q$. Thus

$$\begin{aligned} p &=_F p + \mu;(p' + \tau;q) \\ &=_F p + \mu;(p' + \tau;q) + \mu;q \quad \text{by (T3)} \\ &=_F p + \mu;q. \end{aligned}$$

(ii) $p \stackrel{\tau}{\longrightarrow} p' \stackrel{\mu}{\Longrightarrow} q$. By induction, $p =_F p + \tau;p'$ and $p' =_F p' + \mu;q$. Thus

$$\begin{aligned} p &=_F p + \tau;(p' + \mu;q) \\ &=_F p + \tau;(p' + \mu;q) + p' + \mu;q \quad \text{by (T2)} \\ &=_F p + \tau;p' + p' + \mu;q \\ &=_F p + \tau;p' + \mu;q \quad \text{by (T2)} \\ &=_F p + \mu;q. \end{aligned}$$

This completes the inductive argument. □

To prove the partial completeness result for $\leq_E$ over $FREC_\Sigma \times REC_\Sigma$ we need one further technical result. This is an adaptation of a result originally shown for observational equivalence, [Mil88], and further adapted in [Wal87] to prebisimulations.

Lemma 2.3.5 *For each $p, q \in REC_\Sigma$, $p \lesssim q$ implies $p \lesssim^\star q$ or $\tau;p \lesssim^\star q$ or $p \lesssim^\star \tau;q$.*

Proof: Assume $p, q \in REC_\Sigma$ and $p \lesssim q$. There are three possibilities:

(i) $\exists p' : p \stackrel{\tau}{\longrightarrow} p'$ and $p' \lesssim q$,

(ii) $\exists q' : q \stackrel{\tau}{\longrightarrow} q'$ and $p \lesssim q'$,

(iii) neither (i) nor (ii) holds.

If (i) holds then we show that $p \lesssim^\star \tau;q$. This follows easily by noting that now the move $p \stackrel{\tau}{\longrightarrow} p'$ matches $\tau;q \stackrel{\tau}{\longrightarrow} nil;q \simeq^\star q$ (where $\simeq^\star$ denotes the kernel of $\lesssim^\star$). Moreover, $q\surd\!\!/$ iff $\tau;q\surd\!\!/$.

If (ii) holds then one may show that $\tau;p \lesssim^\star q$. This follows easily from an argument which is based on the following observations:

- $\tau;p \Downarrow \tau$ implies $p \Downarrow \tau$. Then, as $p \lesssim q$, $q \Downarrow \tau$. Moreover, the move $q \stackrel{\tau}{\longrightarrow} q'$ matches $\tau;p \stackrel{\tau}{\longrightarrow} nil;p \simeq^\star p$.
- $\tau;p \Downarrow$ iff $p \Downarrow$. Moreover $\tau;p\surd\!\!/$ iff $p\surd\!\!/$.

If (iii) holds then it is easy to show that $p \lesssim^\star q$. □

The above lemma will be used in the proof of relative completeness to relate processes with respect to $\lesssim$ and $\lesssim^\star$, i. e. in each case in which we know that $p \lesssim q$ we will distinguish three cases:

- $p \lesssim^\star q$,
- $\tau; p \lesssim^\star q$ and
- $p \lesssim^\star \tau; q$.

The proof of completeness will use induction over a binary relation $\prec\!\!\prec$ over $FREC_\Sigma \times REC_\Sigma$. The relation $\prec\!\!\prec$ is defined as follows:

Definition 2.3.4 *For each* $(d,p),(d',p') \in FREC_\Sigma \times REC_\Sigma$, $(d,p) \prec\!\!\prec (d',p')$ *iff*

(i) $\mathtt{dt}(d) < \mathtt{dt}(d')$, *or*
(ii) $\mathtt{dt}(d) = \mathtt{dt}(d')$, $p' \Downarrow$ and $p' \stackrel{\tau}{\Longrightarrow} p$.

To apply induction over $\prec\!\!\prec$ we need to know that $\prec\!\!\prec$ is a *well founded* relation over $FREC_\Sigma \times REC_\Sigma$, i.e. for each $(d,p) \in FREC_\Sigma \times REC_\Sigma$ there does not exist an infinite descending chain $(d,p) \succ\!\!\succ (d_0,p_0) \succ\!\!\succ (d_1,p_1) \ldots$. However, the well-foundedness of $\prec\!\!\prec$ is an easy consequence of the finiteness of $\mathtt{dt}(d)$, $d \in FREC_\Sigma$, the depth of the derivation tree of d, and of the fact that $p \Downarrow$ implies $p \stackrel{\omega}{\not\longrightarrow}$. We have now got all the technical machinery that we need to prove the partial completeness result.

Theorem 2.3.7 (Partial Completeness) *For each* $d \in FREC_\Sigma$, $p \in REC_\Sigma$, $d \lesssim^\star p$ *implies* $[\![d]\!]_E \leq [\![p]\!]_E$.

Proof: First note that the interpretation CI_E satisfies all of the equations in F. As a result $p \leq_F q$ implies $[\![p]\!]_E \leq [\![q]\!]_E$ and so it is sufficient to show that $d \lesssim^\star p$ implies $d \leq_F p$. The proof is by induction over $\prec\!\!\prec$. By Proposition 2.3.3 we may assume that d is a normal form. By definition of normal form, d is either δ or $\sum \mu_i; d_i\{+\Omega\}$.

If $d \equiv \delta$ then $\delta \lesssim^\star p$ implies $p \stackrel{\mu}{\not\longrightarrow}$, for each $\mu \in Act_\tau$, $p \notin \surd$ and $p \Downarrow$. Then, by Proposition 2.3.4, there exists a hnf h such that $h =_{E_r} p$. By the above observations we get that, as $=_{E_r}$ is sound with respect to $\simeq^\star$, it must be the case that $h \equiv \delta$. Hence $p =_{E_r} \delta$. This implies $p =_F \delta$.

Thus we may assume that $d \equiv \sum \mu_i; d_i\{+\Omega\}$. For technical reasons, which will be clear in the remainder of the proof, it will be convenient to isolate the case in which d is of the form $\tau; e$ for some nf e.

- $d \equiv \tau; e$. Since d is a normal form we can assume $d \Downarrow$. The proof of the statement $d \leq_F p$ is divided into two steps:

  (i) We prove that $\tau; e + p \leq_F p$. As $\tau; e \lesssim^\star p$ and $\tau; e \xrightarrow{\tau} nil; e \simeq^\star e$ then, as $e \Downarrow$, there exists p' such that $p \stackrel{\tau}{\Longrightarrow} p'$ and $e \lesssim p'$. By Lemma 2.3.5, $e \lesssim p'$ implies

  $$e \lesssim^\star p \text{ or } \tau; e \lesssim^\star p' \text{ or } e \lesssim^\star \tau; p'.$$

  It is easy to see that:

  $$\begin{array}{lll} (e, p') & \lll (\tau; e, p), & \text{as } \mathtt{dt}(e) < \mathtt{dt}(\tau; e), \\ (\tau; e, p') & \lll (\tau; e, p), & \text{as } p \Downarrow \text{ and } p \stackrel{\tau}{\Longrightarrow} p', \text{ and} \\ (e, \tau; p') & \lll (\tau; e, p) & \text{as } \mathtt{dt}(e) < \mathtt{dt}(\tau; e); \end{array}$$

  thus we may apply induction to obtain

  $$e \leq_F p' \text{ or } \tau; e \leq_F p' \text{ or } e \leq_F \tau; p'.$$

  In each case we obtain, by possibly applying axiom (T1), that $\tau; e \leq_F \tau; p'$. As $p \stackrel{\tau}{\Longrightarrow} p'$, we may apply Proposition 2.3.5 to deduce that $p + \tau; p' =_F p$. Hence $\tau; e + p \leq_F p + \tau; p' =_F p$.

  (ii) We prove $\tau; e \leq_F \tau; e + p$. As $p \Downarrow$, by Proposition 2.3.4 there exists a hnf h such that $p =_{E_{\Gamma}} h$. This obviously implies $p =_F h$. Moreover, as $\tau; e \lesssim^\star p$, h must have the form $\sum \mu_i; p_i$. We show that, for each i, $\tau; e \leq_F \tau; e + \mu_i; p_i$. We distinguish two cases:

  (1) $\mu_i = \tau$. Then $p \xrightarrow{\tau} nil; p_i \simeq^\star p_i$. As $\tau; e \Downarrow \tau$, there exists e' such that $\tau; e \stackrel{\tau}{\Longrightarrow} e'$ and $e' \lesssim p_i$. Once again, by Lemma 2.3.5, $e' \lesssim p_i$ implies

  $$e' \lesssim^\star p_i \text{ or } \tau; e' \lesssim^\star p_i \text{ or } e' \lesssim^\star \tau; p_i.$$

  Moreover, reasoning as in case (i), we may apply induction to obtain

  $$e' \leq_F p_i \text{ or } \tau; e' \leq_F p_i \text{ or } e' \leq_F \tau; p_i.$$

  In each case we get, by possibly applying (T1), that $\tau; e' \leq_F \tau; p_i$. By Proposition 2.3.5, $\tau; e =_F \tau; e + \tau; e' \leq_F \tau; e + \tau; p_i$.

  (2) $\mu_i = a$. We distinguish two subcases:

  (A) if $\tau; e \Downarrow a$ then there exists e' such that $\tau; e \stackrel{a}{\Longrightarrow} e'$ and $e' \lesssim p_i$. By Lemma 2.3.5, $e' \lesssim p_i$ implies

  $$e' \lesssim^\star p_i \text{ or } \tau; e' \lesssim^\star p_i \text{ or } e' \lesssim^\star \tau; p_i.$$

In each case $\mathtt{dt}(e') < \mathtt{dt}(\tau;e)$ and $\mathtt{dt}(\tau;e') < \mathtt{dt}(\tau;e)$; thus we may apply induction to obtain

$$e' \leq_F p_i \text{ or } \tau;e' \leq_F p_i \text{ or } e' \leq_F \tau;p_i.$$

In each case, by possibly applying (T1), $a;e' \leq_F a;p_i$. By Proposition 2.3.5, $\tau;e =_F \tau;e + a;e' \leq_F \tau;e + a;p_i$;

(B) if $\tau;e \Uparrow a$ then, by Fact 2.3.2 and the definition of $=_F$, $\tau;e \Downarrow$ and $\tau;e \Uparrow a$ imply $\tau;e =_F \tau;e + a;\Omega$. This implies that

$$\tau;e =_F \tau;e + a;\Omega \leq_F \tau;e + a;p_i.$$

This completes the proof of the fact that $\tau;e \leq_F \tau;e + \mu_i;p_i$, for each i.

Combining (i) and (ii) we get that $\tau;e \leq_F \tau;e + \sum \mu_i;p_i =_F \tau;e + p \leq_F p$, and this finishes the proof for the case $d \equiv \tau;e$.

- We now consider the general case, $d \equiv \sum \mu_i;d_i\{+\Omega\}$. To show that $d \leq_F p$ we follow the pattern used above.

  (i) We prove, first of all, that $\mu_i;d_i + p \leq_F p$, for each i. As d is a nf and $d \lesssim^\star p$, then $d \xrightarrow{\mu_i} nil;d_i \simeq^\star d_i$ implies that there exists p' such that $p \stackrel{\mu_i}{\Longrightarrow} p'$ and $d_i \lesssim p'$ (note that if $\mu_i = \tau$ then, as d is a nf, $d_i \Downarrow$). By Lemma 2.3.5, $d_i \lesssim p'$ implies

  $$d_i \lesssim^\star p' \text{ or } \tau;d_i \lesssim^\star p' \text{ or } d_i \lesssim^\star \tau;p'.$$

  If $d_i \lesssim^\star p'$ or $d_i \lesssim^\star \tau;p'$ we have that $\mathtt{dt}(d_i) < \mathtt{dt}(d)$; thus we may apply induction to obtain $d_i \leq_F p'$ or $d_i \leq_F \tau;p'$. If $\tau;d_i \lesssim^\star p'$ then we might not be able to apply induction (e.g. consider the case in which $\mathtt{dt}(d) = \mathtt{dt}(\tau;d_i)$ and $p \Uparrow$). However, by the case $d \equiv \tau;e$ examined above, we know that, as $\tau;d_i$ is itself a nf, $\tau;d_i \leq_F p'$. So in each of the three cases, by possibly applying (T1), we obtain $\mu_i;d_i \leq_F \mu_i;p'$. Hence $\mu_i;d_i + p \leq_F \mu_i;p' + p$ and, by Proposition 2.3.5, $p =_F p + \mu_i;p'$. Thus, for each i, $\mu_i;d_i + p \leq_F p$.

  (ii) We now show that $d \leq_F d + p$.

  - If Ω is a summand of d then $d =_F d + \Omega \leq_F d + p$ and we are done.
  - If Ω is not a summand of d then $d \Downarrow$. As $d \lesssim^\star p$ and $d \Downarrow$ we have that $p \Downarrow$. Thus, by Proposition 2.3.4, there exists a hnf $\sum \mu_i;p_i$ such that $p =_F \sum \mu_i;p_i$. As in case (ii) of the part of the proof concerning the case $d \equiv \tau;e$, we can show that, for each i, $d \leq_F d + \mu_i;p_i$. Thus $d \leq_F d + p$.

  Combining (i) and (ii) we get that $d \leq_F d + p \leq_F p$.

This completes the proof. □

2.4 Finite Approximability

This section is entirely devoted to proving that $\sqsubseteq_\omega^{fc}$ is finitely approximable. Using the fact that $\sqsubseteq_\omega^{fc}$ coincides with $\sqsubseteq_\omega^{+}$, it is quite easy to see that it is sufficient to establish that $\sqsubseteq_\omega$ is finitely approximable. This is carried out in two stages. The first consists in a modal characterization of $\sqsubseteq_\omega$; we define a set $\mathcal{L}$ of modal formulae, ϕ, and a satisfaction relation $\models$ with the property that

$$p \sqsubseteq_\omega q \text{ iff } \{\phi \mid p \models \phi\} \subseteq \{\phi \mid q \models \phi\}.$$

This is the topic of §2.4.1 and is a simple modification of similar results in [Mil81], [HM85], [St87], [Ab87b]. In §2.4.2 we show that satisfying a particular modal formula ϕ depends on a finite amount of information. More precisely, if $p \models \phi$ then there is a finite term $d(p, \phi)$ from $FREC_\Sigma$ such that $d(p, \phi) \models \phi$ and $d(p, \phi) \sqsubseteq_\omega p$. Intuitively, $d(p, \phi)$ represents the finite part of p which ensures that p satisfies ϕ. These two results are combined in the final theorem of the chapter, Theorem 2.4.3, which establishes that $\sqsubseteq_\omega$ and $\sqsubseteq_\omega^F$ coincide.

2.4.1 Modal Characterization of $\sqsubseteq_\omega$

We introduce a modal language which is a slight reformulation of the program logics introduced in [Mil81], [HM85], [St87], [Ab87b]. The added atomic formulae will reflect the extra "deadlock structure" which is present in the definition of our transition system semantics for REC_Σ.

Definition 2.4.1 *Let $\mathcal{L}$ be the least class of formulae generated by the following clauses:*

- ∇, Δ, T, $\bot \in \mathcal{L}$,
- ϕ, $\psi \in \mathcal{L}$ *imply* $\phi \wedge \psi$, $\phi \vee \psi \in \mathcal{L}$,
- $\alpha \in Act \cup \{\varepsilon\}$, $\phi \in \mathcal{L}$, *imply* $<\alpha> \phi$, $[\alpha]\phi \in \mathcal{L}$.

The metavariables $\phi, \psi, \varphi \ldots$ will range over $\mathcal{L}$.

The satisfaction relation $\models \subseteq REC_\Sigma \times \mathcal{L}$ is defined as follows:

$$\begin{array}{lllll}
p & \models & T & & \text{always} \\
p & \models & \bot & & \text{never} \\
p & \models & \nabla & \Longleftrightarrow & p \Downarrow \text{ and } p\surd\!\!/ \\
p & \models & \Delta & \Longleftrightarrow & p \Downarrow \text{ and } p \not\in \surd\!\!/ \\
p & \models & \phi \wedge \psi & \Longleftrightarrow & p \models \phi \text{ and } p \models \psi \\
p & \models & \phi \vee \psi & \Longleftrightarrow & p \models \phi \text{ or } p \models \psi \\
p & \models & <\alpha> \phi & \Longleftrightarrow & \exists q : p \stackrel{\alpha}{\Longrightarrow} q \text{ and } q \models \phi \\
p & \models & [\alpha]\phi & \Longleftrightarrow & p \Downarrow \alpha \text{ and } \forall q : p \stackrel{\alpha}{\Longrightarrow} q \; q \models \phi.
\end{array}$$

Here $p \stackrel{\varepsilon}{\Longrightarrow} q$ is used to mean $p \stackrel{\tau}{\Longrightarrow} q$ or $p \equiv q$; moreover, $p \Downarrow \varepsilon$ iff $p \Downarrow$. The set of modal formulae in $\mathcal{L}$ satisfied by a process $p \in REC_\Sigma$, $\mathcal{L}(p)$, is given by:

$$\mathcal{L}(p) =_{def} \{\phi \in \mathcal{L} \mid p \models \phi\}.$$

The *modal depth* of a formula ϕ, $\mathtt{md}(\phi)$, is defined by structural recursion as follows:

$$\mathtt{md}(T) = \mathtt{md}(\bot) = 0$$

$$\mathtt{md}(\nabla) = \mathtt{md}(\Delta) = 1$$

$$\mathtt{md}(\phi \wedge \psi) = \mathtt{md}(\phi \vee \psi) = max\{\mathtt{md}(\phi), \mathtt{md}(\psi)\}$$

$$\mathtt{md}(<\alpha> \phi) = \mathtt{md}([\alpha]\phi) = 1 + \mathtt{md}(\phi).$$

For each ϕ, $\mathtt{md}(\phi)$ measures the maximum depth of the nesting of modalities and atomic formulae Δ and ∇ in ϕ. For example, $\mathtt{md}(<a> ([b]T \vee [c][d]\nabla)) = 4$ and $\mathtt{md}(<a> \Delta \wedge [b]T) = 2$.

Theorem 2.4.1 (Modal Characterization Theorem) *For* $p, q \in REC_\Sigma$, $p \sqsubseteq_\omega q$ *iff* $\mathcal{L}(p) \subseteq \mathcal{L}(q)$.

Proof: Let $\mathcal{L}^{(n)}$ denote the subset of modal formulae $\phi \in \mathcal{L}$ such that $\mathtt{md}(\phi) \leq n$. To prove the "only if" implication it is sufficient to show that, for each $n \in \omega$, $\phi \in \mathcal{L}^{(n)}$,

$$p \sqsubseteq_n q \text{ and } p \models \phi \text{ imply } q \models \phi.$$

We proceed by induction on n. The claim is trivial for $n = 0$. Assume $p \sqsubseteq_{n+1} q$; we show that, for each $\phi \in \mathcal{L}^{(n+1)}$, $p \models \phi$ implies $q \models \phi$. The proof proceeds by structural induction over ϕ. We consider only two cases and leave the others to the reader.

$\phi \equiv \nabla$. Obviously $\nabla \in \mathcal{L}^{(n+1)}$. Assume $p \models \nabla$. By the definition of $\models$, $p \models \nabla$ iff $p \Downarrow$ and $p\surd\!\!/$. As $p \sqsubseteq_{n+1} q$, we have that $q \Downarrow$ and $q\surd\!\!/$. Thus $q \models \nabla$.

$\phi \equiv [\alpha]\psi$. Assume $p \models [\alpha]\psi$. Then, by the definition of $\models$, $p \Downarrow \alpha$ and, for each p' such that $p \stackrel{\alpha}{\Longrightarrow} p'$, $p' \models \psi$. As $p \sqsubseteq_{n+1} q$, $p \Downarrow \alpha$ implies $q \Downarrow \alpha$. Moreover, it is easy to see that $q \stackrel{\alpha}{\Longrightarrow} q'$ implies

$$\exists\, p' : p \stackrel{\alpha}{\Longrightarrow} p' \text{ and } p' \sqsubseteq_n q'.$$

By the outer inductive hypothesis, $q' \models \psi$, for each q' such that $q \stackrel{\alpha}{\Longrightarrow} q'$. Thus $q \models [\alpha]\psi$.

We now show the converse implication. In fact, we will show a stronger result which depends on the sort finiteness of the operational semantics for REC_Σ. Assume $p, q \in REC_\Sigma$. Let A be a finite subset of $Act \cup \{\varepsilon\}$ such that $\varepsilon \in A$ and $(Sort(p) \cup Sort(q)) - \{\tau\} \subseteq A$. Such an A exists by the sort finiteness of our transition system semantics for REC_Σ. For each $n \in \omega$, let $\mathcal{L}^{(A,n)}$ denote the subset of modal formulae ϕ over A of modal depth at most n. By induction on n we prove that

$$p \not\sqsubseteq_n q \text{ implies } \exists \phi \in \mathcal{L}^{(A,n)} : p \models \phi \text{ and } q \not\models \phi.$$

The base case is vacuous. Assume $p \not\sqsubseteq_{n+1} q$. Then one of the following cases occurs:

(1) $p \stackrel{\mu}{\Longrightarrow} p'$ and, for all q', $q \stackrel{\hat{\mu}}{\Longrightarrow} q'$ implies $p' \not\sqsubseteq_n q'$. By the inductive hypothesis, for each q' such that $q \stackrel{\hat{\mu}}{\Longrightarrow} q'$ there exists $\phi_{q'} \in \mathcal{L}^{(A,n)}$ such that $p \models \phi_{q'}$ and $q \not\models \phi_{q'}$. The set

$$\Gamma = \{\phi_{q'} \mid q \stackrel{\hat{\mu}}{\Longrightarrow} q'\}$$

might be infinite, e. g. if $q \Uparrow \mu$; however, as A is a finite set, $\mathcal{L}^{(A,n)}$ is finite up to logical equivalence, $\equiv_\lambda$. Hence $\Gamma/ \equiv_\lambda$ is finite as well and there exists a formula $\psi \in \mathcal{L}^{(A,n)}$ which is logically equivalent to $\bigwedge \Gamma$. Take $\phi = <\hat{\mu}> \psi$. Then $\mathtt{md}(\phi) \leq n+1$ and $p \models \phi$. On the other hand $q \not\models \phi$.

(2) $p \Downarrow \alpha$ and $q \Uparrow \alpha$. We distinguish two cases: if $q \stackrel{\alpha}{\Longrightarrow}$ then $[\alpha]T \in \mathcal{L}^{(A,n+1)}$ and $p \models [\alpha]T$ while $q \not\models [\alpha]T$. Otherwise it must be the case that $q \Uparrow$. Hence $p \models [\varepsilon]T$ while $q \not\models [\varepsilon]T$.

(3) $p \Downarrow \mu$, $q \Downarrow \mu$, $q \stackrel{\mu}{\Longrightarrow} q'$ and, for all p', $p \stackrel{\hat{\mu}}{\Longrightarrow} p'$ implies $p' \not\sqsubseteq_n q'$. By the inductive hypothesis, for each p' such that $p \stackrel{\hat{\mu}}{\Longrightarrow} p'$ there exists $\phi_{p'} \in \mathcal{L}^{(A,n)}$ such that $p' \models \phi_{p'}$ and $q' \not\models \phi_{p'}$. The set of formulae $\Gamma = \{\phi_{p'} \mid p \stackrel{\hat{\mu}}{\Longrightarrow} p'\}$ might be infinite, however, reasoning as above, we may deduce the existence of a formula $\psi \in \mathcal{L}^{(A,n)}$ which is logically equivalent to $\bigvee \Gamma$. Take

$$\phi = [\hat{\mu}]\psi.$$

It is easy to see that $\mathtt{md}(\phi) \leq n+1$ and $p \models \phi$. On the other hand, $q \not\models \phi$.

(4) $p \Downarrow$ and $\neg(p \surd\!\!\!/ \text{ iff } q \surd\!\!\!/)$. We may assume, wlog, that $q \Downarrow$, otherwise case (2) applies. There are two subcases to examine:

(a) $p \Downarrow, q \Downarrow, p\surd\!\!\surd$ and $q \notin \surd\!\!\surd$. Then $p \models \nabla$ whilst $q \not\models \nabla$.

(b) $p \Downarrow, q \Downarrow, p \notin \surd\!\!\surd$ and $q\surd\!\!\surd$. Then $p \models \Delta$ whilst $q \not\models \Delta$.

This completes the proof. □

2.4.2 Finitary Properties

Let us recall that the finitary part of the preorder $\sqsubseteq_\omega$, $\sqsubseteq_\omega^F$, is defined as follows:

$$p \sqsubseteq_\omega^F q \text{ iff, } \forall d \in FREC_\Sigma,\ d \sqsubseteq_\omega p \Longrightarrow d \sqsubseteq_\omega q.$$

As an easy consequence of Theorem 2.3.2 we have that $\sqsubseteq_\omega^F = \sqsubseteq^F$. Moreover, it is easy to establish that $p \sqsubseteq_\omega q$ implies $p \sqsubseteq_\omega^F q$. We now show that the converse implication also holds. As stated at the beginning of §2.4, the key point of the proof of the inclusion $\sqsubseteq_\omega^F \subseteq \sqsubseteq_\omega$ will be a construction that, given $p \in REC_\Sigma$ and $\phi \in \mathcal{L}(p)$, will generate a finite process $d(p,\phi)$ such that $d(p,\phi) \models \phi$ and $d(p,\phi) \sqsubseteq p$. In order to give the construction of $d(p,\phi)$ we need a technical definition which uses the precongruence $\leq_{Er}$ defined immediately before Proposition 2.3.4:

Definition 2.4.2 *Let $p \in REC_\Sigma$, $d \in FREC_\Sigma$ and suppose that $d \leq_{Er} p$. It follows that $CI_E[\![d]\!] \leq CI_E[\![p]\!]$. Since CI_E is an algebraic cpo, there exists some $n \geq 0$ such that $CI_E[\![d]\!] \leq CI_E[\![p^n]\!]$. Since both are finite terms, this implies $d \leq_E p^n$ and therefore $d \leq_{Er} p^n$. So, for d, $e \in FREC_\Sigma$ such that $d \leq_{Er} p$ and $e \leq_{Er} p$, let $d \vee e$ denote the least principal approximation p^k such that $d \leq_{Er} p^k$ and $e \leq_{Er} p^k$.*

This operator will be used in the construction of $d(p,\phi)$. In what follows we will consider $\mathcal{L}$ modulo logical equivalence. Note that, because of this assumption and the law

$$[\alpha] \bigwedge_{i \in I} \phi_i = \bigwedge_{i \in I} [\alpha]\phi_i,$$

when considering formulae of the form $[\alpha]\phi$ we can restrict ourselves to the cases in which ϕ is either T or has the form $\bigvee_{i\in I} \phi_i$, where each ϕ_i is either of the form $<\beta>\psi$ or $[\beta]\psi$, for some $\beta \in Act \cup \{\varepsilon\}$ and $\psi \in \mathcal{L}$.

The construction $d(-,-) : \{(p,\phi) \mid p \models \phi\} \longrightarrow FREC_\Sigma$ will be given by induction over the relation $\ll \subseteq (\mathcal{L} \times REC_\Sigma)^2$ defined as follows:

$$(\phi, q) \ll (\psi, p) \iff \begin{cases} (i)\ \mathtt{ht}(\phi) < \mathtt{ht}(\psi), \text{ or} \\ (ii)\ \mathtt{ht}(\phi) = \mathtt{ht}(\psi),\ p \Downarrow \text{ and } p \overset{\tau}{\Longrightarrow} q, \end{cases}$$

where the *height* of a formula ϕ, $\mathtt{ht}(\phi)$, is easily defined by structural recursion on ϕ, [Ab87b]. Of course, for the following inductive construction to make sense we have to ensure that $\ll$ is a well founded relation. However, this follows from the fact that $\mathtt{ht}(\phi)$ is finite for each $\phi \in \mathcal{L}$ and the fact that $p \Downarrow$ implies that $\{q \mid p \stackrel{\tau}{\Longrightarrow} q\}$ is finite. We can now state and prove the main theorem of this section.

Theorem 2.4.2 *For each $p \in REC_\Sigma$, $\phi \in \mathcal{L}(p)$, there exists a finite process $d(p, \phi)$ such that:*

(i) *$d(p,\phi) \models \phi$, and*

(ii) *$d(p,\phi) \leq_{Er} p$.*

Proof: The proof of the theorem is constructive. By induction on the relation $\ll$ we construct, for each $p \in REC_\Sigma$ and $\phi \in \mathcal{L}(p)$, a finite process which meets the statement of the theorem. We proceed by structural recursion on the formula ϕ.

$p \models \phi \equiv T$. Then $d(p,T) \equiv \Omega$.

$\phi \equiv \bot$. Vacuous.

$p \models \phi \equiv \nabla$. Recall that, by the definition of $\models$, $p \models \nabla$ iff $p \Downarrow$ and $p \surd\!\!\!/$. As $p \Downarrow$, $p =_{Er} h(p)$ for some hnf $h(p)$. By the soundness of $=_{Er}$, $p \simeq^{\star} h(p)$. Moreover, as $p \surd\!\!\!/$, $h(p) \equiv \sum_{i \in I} \tau; p_i + \sum_{j \in J} a_j; p_j$. As $\lesssim^{\star} \subseteq \sqsubseteq_{\omega}$, by the modal characterization theorem

$$\sum_{i \in I} \tau; p_i + \sum_{j \in J} a_j; p_j \models \nabla.$$

Take $d(p,\nabla) \equiv \sum_{i \in I} \tau; d(p_i, \nabla) + \sum_{j \in J} a_j; \Omega$.

- We show that $d(p,\nabla) \models \nabla$. By the inductive hypothesis, $d(p_i,\nabla) \models \nabla$, for each $i \in I$. This implies that $d(p_i,\nabla) \Downarrow$ and $d(p_i,\nabla) \surd\!\!\!/$, for each $i \in I$. Thus $d(p,\nabla) \Downarrow$ and $d(p,\nabla) \surd\!\!\!/$ (note that, as $p \models \nabla$, $I = \emptyset$ implies $J = \emptyset$).
- By induction $d(p_i,\nabla) \leq_{Er} p_i$, for each $i \in I$, from which it follows that $d(p,\nabla) \leq_{Er} h(p) =_{Er} p$.

$p \models \phi \equiv \Delta$. By the definition of $\models$, $p \models \Delta$ iff $p \Downarrow$ and $p \notin \surd\!\!\!/$. As $p \Downarrow$, we may assume, wlog, that p has either the form δ or $\sum_{i \in I} \tau; p_i + \sum_{j \in J} a_j; p_j$. If p is δ then $d(p,\Delta) \equiv \delta$.

If p is $\sum_{i \in I} \tau; p_i + \sum_{j \in J} a_j; p_j$ then partition I into $I_1 =_{def} \{i \in I \mid p_i \surd\!\!\!/\}$ and $I_2 =_{def} \{i \in I \mid p_i \notin \surd\!\!\!/\}$. Note that, as $p \notin \surd\!\!\!/$, $I \neq \emptyset$ implies $I_2 \neq \emptyset$. Then

$$d(p,\Delta) \equiv \sum_{i \in I_1} \tau; d(p_i,\nabla) + \sum_{i \in I_2} \tau; d(p_i,\Delta) + \sum_{j \in J} a_j; \Omega.$$

- We show that $d(p,\Delta) \models \Delta$. The claim is trivial when $I = \emptyset$. Assume now $I \neq \emptyset$. Then $I_2 \neq \emptyset$; thus $d(p,\Delta) \xrightarrow{\tau} nil; d(p_k,\Delta)$, for some $k \in I_2$. By the inductive hypothesis, $d(p_k,\Delta) \models \Delta$. Thus $d(p_k,\Delta) \notin \surd\!\!\!/$. It is easy to see that $d(p,\Delta) \Downarrow$; in fact, by the inductive hypothesis, for each $i \in I_1 \cup I_2$, $d(p_i,\star) \Downarrow$, $\star \in \{\Delta,\nabla\}$. Thus $d(p,\Delta) \models \Delta$.

- To prove that $d(p,\Delta) \leq_{Er} p$ it is sufficient to note that, by induction, $d(p_i,\nabla) \leq_{Er} p_i$, for each $i \in I_1$, and $d(p_i,\Delta) \leq_{Er} p_i$, for each $i \in I_2$.

$p \models \phi \equiv \phi_1 \vee \phi_2$. By the definition of $\models$, $p \models \phi_1 \vee \phi_2$ iff $p \models \phi_1$ or $p \models \phi_2$. Assume, wlog, that $p \models \phi_1$. Then $d(p,\phi) \equiv d(p,\phi_1)$. Both the statements of the theorem then follow by induction.

$p \models \phi_1 \wedge \phi_2$. Then $p \models \phi_1$ and $p \models \phi_2$. Take $d(p,\phi) \equiv d(p,\phi_1) \vee d(p,\phi_2)$. We show that $d(p,\phi) \models \phi$. By the inductive hypothesis, $d(p,\phi_i) \models \phi_i$, $i = 1,2$. By the definition of $\vee$, $d(p,\phi_i) \leq_{Er} d(p,\phi_1) \vee d(p,\phi_2)$, $i = 1,2$. As $\sqsubseteq$ and $\sqsubseteq_\omega$ coincide over $FREC_\Sigma \times REC_\Sigma$ by Theorem 2.3.2, $d(p,\phi_i) \sqsubseteq_\omega d(p,\phi_1) \vee d(p,\phi_2)$, $i = 1,2$. By the modal characterization theorem, $d(p,\phi_1) \vee d(p,\phi_2) \models \phi_i$, $i = 1,2$. Hence $d(p,\phi) \models \phi_1 \wedge \phi_2$.

$p \models <\alpha> \phi$. By the definition of $\models$, $p \models <\alpha> \phi$ iff there exists q such that $p \stackrel{\alpha}{\Longrightarrow} q$ and $q \models \phi$. Then:

- if $\alpha = \varepsilon$ then we distinguish two possibilities depending on whether $p \equiv q$ or $p \stackrel{\tau}{\Longrightarrow} q$. If $p \equiv q$ then $d(p, <\varepsilon> \phi) \equiv d(p,\phi)$. Otherwise, we set $d(p, <\varepsilon> \phi) \equiv \tau; d(q,\phi) + \Omega$;
- if $\alpha = a$ then $d(p, <a> \phi) \equiv a; d(q,\phi) + \Omega$.

In both cases, it is easy to see that both the statements of the theorem are met by $d(p, <\alpha> \phi)$. The details are omitted.

$p \models [\varepsilon]\phi$. By the definition of $\models$, $p \models [\varepsilon]\phi$ iff $p \Downarrow$ and, for each q such that $p \stackrel{\varepsilon}{\Longrightarrow} q$, $q \models \phi$. As $p \Downarrow$, we may assume, wlog, that p is a hnf. If p is δ then $d(p,[\varepsilon]\phi) \equiv \delta$ and both the statements are trivially seen to hold.

Assume now that $p \equiv \sum_{i \in I} \tau; p_i + \sum_{j \in J} a_j; p_j$. As previously remarked, we may assume that up to logical equivalence ϕ is either T or is of the form $\bigvee_{h \in H} \phi_h$, where each ϕ_h is of the form $<\alpha> \psi$ or $[\alpha]\psi$, for some $\alpha \in Act \cup \{\varepsilon\}$ and $\psi \in \mathcal{L}$.

- If ϕ is T then $d(p,[\varepsilon]T) \equiv \sum_{i \in I} \tau; d(p_i,[\varepsilon]T) + \sum_{j \in J} a_j; \Omega$. To see that $d(p,[\varepsilon]T) \models [\varepsilon]T$ it is sufficient to prove that $d(p,[\varepsilon]T) \Downarrow$. This follows from the fact that, by the inductive hypothesis, $d(p_i,[\varepsilon]T) \models [\varepsilon]T$ and thus $d(p_i,[\varepsilon]T) \Downarrow$, for each $i \in I$. The proof of the fact that $d(p,[\varepsilon]T) \leq_{Er} p$ is routine and is omitted.

- If $\phi \equiv \bigvee_{h \in H} \phi_h$ then $p \models \bigvee_{h \in H} \phi_h$. This is because there exists $k \in H$ such that $p \models \phi_k$. We proceed by analyzing the form of ϕ_k:

    $\phi_k \equiv [a]\psi$. Then $p \Downarrow a$ and, for each q such that $p \stackrel{a}{\Rightarrow} q$, $q \models \psi$. Take $d(p, [\varepsilon]\phi) \equiv d(p, [a]\psi)$. First of all, note that $d(p, [a]\psi) \leq_{Er} p$ follows immediately by the inductive hypothesis. We now show that $d(p, [a]\psi) \models [\varepsilon] \bigvee_{h \in H} \phi_h$. Of course $d(p, [a]\psi) \Downarrow$ as, by the inductive hypothesis, $d(p, [a]\psi) \models [a]\psi$. Assume now that $d(p, [a]\psi) \stackrel{\varepsilon}{\Rightarrow} x$. If $x \equiv d(p, [a]\psi)$ then, by the inductive hypothesis, $d(p, [a]\psi) \models [a]\psi = \phi_k$. Hence $d(p, [a]\psi) \models \bigvee_{h \in H} \phi_h$. If $d(p, [a]\psi) \stackrel{\tau}{\Rightarrow} x$ then it is easy to see that $x \models [a]\psi$ as well. Thus $x \models \bigvee_{h \in H} \phi_h$. This establishes that $d(p, [a]\psi) \models [\varepsilon] \bigvee_{h \in H} \phi_h$.

    $\phi_k \equiv <a> \psi$. Then $p \models <a> \psi$ iff there exists $j_1 \in J$ such that $a_{j_1} = a$ and $p_{j_1} \stackrel{\varepsilon}{\Rightarrow} q \models \psi$, for some q. Take

    $$d(p, \phi) \equiv \sum_{i \in I} \tau; d(p_i, \phi) + \sum_{j: a_j \neq a} a_j; \Omega + a; d(p_{j_1}, <\varepsilon> \psi) + a; \Omega.$$

    Note that, by construction and the inductive hypothesis, $d(p, \phi) \leq_{Er} p$. We show that $d(p, \phi) \models [\varepsilon] \bigvee_{h \in H} \phi_h$. Obviously $d(p, \phi) \Downarrow$ as, by the inductive hypothesis, $d(p_i, \phi) \Downarrow$, for each $i \in I$. Assume that $d(p, \phi) \stackrel{\varepsilon}{\Rightarrow} x$. Then there are two cases to examine:

    (A) $x \equiv d(p, \phi)$. Then $d(p, \phi) \stackrel{a}{\rightarrow} nil; d(p_{j_1}, <\varepsilon> \psi)$. As by the inductive hypothesis $d(p_{j_1}, <\varepsilon> \psi) \models <\varepsilon> \psi$, we have that $d(p, \phi) \models <a> \psi$ and this implies $d(p, \phi) \models \bigvee_{h \in H} \phi_h$.

    (B) $d(p_i, \phi) \stackrel{\varepsilon}{\Rightarrow} x$ for some $i \in I$. By the inductive hypothesis, $d(p_i, \phi) \models \phi$ and this implies $x \models \bigvee_{h \in H} \phi_h$.

    Hence $d(p, \phi) \models \phi$.

    $\phi_k \equiv [\varepsilon]\psi$. Take $d(p, \phi) \equiv d(p, [\varepsilon]\psi)$. Proving that $d(p, [\varepsilon]\psi)$ meets the statement of the theorem is done exactly as in the case $\phi_k \equiv [a]\psi$.

    $\phi_k \equiv <\varepsilon> \psi$. Take

    $$d(p, \phi) \equiv d(p, <\varepsilon> \psi) \vee (\sum_{i \in I} \tau; d(p_i, \phi) + \sum_{j \in J} a_j; \Omega).$$

    Note that, by the inductive hypothesis, $d(p, <\varepsilon> \psi) \leq_{Er} p$ and, as $d(p_i, \phi) \leq_{Er} p_i$ $(i \in I)$, we also have that $\sum_{i \in I} \tau; d(p_i, \phi) + \sum_{j \in J} a_j; \Omega \leq_{Er} p$. Thus, by the definition of $\vee$, $d(p, \phi) \leq_{Er} p$ and we have checked the second part of the statement of the theorem.

    We are left to show that $d(p, \phi) \models \phi$. By the inductive hypothesis, $d(p_i, \phi) \models \phi$, for each $i \in I$. Thus, for each $i \in I$, $d(p_i, \phi) \Downarrow$. By the definition of $\Downarrow$, this implies that $(\sum_{i \in I} \tau; d(p_i, \phi) + \sum_{j \in J} a_j; \Omega) \Downarrow$ and therefore that $d(p, \phi) \Downarrow$.

Moreover, $d(p, <\varepsilon> \psi) \equiv d(p, \phi_k) \models \phi_k$, by the inductive hypothesis. As $d(p, \phi_k) \sqsubseteq_\omega d(p, \phi)$, by the modal characterization theorem we get that $d(p, \phi) \models \phi_k$. Hence, by the definition of $\models$, $d(p, \phi) \models \bigvee_{h \in H} \phi_h$.

Assume now that $d(p, \phi) \stackrel{\tau}{\Longrightarrow} y$. Then, as $\sum_{i \in I} \tau; d(p_i, \phi) + \sum_{j \in J} a_j; \Omega \sqsubseteq d(p, \phi)$, there exists some $i \in I$ and x such that $d(p_i, \phi) \stackrel{\varepsilon}{\Longrightarrow} x$ and $x \sqsubseteq y$. As $d(p_i, \phi) \models [\varepsilon] \bigvee_{h \in H} \phi_h$ we have that $x \models \bigvee_{h \in H} \phi_h$. By the modal characterization theorem, the finiteness of x and y and Theorem 2.3.2, this implies $y \models \bigvee_{h \in H} \phi_h$. Thus we have shown that $d(p, \phi) \models \phi$.

The proof of the case $\phi \equiv [\varepsilon] \bigvee_{h \in H} \phi_h$ is thus complete.

$p \models [a]\phi$. As $p \models [a]\phi$ we have that $p \Downarrow a$. This implies that $p \Downarrow$ and thus we may assume, wlog, that p is a hnf. If $p \equiv \delta$ then $d(p, [a]\phi) \equiv \delta$.

If $p \equiv \sum_{i \in I} \tau; p_i + \sum_{j \in J} a_j; p_j$ then

$$d(p, [a]\phi) \equiv \sum_{i \in I} \tau; d(p_i, [a]\phi) + \sum_{j : a_j = a} a; d(p_j, [\varepsilon]\phi) + \sum_{j : a_j \neq a} a_j; \Omega.$$

Both the statements of the theorem are easily seen to hold when $d(p, [a]\phi) \equiv \delta$. Assume now that the other case occurs. First of all note that $d(p, [a]\phi)$ is well defined as

- $p \models [a]\phi$ implies $p_i \models [a]\phi$, for each $i \in I$, and
- $p \models [a]\phi$ implies $p_j \models [\varepsilon]\phi$, for each $j \in J$ such that $a_j = a$.

By the inductive hypothesis, $d(p_i, [a]\phi) \models [a]\phi$, for each $i \in I$, and $d(p_j, [\varepsilon]\phi) \models [\varepsilon]\phi$, for each $j \in J$ such that $a_j = a$. Thus, by the definition of $\models$, $d(p_i, [a]\phi) \Downarrow a$ and $d(p_j, [\varepsilon]\phi) \Downarrow$. It is easy to see that this implies $d(p, [a]\phi) \Downarrow a$. Assume now that $d(p, [a]\phi) \stackrel{a}{\Longrightarrow} x$. Then either there exists $i \in I$ such that $d(p_i, [a]\phi) \stackrel{a}{\Longrightarrow} x$ or, for some j such that $a_j = a$, $d(p_j, [\varepsilon]\phi) \stackrel{\varepsilon}{\Longrightarrow} x$. In both cases, by the inductive hypothesis and the definition of $\models$, we get that $x \models \phi$. We have thus shown that $d(p, [a]\phi) \models [a]\phi$.

Finally note that, by the inductive hypothesis, $d(p_i, [a]\phi) \leq_{Er} p_i$ and $d(p_j, [\varepsilon]\phi) \leq_{Er} p_j$, for each $i \in I$ and j such that $a_j = a$. Therefore, by construction, $d(p, [a]\phi) \leq_{Er} p$. This completes the proof of the theorem. □

Example 2.4.1 *As an example of application of the construction of $d(p, \phi)$ given in the above proof, we shall give $d(p, \phi)$ for $p \equiv rec\, x.\, (x; a + a)$ and $\phi \equiv \phi_1 \wedge \phi_2$, with $\phi_1 \equiv <a> [a] \perp$ and $\phi_2 \equiv <a><a> T$. It is easy to see that $p \models \phi$ as*

(a) $p \models \phi_1$ *because* $p \stackrel{a}{\longrightarrow} nil \models [a] \perp$, *and*

(b) $p \models \phi_2$ *because* $p \stackrel{a}{\longrightarrow} nil; a \models <a> T$.

By the construction of the above theorem, $d(p, \phi_1) \equiv a; nil + \Omega$ *and* $d(p, \phi_2) \equiv a; (a; \Omega + \Omega) + \Omega$. *Hence*

$$d(p, \phi) \equiv d(p, \phi_1) \vee d(p, \phi_2) = p^2 =_{Er} \Omega + a + a; a. \quad \square$$

After this rather delicate and lengthy proof we have all the technical machinery that is needed to prove that the preorder $\precsim_\omega$ is finitely approximable over REC_Σ.

Theorem 2.4.3 *For each* $p, q \in REC_\Sigma$, $p \precsim_\omega q$ *iff* $p \precsim_\omega^F q$.

Proof: We have already recalled that $p \precsim_\omega q$ implies $p \precsim_\omega^F q$. We now show that the converse implication also holds. Assume that $p \precsim_\omega^F q$ and $p \models \phi$, $\phi \in \mathcal{L}$. Then, by Theorem 2.4.2, $d(p, \phi) \models \phi$ and $d(p, \phi) \leq_{Er} p$. By the soundness of $\leq_{Er}$ with respect to $\precsim_\omega$, $d(p, \phi) \precsim_\omega p$.

As $p \precsim_\omega^F q$, $d(p, \phi) \precsim_\omega q$. As, by Theorem 2.4.2, $d(p, \phi) \models \phi$, by the modal characterization theorem $q \models \phi$. Thus $\mathcal{L}(p) \subseteq \mathcal{L}(q)$ and, by the modal characterization theorem, this implies that $p \precsim_\omega q$. $\square$

Because of the coincidence of $\precsim$ and $\precsim_\omega$ over $FREC_\Sigma \times REC_\Sigma$ we have that $\precsim_\omega$ is indeed the finitary part of the prebisimulation preorder $\precsim$, $\precsim^F$.

2.5 Concluding Remarks

In this chapter we have developed a semantic theory for a process algebra which incorporates some explicit representation of successful termination, deadlock and divergence. The process algebra that we have considered has been endowed with both an operational and a denotational semantics and the two semantic views of processes have been shown to agree. Namely, we have shown that the denotational model that we have proposed, the initial continuous algebra which satisfies a set of equations CI_E, is fully abstract with respect to a natural operational preorder over the language. The proof of the full abstraction theorem relies on several results of independent interest; namely the finite approximability of the behavioural preorder and a partial completeness result for the set of inequations E with respect to the preorder. The proof of the finite approximability of the behavioural preorder is one of the main novelties of the chapter; it relies on a characterization of the behavioural preorder in terms of a modal logic and makes a fundamental use of a novel construction which produces, for each term p and modal formula ϕ satisfied by p, a finite approximant of p, $d(p, \phi)$, which satisfies ϕ. We believe that the pattern followed in the proof of Theorem 2.4.3 provides a general technique to establish the finite approximability of bisimulation-like preorders which afford logical characterizations in terms of modal logics allowing finite conjunctions and disjunctions only.

As pointed out in [H81], [H88a], our choice of a denotational semantics for the language studied in this chapter gives us a complete axiomatic proof system (albeit a non recursively enumerable one) for closed terms of the language. Moreover, as our denotational model is based upon the well known theory of algebraic cpo's, rather than metric spaces as in [BK82], we may obtain effective proof systems for the language by using induction rules such as Scott Induction and Fixed-Point Induction, [LS87]. Other advantages of using the theory of cpo's rather than metric spaces are that all of the usual operators found in process algebras may be readily interpreted, as no restriction need be placed on the recursive definitions allowed in the language REC_{Σ}, and that features like silent actions and encapsulation/abstraction operators may be smoothly dealt with within it. On the other hand, as pointed out in §2.1, using metric spaces we can only readily interpret operators which are *contractive* and this requirement imposes restrictions on their applicability. For instance, it is well-known that unguarded recursive definitions give rise to operators which are not contractive. Moreover, essential features of process algebras like silent actions and abstraction/encapsulation operators have never been dealt with satisfactorily in this framework.

The language we have considered in this chapter incorporates features from **CCS** and **ACP**. It extends **ACP** by allowing an explicit representation of successful termination and divergence; moreover, our language allows for general recursive definitions. The auxiliary operators which **ACP** uses to axiomatize $|$ (namely, $\not|$, for *left-merge*, and $|_c$, for the *communication merge*) could be added to the language without affecting our results. In this presentation, we have omitted the treatment of these operators for reasons of simplicity and because we shall not have much use for them in the remainder of this thesis. The only exception will be Chapter 3, where the left-merge operator is used for axiomatization purposes over a simple sublanguage of REC_{Σ}. The language REC_{Σ} extends **CCS** as it allows general sequential composition and an explicit representation of deadlock (as opposed to successful termination). However, the signature of **CCS** contains a family of *relabelling operators* $\_[R]$, where $R : Act_{\tau} \longrightarrow Act_{\tau}$ is a function such that $R(\bar{a}) = \overline{R(a)}$ and $R(\tau) = \tau$. The introduction of such an operator in the signature of our language would cause some problems. To see this, we recall that our results about the finite approximability of the behavioural preorder $\sqsubseteq^{c}_{\omega}$ depend on the sort-finiteness of our transition system semantics for the language (see §2.4). However, if Act is infinite this is no longer the case. To show this, consider an enumeration $\{a_0, a_1, \ldots, a_i, \ldots\}$ of the set of observable actions Act. Using the enumeration of Act, we may define a relabelling S such that $S(a_i) = a_{i+1}$, for each $i \in \omega$. Take the process p defined as follows, [Ab87b]:

$$p \equiv rec\, x.\ a_0 + x[S].$$

Then it is easy to see that the unguarded recursive definition and the generality of S give rise to a process which is not sort-finite. In fact, $p \xrightarrow{a_i}$ for each $i \in \omega$. As a consequence of these observations, our behavioural preorder would not be finitely approximable and CI_E would not be fully abstract with respect to it.

However, it may be argued that one rarely, if ever, needs relabelling operators of such a generality. In practice, relabelling functions are usually assumed to be constant on all but finitely many actions in Act_τ. If we allow only this kind of relabellings in our language then the resulting transition system semantics will again be sort-finite, [Ab87a,b], and thus all of the results of this chapter will carry through to this extended language.

An interesting point to note is that most of the technical analysis of our operational preorder $\sqsubseteq^c_\omega$ has been carried out by using $\lesssim$ and $\lesssim^\star$. As $\lesssim$ is technically simpler than $\sqsubseteq$, it might be that some of the results of the chapter could have been obtained in a simpler way by using $\lesssim^c_\omega$ as behavioural preorder. However, this is not the case. It turns out that no denotational model of the form $CI_{E'}$, for any set of equations E', can be fully abstract with respect to $\lesssim^c_\omega$. This is because in such a model all the syntactically finite terms, i. e. terms from $FREC_\Sigma$, are interpreted as semantically finite elements. That is if $d \in FREC_\Sigma$ and $d \leq_{CI_{E'}} p$ then, for some finite approximation of p, p^n, $d \leq_{CI_{E'}} p^n$. This property does not hold of $\lesssim^c_\omega$ and so $\lesssim^c_\omega$ can coincide with $\leq_{CI_{E'}}$ for no set of equations E'. As a counterexample, consider the two synchronization trees:

$$\begin{aligned} d &\equiv b; a + \Omega \\ p &\equiv \sum_{k \geq 1} b; q(k) + \Omega, \end{aligned}$$

where

$$\begin{aligned} q(1) &\equiv a + \tau; a; c \\ q(k+1) &\equiv a + \tau; q(k), \ \ k \geq 1. \end{aligned}$$

Note that, for each $k \geq 1$, $a \lesssim_k q(k)$ and therefore $d \lesssim_\omega p$. The finite approximations to p are all of the form

$$p^m \equiv \sum_{1 \leq k \leq m} b; q(k) + \Omega.$$

However, for each $m \geq 1$, $d \not\lesssim_\omega p^m$. In fact, $d \not\lesssim_{m+3} p^m$ because, for each k, $a \not\lesssim_{k+2} q(k)$.

We end this section with a brief comparison with related work. Several term model constructions, [Mil77], for **CCS** and **SCCS**-like languages have been proposed in the literature. See for example [HP80], [H81]. In each of these papers, a denotational semantics is given to the languages considered by means of the initial continuous algebra which satisfies a set of equations E. The denotational model is then shown to be fully abstract with respect to a behavioural preorder. In [DH84] the authors show how, for the *testing equivalences* they introduce, the denotational models have a natural representation in terms of a particular class of trees, the *acceptance trees* of [H85]. In [Ab87b], the author takes a language-independent standpoint and analyzes the general relationships between strong prebisimulation, $\sqsubseteq$, over transition systems and its finitary part, $\sqsubseteq^F$. The author also shows how his general results may be used to obtain a fully abstract model with respect to (the finitary part of) strong prebisimulation over a version of **SCCS**, [Mil83], with only finite summations and relates his model to the one in [H81]. In [Wal87] a behavioural relation similar to $\lesssim$ is studied and applied to **CCS**; complete axiomatizations are given for finite and regular processes. In many ways the work in the present chapter may be considered as an extension of this work, employing ideas from [HP80]. It provides the first comprehensive treatment of a weak version of prebisimulation and, in addition, it establishes a mathematical setting within which the notions of termination, divergence and deadlock may be compared and contrasted. Similar motivations are at the heart of [BKO87]. There the authors present several axiomatic systems to reason about successful termination, deadlock and divergence in the theories of both bisimulation and failures equivalence, [BHR84]. Models for the equational theories are exhibited, thus proving their logical consistency. Apart from a systematic analysis of axiom systems and semantic models dealing with the notions of abstraction and divergence, the paper presents a new failure semantics which allows *fair abstraction of unstable divergence.* This semantics does not always consider divergence as catastrophic, as it is done in, e.g., [BHR84], and a weak form of *Koomen's Fair Abstraction Rule*, [BBK87], holds for it. The theory presented in [BKO87], however, only deals with a language without parallel features and no completeness result, relating the axiomatic systems and the equivalences presented in the paper, is shown. In [Rou85], a modal logic similar to the one employed in §2.4 is used to construct an information system [Sco82] which generates a complete partial order of synchronization trees, [Mil80], $\mathcal{P}_c(T_\Sigma)$. The elements of $\mathcal{P}_c(T_\Sigma)$, called *forests*, are sets of synchronization trees closed with respect to strong observational equivalence [Mil80] and a suitable metric, [GR83]. Some operations, among which a general notion of sequential composition dealing with deadlock and successful termination, are defined over $\mathcal{P}_c(T_\Sigma)$ and used to give a denotational semantics for a **CSP**-like language. However, the paper, being mostly concerned with a study of the mathematical properties of the space $\mathcal{P}_c(T_\Sigma)$, does not attempt an operational justification of the denotational semantics or an equational characterization of the congruence induced by it.

The dichotomy deadlock/successful termination has been dealt with in a different fashion in **CSP** [BHR84], [Hoare85] and the latest papers on **ACP** [BG87a,b]. Both these process algebras introduce an explicit constant standing for successful termination, SKIP in **CSP** and ε in **ACP**. These constants obey the following operational rules:

- SKIP $\xrightarrow{\checkmark}$ STOP, and

- $\varepsilon \xrightarrow{\surd} \delta$,

where STOP and δ are the constants used to denote deadlock in **CSP** and **ACP**, respectively. The intuition captured by the above-given rules is that successful termination is an action in the behaviour of a process, the action processes perform when they terminate. On the other hand, a deadlocked process like STOP or δ is one that cannot perform any move, not even a successful termination one. This is reflected in the equational laws satisfied by, e.g., ε in **ACP**. For instance, in the equational theory of **ACP** with the empty process ε, the equation

$$\delta + \varepsilon = \varepsilon$$

replaces our $\delta + nil = \delta$. Indeed, in that theory δ always gets cancelled in a sum context, i.e. the equation

$$x + \delta = x$$

holds without any conditions on x. However, the equation

$$x + \varepsilon = x$$

no longer holds (contrary to what happens for our *nil*). As pointed out in §2.1, in the theory of **ACP**, these equations express some a priori considerations about the properties that concurrent, communicating systems are expected to have and are used to describe the intended semantics of processes. The consistency of such axiomatic descriptions of the semantics of processes is then shown by exhibiting models for the axioms. In this chapter, following Milner [Mil80], we have taken the view that operational semantics should be the touchstone for assessing mathematical models of concurrent processes. In this approach, operational semantics is used as a framework within which different intuitions about the behaviour of processes may be expressed and compared. Equational theories, for example complete axiomatizations of some notion of behavioural equivalence or preorder, are then derived from and justified by the operational description of processes. An operational description of the semantics of processes allows us to discuss different intuitions about successful termination and deadlock. The approach we have followed in this chapter is based upon the intuition that both deadlocked processes and successfully terminated ones do not perform any move and that the only way of behaviourally distinguishing them is to observe their behaviour in contexts built using sequential composition. However, we can revise our framework in at least two ways so as to give an operational understanding to the **ACP** theory of ε. One involves changing the interpretation of the termination predicate $\surd$. Following the intuition underlying the **ACP** treatment of the successfully terminated process ε, $p\surd$ may be read as *p has a termination possibility* or *p may terminate*, as it is done in [BG87b]. A termination predicate which is more in line with the **ACP** theory may then be defined by changing rule (iii) of definition 2.2.2 to

$$p\surd \text{ implies } (p+q)\surd \text{ and } (q+p)\surd.$$

Another possibility involves the introduction of a special action, $\surd$, and defining ε to be $\surd;\delta$. In both cases we would obtain the **ACP** laws for ε. Alternatively, we could revise the language by replacing *nil* with ε. Our results carry through to the revised language after simple modifications to the operational semantics, the set of equations E, the behavioural preorder and the modal logic considered in §2.4. These changes are needed in order to take into account the different nature between ε and *nil*. This shows that the proof techniques employed in this chapter to prove our full-abstraction result are indeed quite general and easily adapted to capture different intuitive notions of successful termination in a language.

The semantic theory presented in this chapter for the language REC_Σ has been based upon the interleaving assumption that processes evolve by performing actions which are *atomic*. The following three chapters will be devoted to the semantic study of process algebras for which the assumption of atomicity of action occurrences does not lead, in general, to suitable semantic theories. In particular, we shall investigate process algebras which incorporate an operator for the refinement of actions by processes, thus allowing us to change the level of atomicity of actions.

The languages which will be considered in the future chapters will all be "sublanguages" of REC_Σ. In particular, apart from the use of the left-merge operator in Chapter 3, the treatment of action refinement in Chapters 3-5 will be based on subsets of the set of finite processes $FREC_\Sigma$. In Chapter 6 we shall instead consider a **CCS**-like sublanguage of REC_Σ in which action-prefixing will replace general sequential composition. This language will not contain a restriction operator and its parallel composition operator will not permit communication between concurrent processes.

Chapter 3

Action Refinement for a Simple Language

3.1 Introductory Remarks

As pointed out in Chapter 1, *Process Algebras* like **CCS** [Mil80,89], **CSP** [Hoare85] and **ACP** [BK85] are some of the many languages which have been devised for describing reactive systems. They consist of a set of combinators for constructing new systems from existing ones together with a mechanism for recursive definitions and are normally parameterized with respect to a predefined set of actions. For example, if a, b, c are actions then P, defined by

$$P \Leftarrow (a; P) + (b; c; P),$$

describes a reactive system which can either perform an a-action and proceed as before or a b-action followed by a c-action and then proceed as before.

These languages are designed to describe not only actual systems but also their specifications. It follows that a very important component for these languages is a notion of equivalence between descriptions: one description might be a specification, say SPEC, and another, SYS, the description of an actual implementation and to say that they are equivalent means that SYS is a correct implementation of SPEC; for both describe essentially the same behaviour but at different levels of *abstraction* or *refinement*.

A variety of equivalences have been proposed, [Mil80], [HM85], [Hoare85], [DH84], and methods for proving pairs of descriptions equivalent. Although the resulting theories are very different, speaking in general terms they have much in common. They all allow abstraction from internal actions and, in addition, the equivalences in the latter two references allow descriptions to be equivalent if they both describe the same behaviour *modulo nondeterminism*.

We would like to develop a language for reactive systems and a related equivalence which would support a form of abstraction/refinement where actions may be refined to processes or dually processes may be abstracted to actions. For example, as pointed out in §1.2, at a very high level of abstraction a system might be described by:

$$SPEC \Leftarrow input; output; SPEC.$$

At a more refined level it may be that the actions of input and output are rather complicated and are carried out by processes P and Q, respectively. So SPEC could be refined to:

$$SYS \Leftarrow P; Q; SYS.$$

This form of abstraction/refinement is not supported by any existing Process Algebra and it is the topic of the present chapter. More specifically we consider a very simple (even minimal) process algebra, add a new combinator for refining an action by a process and address the question of an appropriate equivalence for the augmented language. The main result of the chapter is that, at least for the simple language we consider, an adequate equivalence can be defined in a very intuitive manner and moreover can be axiomatised in much the same way as the standard behavioural equivalences, [HM85], [DH84].

We now give a brief outline of the remainder of the chapter. In §3.2 we define our language, essentially a simple subset of the one considered in the previous chapter, which is deliberately designed to be the simplest possible in which the questions we address are interesting. We also give a standard operational semantics in terms of a labelled transition system, [Pl81]. This is the starting point for most of the existing behavioural equivalences. We give one example, strong observational equivalence $\sim$, and show that it is not adequate for our language. The basic problem is that two equivalent terms can have different effects when used as part of a larger system or context. It means that this equivalence will not support a compositional proof method. Nevertheless it is an intuitive equivalence and we argue that an appropriate equivalence for our language is the largest context-preserving equivalence contained in $\sim$, $\sim^c$. We show that this relation $\sim^c$ is intuitively and mathematically tractable; the former by showing that it coincides with timed-equivalence, [H88c], and the latter by showing that it can be equationally characterized.

Timed-equivalence is the subject of §3.3. If an action can be refined by a process, we can no longer consider actions as being atomic events. A minimal consequence is that actions have distinct beginnings and endings. This is the intuition underlying timed-equivalence, which we denote by $\sim_t$. It is strong observational equivalence but defined using subactions, beginnings and endings, $S(a)$ and $F(a)$ for each action a.

We prove that $\sim_t$ and $\sim^c$ coincide. However, the proof of $\sim_t \subseteq \sim^c$ relies on an equational characterization of $\sim_t$. This equational characterization is given in detail in §3.5 as it is independent of the remainder of the chapter. One consequence of the fact that $\sim_t$ and $\sim^c$ coincide is that $\sim_t$ is preserved by action-refinements, i.e. $p \sim_t q$ implies that $p[a \rightsquigarrow r] \sim_t q[a \rightsquigarrow r]$, for each process r. However, the proof is very indirect as it relies on the completeness of a set of equations for $\sim_t$ and the particular form these equations take. It would be preferable to have an operational or behavioural proof of this fact as we are unlikely to have complete equational theories of the appropriate form for more complicated languages. (See Chapter 4, where the work presented in this chapter is generalized to a much richer language)

With this in mind, in §3.4 we introduce a variation of timed-equivalence called *refine-equivalence*, $\sim_r$. This is also a strong observational equivalence defined using the sub-actions $S(a)$ and $F(a)$, but the extra ingredient is that the beginnings and endings of actions must be properly matched. In many ways this is a more intuitive formulation of strong observational equivalence for non-atomic actions, although the formal definition is somewhat more complicated. We show that, for our simple language, $\sim_t$ and $\sim_r$ coincide and we also give a purely behavioural proof of the fact that $\sim_r$ is preserved by action-refinements.

Again, we end this chapter with a section of concluding remarks.

3.2 The Language

In this section we will present the language used in this chapter to discuss the problems mentioned in the introduction. The language we use is a Process Algebra, in the style of **CCS** [Mil80], **CSP** [Hoare85] and **ACP** [BK85], essentially a simple subset of the one presented in the previous chapter, equipped with a new combinator for refining an action by a process. The set of combinators we consider is a minimal one in which the issue we want to address in this chapter is interesting. For instance, it lacks a notion of communication between parallel agents and a facility for defining recursive agents. However, its set of combinators constitutes the core of any process algebra and an understanding of the theory we aim at developing for them will prove to be useful in extensions to more complex algebras which will be presented in Chapter 4.

This section is organized as follows: in §3.2.1 we briefly review two standard means of defining the operational semantics of concurrent agents, Labelled Transition Systems (LTS), [Kel76], and bisimulation, [Pa81]. The notion of (strong) bisimulation will be used many times in the remainder of the chapter for many different LTS's and we will often refer the reader to the general definitions given in this section.

In §3.2.2 we present our basic language **P** and its operational semantics. The operational rules will be used to define a standard observational semantics for **P** by means of the strong bisimulation technique outlined in §3.2.1.

In §3.2.3 we enrich **P** with a simple refinement operator $[a \rightsquigarrow q]$ and the resulting language will be called $\mathbf{P}_{ext}$. This operator allows the refinement of an action by a process. The operational semantics for this language will be defined in two ways: the first reduces a process $p \in \mathbf{P}_{ext}$ to a process $\mathbf{red}(p) \in \mathbf{P}$ by essentially considering the new operator as a syntactic substitution. The operational behaviour of p is then indirectly defined by that of $\mathbf{red}(p)$. The second defines a set of transition relations which directly give the operational semantics for $p \in \mathbf{P}_{ext}$.

The associated strong bisimulations will be shown to coincide on $\mathbf{P}_{ext}$. We will denote them by $\sim$.

However, as we will show by means of examples, it turns out that $\sim$ is not an adequate semantic equivalence for $\mathbf{P}_{ext}$. In fact, it is not a congruence with respect to action refinement. This means that $\sim$ would not support compositional proof techniques. As it is standard practice we consider the largest congruence, with respect to $\mathbf{P}_{ext}$, contained in $\sim$. We will denote it by $\sim^c$. The remainder of the chapter will be devoted to studying the properties of $\sim^c$.

3.2.1 Labelled Transition Systems and Bisimulation

A standard way of defining the operational semantics of concurrent processes is the so-called *two-step approach.* This approach, advocated by Milner in [Mil80] and most of his subsequent work, is based on associating an operational semantics to processes following Plotkin's *Structured Operational Semantics* (SOS), [Pl81], and then abstracting from unwanted details on the way processes evolve using a behavioural equivalence which equates processes which cannot be told apart by means of external observation of their behaviour. A semantic process is then taken to be a congruence class of syntactic objects (terms). In this section we briefly review these ideas and refer to the quoted references for more details.

A standard notion which is used in defining the operational semantics of concurrent processes using Plotkin's SOS is that of *Labelled Transition System* (LTS), [Kel76].

Informally an LTS is based on a notion of global state and of transition from one (global) state to the other. The transition relation is usually parameterized on a set of atomic actions **Act** and several interpretations of the relations $\xrightarrow{\alpha}$, $\alpha \in \mathbf{Act}$, are possible. A standard one is the following, [Mil80]:

$p \xrightarrow{\alpha} p'$ iff p may perform the action α and become p' in doing so.

Formally the definition of labelled transition system is the following one:

Definition 3.2.1 *A* Labelled Transition System *(LTS) is a triple* $(\mathbf{St}, \mathbf{Act}, \{\xrightarrow{\alpha} | \alpha \in \mathbf{Act}\})$, *where:*

- $\mathbf{St}$ *is a set of* states,
- $\mathbf{Act}$ *is a set of* actions,
- *for each* $\alpha \in \mathbf{Act}$, $\xrightarrow{\alpha} \subseteq \mathbf{St}^2$ *is called a* transition relation.

The second step of the operational approach to the semantics of concurrent processes is the *abstraction from unwanted details.* This is achieved by factorizing the set of processes via one of the observational equivalences proposed in the literature, [Mil80,89], [DH84], [BHR84]. The essence of these equivalences is that they are based upon the idea of *observing* a process, which is usually taken to mean communicating with it. The idea is that processes are equal iff they are indistinguishable in any experiment based upon observation. One of the most extensively studied approaches to the definition of observational equivalences for processes is the one based on Park's notion of *bisimulation,* [Pa81]. This way of defining observational equivalences on processes will be frequently used in this chapter. For this reason we now give the formal definition of bisimulation for LTS and will refer to it every time we define an equivalence in this style.

Definition 3.2.2 *Let* $(\mathbf{St}, \mathbf{Act}, \{\xrightarrow{\alpha} | \alpha \in \mathbf{Act}\})$ *be an LTS.*

A binary relation $\mathcal{R}$ *on* $\mathbf{St}$ *is called a* bisimulation *if for each* $(s_1, s_2) \in \mathcal{R}$, $\alpha \in \mathbf{Act}$ *the following clauses hold:*

(i) $s_1 \xrightarrow{\alpha} s_1' \Rightarrow \exists s_2' : s_2 \xrightarrow{\alpha} s_2'$ and $(s_1', s_2') \in \mathcal{R}$,

(ii) $s_2 \xrightarrow{\alpha} s_2' \Rightarrow \exists s_1' : s_1 \xrightarrow{\alpha} s_1'$ and $(s_1', s_2') \in \mathcal{R}$.

The following result is then standard.

Proposition 3.2.1 *Let* $\sim = \bigcup\{\mathcal{R} \subseteq \mathbf{St}^2 | \mathcal{R}$ is a bisimulation $\}$. *Then:*

(i) $\sim$ *is the maximum bisimulation.*

(ii) $\sim$ *is an equivalence relation on* **St**.

In the remainder of the chapter this technique for defining the operational semantics of concurrent processes will be used several times. We will introduce simple languages $\mathcal{L}$ and will define the operational semantics of the terms by induction upon their structure. This will associate an LTS to each term $t \in \mathcal{L}$ and thus to $\mathcal{L}$ itself. Consequently we will apply the notion of bisimulation, described above for general LTS's, to obtain a standard observational equivalence.

3.2.2 A Basic Language

The language we will consider, which is closely related to the language PA investigated by Bergstra and Klop in several papers in the literature, see e.g. [BK84,88], will be parameterized over a set of actions **A**. The set of actions **A** will be a subset of the set of processes. Actions, together with the terminated process *nil*, will be the constants of our algebra of processes. The process combinators used to build new systems from existing ones will be the following:

- + for *non-deterministic choice*,
- ; for *sequential composition*,
- $|$ for *parallel composition* without communication, and
- $|\!\!/$ for *left-merge*, [BK84], [H88c], [CH89]. Intuitively, for processes p and q, $p|\!\!/q$ will denote a process which behaves like the parallel composition of p and q, but with the restriction that its first move has to originate from p.

Apart from the presence of the left-merge operator, which will play an important rôle in the axiomatization presented in §3.5, the language considered in this chapter is a simple sublanguage of the set of finite processes $FREC_\Sigma$ used in Chapter 2. Due to the absence of a restriction operator from the signature for processes, we can restrict ourselves to consider only one form of termination and thus, for the sake of simplicity, we will not have an explicit constant for deadlock, like δ in Chapter 2, in the language. For the sake of clarity, we now give the formal definition of the language considered in this chapter.

Definition 3.2.3 *Let* **A** *be an uninterpreted set of actions.* **A** *will be ranged over by* $a, b, a', b', \ldots$.

For each $n \in \mathbf{N}$ *let* Σ'^n *be defined as follows:*

(i) $\Sigma'^0 = \{nil\} \cup \mathbf{A}$,

(ii) $\Sigma'^2 = \{+, ;, |, \not|\}$,

(iii) $\Sigma'^n = \emptyset \ \forall n \notin \{0, 2\}$.

The signature Σ' is defined as $\Sigma' =_{def} \bigcup_{n \geq 0} \Sigma'^n$. Let $\mathbf{P}$ denote the word algebra over Σ', $\mathbf{T}_{\Sigma'}$.

Terms in $\mathbf{P}$ will be written considering $+, ;, |$ and $\not|$ as infix operators. $\mathbf{P}$ will be ranged over by $p, q, q', q_1 \ldots$ and it is given the structure of a labelled transition system by defining a standard operational semantics for it in Plotkin's SOS style [Mil80], [Pl81]. Due to the fact that we are considering general sequential composition rather than action-prefixing, we need a termination predicate on $\mathbf{P}$, which is defined following our treatment of termination in Chapter 2. (Another alternative is shown in [BHR84]) The definition of the termination predicate for $\mathbf{P}$ follows the lines of the corresponding one given in Definition 2.2.2 for the language REC_Σ. The only new case is that dealing with the left-merge operator. Intuitively, as each initial move of $p \not| q$ has to originate from p, $p \not| q$ is terminated if, and only if, p is terminated.

Definition 3.2.4 *Let $\surd$ be the least set which satisfies:*

- $nil \in \surd$
- $q_1 \in \surd \Rightarrow \forall q \in \mathbf{P}\ (q_1 \not| q) \in \surd$
- $q_1, q_2 \in \surd \Rightarrow q_1 + q_2,\ q_1 ; q_2,\ q_1 | q_2 \in \surd$.

Notation 3.2.1 *We will often write $q\surd$ for $q \in \surd$.*

Intuitively, for each $q \in \mathbf{P}$, $q\surd$ iff q is a terminated process, which will mean, as it will become clear after the next definition, that q cannot perform any transition.

The definition of the transition relation $\xrightarrow{a}$, $a \in \mathbf{A}$, given below is based on Definition 2.2.4. Again, the only new defining clause for $\xrightarrow{a}$ deals with the left-merge operator. It captures the intuition that the first transition of a process of the form $p \not| q$ must be due to process p.

Definition 3.2.5 *For each $a \in \mathbf{A}$, $\xrightarrow{a}$ is the least binary relation on $\mathbf{P}$ which satisfies the following axiom and rules:*

(1) $a \xrightarrow{a} nil$

(2) $q_1 \xrightarrow{a} q' \Rightarrow q_1 + q_2 \xrightarrow{a} q',\ q_2 + q_1 \xrightarrow{a} q'$

(3) $q_1 \xrightarrow{a} q' \Rightarrow q_1; q_2 \xrightarrow{a} q'; q_2$

(4) $q_1\surd,\ q_2 \xrightarrow{a} q' \Rightarrow q_1; q_2 \xrightarrow{a} q'$

(5) $q_1 \xrightarrow{a} q' \Rightarrow q_1|q_2 \xrightarrow{a} q'|q_2,\ q_2|q_1 \xrightarrow{a} q_2|q'$

(6) $q_1 \xrightarrow{a} q' \Rightarrow q_1 \not| q_2 \xrightarrow{a} q'|q_2$.

We may now define a standard observational equivalence on **P** using the notion of strong bisimulation presented in the previous section. The equivalence relation generated in this way for the language **P** will be denoted by $\sim$.

Proposition 3.2.2 *$\sim$ is a Σ'-congruence.*

It is natural to require that semantic equivalence relations be congruences. This means that whenever we have two equivalent specifications we may place either of them into a larger system (or *context*) obtaining equivalent behaviours.

An important feature of the theory of $\sim$ is that $|$ is not a primitive operator, at least for finite processes. In fact, every finite parallel process has an equivalent purely non-deterministic counterpart.

This is illustrated by the following standard example.

Example 3.2.1 *Let $q_1 = a|b$ and $q_2 = (a;b) + (b;a)$. Then*

$$\mathcal{R} = \{(q_1, q_2), (a|nil, nil; a), (nil|b, nil; b), (nil|nil, nil)\}$$

is a bisimulation. Hence $q_1 \sim q_2$.

Another important feature of strong bisimulation equivalence is that its theory has several complete axiomatizations over different languages, [HM85], [H88c], [BK85]. This shows that this equivalence is mathematically tractable. In what follows we would like to show the same property for a language based on **P** enriched with a feature for action-refinement.

3.2.3 The Extended Language

In this section we introduce the language whose semantic properties will be investigated in the remainder of this chapter. The language, which we call the *extended process language*, $\mathbf{P}_{ext}$, is based on the signature Σ' enriched with a combinator, $[a \rightsquigarrow p]$, which allows the refinement of an action by a process p. Formally:

Definition 3.2.6 *The* extended process language $\mathbf{P}_{ext}$ *is the language generated by the following grammar:*

$$p ::= nil \mid a \mid p + p \mid p; p \mid p|p \mid p\not| p \mid p[a \rightsquigarrow p]$$

where $a \in \mathbf{A}$.

$\mathbf{P}_{ext}$ *will be ranged over, with abuse of notation, by* $p, q, r, p', p_1, \ldots$.

Example 3.2.2 $p = (a|b)[a \rightsquigarrow a_s; a_f]$ *stands for a process in which action* a *has been refined to the sequential composition of two subactions* a_s *and* a_f.

With regard to the simple process presented in the above example, operationally we would expect it to be able to perform any shuffle of the strings $a_s a_f$ and b. We have at least two ways of capturing the operational behaviour of the processes definable in $\mathbf{P}_{ext}$:

- we might regard $p[a \rightsquigarrow p']$ as indicating the syntactical substitution of process p' for any occurrence of a in p. Iterating this procedure we could translate or reduce every process in $\mathbf{P}_{ext}$ into a process in $\mathbf{P}$. The operational behaviour of $p[a \rightsquigarrow p']$ would then be determined by that of its translation according to the operational semantics of $\mathbf{P}$-processes;
- we might explicitly define an operational semantics for $\mathbf{P}_{ext}$.

We will follow both lines and show that the resulting semantic equivalences coincide. Having done so, in the following sections of this chapter we will use the operational semantics of $\mathbf{P}_{ext}$-processes given via the reduction to $\mathbf{P}$. This method, although it is maybe less elegant, will be technically more tractable.

Definition 3.2.7

(1) *The* reduction function, $\mathbf{red} : \mathbf{P}_{ext} \to \mathbf{P}$, *is defined by structural induction as follows:*

(i) $\mathbf{red}(nil) = nil$

(ii) $\mathbf{red}(a) = a$

(iii) $\mathbf{red}(p_1 \; op \; p_2) = \mathbf{red}(p_1) \; op \; \mathbf{red}(p_2)$, $op \in \{+, ;, |, \not|\}$

(iv) $\mathbf{red}(p_1[a \rightsquigarrow p_2]) = \mathbf{red}(p_1)[a/\mathbf{red}(p_2)]$, *where* $[a/\mathbf{red}(p_2)]$ *denotes the syntactic substitution of* $\mathbf{red}(p_2)$ *for each occurrence of* a *in* $\mathbf{red}(p_1)$.

(2) *A* refinement function $\rho : \mathbf{A} \rightarrow \mathbf{P}_{ext}$ *may be used to generalize the construct* $[a \rightsquigarrow p]$. *The language obtained in this way will be denoted by* $\mathbf{P}_\rho$. *The reduction function for this language will be defined as in (1) with the following clause in place of clause* (iv)*:*

$$\mathbf{red}(p[\rho]) = \mathbf{red}(p)\rho_r$$

where $\rho_r : \mathbf{A} \rightarrow \mathbf{P}$ *is the syntactic substitution defined by:*

$$\forall a \in \mathbf{A}\ \rho_r(a) = \mathbf{red}(\rho(a)).$$

Intuitively, a refinement function ρ may be seen as the simultaneous refinement of each action $a \in \mathbf{A}$ by the process $\rho(a)$. It is interesting to note that it is essential to consider refinement functions ρ whose codomain is $\mathbf{P}_{ext}$ rather than $\mathbf{P}_\rho$. In fact, had we allowed refinement maps $\rho : \mathbf{A} \rightarrow \mathbf{P}_\rho$, the reduction function $\mathbf{red}(\cdot)$ would not be properly defined for refinements such that $\rho(a) = a[\rho]$.

Fact 3.2.1

(i) $\forall p \in \mathbf{P}_{ext}(\mathbf{P}_\rho)\ \mathbf{red}(p) \in \mathbf{P}$.

(ii) $\forall q \in \mathbf{P}\ \ \mathbf{red}(q) = q$.

With the above notion of reduction each process $p \in \mathbf{P}_{ext}(\mathbf{P}_\rho)$ inherits an operational semantics from its reduction. For example:

$$\mathbf{red}(\ (a|b)[a \rightsquigarrow a_s; a_f]\ ) = (a_s; a_f)|b,$$

and the transitions of such a $\mathbf{P}$-process can be easily determined using the operational rules given in the previous section.

Moreover, $\mathbf{P}_{ext}(\mathbf{P}_\rho)$ inherits a notion of observational equivalence for processes from $\mathbf{P}$ via $\mathbf{red}(\cdot)$. Two processes $p_1, p_2 \in \mathbf{P}_{ext}(p_1, p_2 \in \mathbf{P}_\rho)$ will be equivalent iff their reductions are equivalent with respect to $\sim$. From now on we will concentrate on $\mathbf{P}_\rho$ as $\mathbf{P}_{ext}$ is a "sublanguage" of $\mathbf{P}_\rho$.

Definition 3.2.8 $\forall p_1, p_2 \in \mathbf{P}_\rho\ \ p_1 \sim_1 p_2 \Leftrightarrow \mathbf{red}(p_1) \sim \mathbf{red}(p_2)$.

We will now define an explicit operational semantics for $\mathbf{P}_\rho$ from which a standard observational equivalence is obtained.

First of all we define the termination predicate $\surd$ for the extended language.

Definition 3.2.9

(i) *For each* $p \in \mathbf{P}_\rho$ *define* $\mathbf{Ter}(p) \subseteq \mathbf{A}$, *the set of actions of* p *which have to be mapped to terminated processes to force* p *to terminate, by structural induction as follows:*

- $\mathbf{Ter}(nil) = \emptyset$
- $\mathbf{Ter}(a) = \{a\}$
- $\mathbf{Ter}(p_1 \not| p_2) = \mathbf{Ter}(p_1)$
- $\mathbf{Ter}(p_1\ op\ p_2) = \mathbf{Ter}(p_1) \cup \mathbf{Ter}(p_2)$, *for* $op \in \{+, ;, |\}$
- $\mathbf{Ter}(p[\rho]) = \bigcup_{a \in \mathbf{Ter}(p)} \mathbf{Ter}(\rho(a))$.

(ii) *Let* $\surd$ *be the least subset of* $\mathbf{P}_\rho$ *which satisfies:*

- $nil \in \surd$
- $p_1 \in \surd \Rightarrow p_1 \not| p_2 \in \surd$
- $p_1, p_2 \in \surd \Rightarrow p_1\ op\ p_2 \in \surd$, *where* $op \in \{+, ;, |\}$
- $\forall a \in \mathbf{Ter}(p)\ \rho(a) \in \surd \Rightarrow p[\rho] \in \surd$.

In what follows we will write $p\surd$ *iff* $p \in \surd$.

Notation 3.2.2 *Given two refinement functions* $\rho, \rho' : \mathbf{A} \to \mathbf{P}_{ext}$, *we write* $\rho \circ \rho'$ *for the refinement function defined as follows:*

$$\forall a \in \mathbf{A}\ (\rho \circ \rho')(a) = \rho'(a)\rho,$$

the application of the syntactic substitution ρ *to each* $\rho'(a)$.

One can prove by induction on p *that* $\mathbf{red}(p\rho) = \mathbf{red}(p)\rho_r$, *from which it follows that* $(\rho \circ \rho')_r = \rho_r \circ \rho'_r$.

We can now define an explicit operational semantics for the language $\mathbf{P}_\rho$ (and hence for its sublanguage $\mathbf{P}_{ext}$).

Definition 3.2.10 *For each* $a \in \mathbf{A}$ *let* $\xrightarrow{a}$ *be the least binary relation on* $\mathbf{P}_\rho$ *which satisfies the following axiom and rules:*

(1) $a \xrightarrow{a} nil$

(2) $p_1 \xrightarrow{a} p'_1 \Rightarrow p_1 + p_2 \xrightarrow{a} p'_1,\ p_2 + p_1 \xrightarrow{a} p'_1$

(3) $p_1 \xrightarrow{a} p_1' \Rightarrow p_1; p_2 \xrightarrow{a} p_1'; p_2$

(4) $p_1\surd, p_2 \xrightarrow{a} p' \Rightarrow p_1; p_2 \xrightarrow{a} p'$

(5) $p_1 \xrightarrow{a} p_1' \Rightarrow p_1 | p_2 \xrightarrow{a} p_1' | p_2,\ p_2 | p_1 \xrightarrow{a} p_2 | p_1'$

(6) $p_1 \xrightarrow{a} p_1' \Rightarrow p_1 \lfloor\!\lfloor p_2 \xrightarrow{a} p_1' | p_2$

(7) $\rho(b) \xrightarrow{a} p' \Rightarrow b[\rho] \xrightarrow{a} p'$

(8) $p_1[\rho] \xrightarrow{a} p' \Rightarrow (p_1 + p_2)[\rho] \xrightarrow{a} p', (p_2 + p_1)[\rho] \xrightarrow{a} p'$

(9) $p_1[\rho] \xrightarrow{a} p' \Rightarrow (p_1; p_2)[\rho] \xrightarrow{a} p'; (p_2[\rho])$

(10) $p_1[\rho]\surd, p_2[\rho] \xrightarrow{a} p' \Rightarrow (p_1; p_2)[\rho] \xrightarrow{a} p'$

(11) $p_1[\rho] \xrightarrow{a} p_1' \Rightarrow (p_1 | p_2)[\rho] \xrightarrow{a} p_1' | (p_2[\rho]), (p_2 | p_1)[\rho] \xrightarrow{a} (p_2[\rho]) | p_1'$

(12) $p_1[\rho] \xrightarrow{a} p_1' \Rightarrow (p_1 \lfloor\!\lfloor p_2)[\rho] \xrightarrow{a} p_1' | (p_2[\rho])$

(13) $p[\rho' \circ \rho] \xrightarrow{a} p' \Rightarrow (p[\rho])[\rho'] \xrightarrow{a} p'$.

Using this operational semantics we can define a standard observational equivalence on $\mathbf{P}_\rho$ using the strong bisimulation technique as we have done for $\mathbf{P}$. The resulting equivalence relation on $\mathbf{P}_\rho$ will be denoted by $\sim_2$. We will now show that $\sim_1$ and $\sim_2$ coincide on $\mathbf{P}_\rho$. To do so we will need to relate the moves of a process $p \in \mathbf{P}_\rho$ with those of its reduction $\mathbf{red}(p) \in \mathbf{P}$.

Lemma 3.2.1 *The following statements hold:*

(i) $\forall p \in \mathbf{P}_\rho\ \mathbf{Ter}(p) = \mathbf{Ter}(\mathbf{red}(p))$.

(ii) $\forall p \in \mathbf{P}_\rho\ p\surd \Leftrightarrow \mathbf{red}(p)\surd$.

Proof: Both the statements can be shown by induction on the structure of p. The only interesting case is $p = p_1[\rho]$.

(i) $\mathbf{Ter}(p_1[\rho]) = \bigcup_{a \in \mathbf{Ter}(p_1)} \mathbf{Ter}(\rho(a))$, by definition.

This is equal to $\bigcup_{a \in \mathbf{Ter}(\mathbf{red}(p_1))} \mathbf{Ter}(\mathbf{red}(\rho(a)))$, by the inductive hypothesis, which by definition is equal to $\mathbf{Ter}(\mathbf{red}(p_1)\rho_r)$.

(ii) $p_1[\rho]\surd \Leftrightarrow \forall a \in \mathbf{Ter}(p_1)\ \rho(a)\surd$, by definition.

By inductive hypothesis, this is equivalent to

$$\forall a \in \mathbf{Ter}(p_1)\ \mathbf{red}(\rho(a))\surd.$$

By clause (i) of the lemma, this is equivalent to

$$\forall a \in \mathbf{Ter}(\mathbf{red}(p_1))\ \mathbf{red}(\rho(a))\surd.$$

This is in turn equivalent to $(\mathbf{red}(p_1)\rho_r)\surd$ and $\mathbf{red}(p_1[\rho])\surd$.

This completes the proof of the lemma. □

We can now relate the moves of a process $p \in \mathbf{P}_\rho$ with those of its reduction $\mathbf{red}(p)$.

Lemma 3.2.2 *For each $p \in \mathbf{P}_\rho$ the following statements hold:*

(i) $p \stackrel{a}{\longrightarrow\!\!\triangleright} p' \Rightarrow \mathbf{red}(p) \stackrel{a}{\longrightarrow} \mathbf{red}(p')$;

(ii) $\mathbf{red}(p) \stackrel{a}{\longrightarrow} x \Rightarrow \exists p' : p \stackrel{a}{\longrightarrow\!\!\triangleright} p'$ and $\mathbf{red}(p') = x$.

Proof: We prove both statements by induction on the length n of the proof of $p \stackrel{a}{\longrightarrow\!\!\triangleright} p'$ and $\mathbf{red}(p) \stackrel{a}{\longrightarrow} x$, respectively.

We proceed by examining the structure of p and sketch the details of two cases of the proof of statement (i) leaving the others to the reader.

$p = p_1;p_2$ Assume $p_1;p_2 \stackrel{a}{\longrightarrow\!\!\triangleright} p'$. There are two subcases to examine:

(a) $p_1 \stackrel{a}{\longrightarrow\!\!\triangleright} p'_1$ and $p' = p'_1;p_2$.

The proof of the derivation $p_1 \stackrel{a}{\longrightarrow\!\!\triangleright} p'_1$ has length $n-1$. Hence, by inductive hypothesis, $\mathbf{red}(p_1) \stackrel{a}{\longrightarrow} \mathbf{red}(p'_1)$.

By the operational semantics for $\mathbf{P}$,

$$\mathbf{red}(p_1;p_2) = \mathbf{red}(p_1);\mathbf{red}(p_2) \stackrel{a}{\longrightarrow} \mathbf{red}(p'_1);\mathbf{red}(p_2) = \mathbf{red}(p'_1;p_2).$$

(b) $p_1\surd$ and $p_2 \stackrel{a}{\longrightarrow\!\!\triangleright} p'$.

The proof of the derivation $p_2 \stackrel{a}{\longrightarrow\!\!\triangleright} p'$ has length $n-1$. Hence, by inductive hypothesis, $\mathbf{red}(p_2) \stackrel{a}{\longrightarrow} \mathbf{red}(p')$.

By the above lemma, $p_1\surd \Leftrightarrow \mathbf{red}(p_1)\surd_{\mathbf{P}}$.

Hence, by the operational semantics for $\mathbf{P}$,

$$\mathbf{red}(p_1;p_2) = \mathbf{red}(p_1);\mathbf{red}(p_2) \stackrel{a}{\longrightarrow} \mathbf{red}(p').$$

$p = p_1[\rho]$ We now examine the structure of p_1 concentrating on showing the claim for two subcases.

$p_1 = b$ Then $\rho(b) \overset{a}{\longrightarrow\!\!\triangleright} p'$. The proof of this derivation has length $n-1$. Hence, by inductive hypothesis, $\mathbf{red}(\rho(b)) \overset{a}{\longrightarrow} \mathbf{red}(p')$.

This implies $\mathbf{red}(b[\rho]) = b\rho_r \overset{a}{\longrightarrow} \mathbf{red}(p')$.

$p_1 = p_2[\rho']$ Assume $(p_2[\rho'])[\rho] \overset{a}{\longrightarrow\!\!\triangleright} p'$.

Then $p_2[\rho \circ \rho'] \overset{a}{\longrightarrow\!\!\triangleright} p'$.

The length of the proof of this derivation is less than n. Hence, by inductive hypothesis, $\mathbf{red}(p_2[\rho \circ \rho']) \overset{a}{\longrightarrow} \mathbf{red}(p')$.

Moreover, by definition of $\mathbf{red}(\cdot)$,

$$\mathbf{red}(p_2[\rho \circ \rho']) = \mathbf{red}(p_2)(\rho \circ \rho')_r.$$

By a previous observation (notation 3.2.2), $\mathbf{red}(p_2)(\rho \circ \rho')_r = \mathbf{red}(p_2)(\rho_r \circ \rho'_r)$.

It follows that,

$$\mathbf{red}(p_2)(\rho_r \circ \rho'_r) = (\mathbf{red}(p_2)\rho'_r)\rho_r = \mathbf{red}((p_2[\rho'])[\rho]).$$

This completes the inductive argument for statement (i). The proof of statement (ii) follows entirely similar lines. □

We will now show that the equivalences $\sim_1$ and $\sim_2$ coincide on $\mathbf{P}_\rho$ (and thus on $\mathbf{P}_{ext}$). This is stated by the following theorem.

Theorem 3.2.1 $\forall p_1, p_2 \in \mathbf{P}_\rho\ p_1 \sim_1 p_2 \Leftrightarrow p_1 \sim_2 p_2$.

Proof: We will show that for each $p_1, p_2 \in \mathbf{P}_\rho$,

$$p_1 \sim_2 p_2 \Leftrightarrow \mathbf{red}(p_1) \sim \mathbf{red}(p_2).$$

(Only if) Consider the relation $\mathcal{R} \subseteq \mathbf{P}^2$ defined as follows:

$$\mathcal{R} =_{def} \{(\mathbf{red}(p_1), \mathbf{red}(p_2)) \mid p_1 \sim_2 p_2\}.$$

We show that $\mathcal{R}$ is a $\mathbf{P}$-bisimulation. By symmetry it is sufficient to show that

$$\mathbf{red}(p_1) \overset{a}{\longrightarrow} x \Rightarrow \exists y : \mathbf{red}(p_2) \overset{a}{\longrightarrow} y \text{ and } (x, y) \in \mathcal{R}.$$

Assume $\mathbf{red}(p_1) \xrightarrow{a} x$. Then, by the above lemma,

$$\exists p_1' : p_1 \xrightarrow{a}\!\!\triangleright\, p_1' \text{ and } \mathbf{red}(p_1') = x.$$

As $p_1 \sim_2 p_2$, there exists p_2' such that

$$p_2 \xrightarrow{a}\!\!\triangleright\, p_2' \text{ and } p_1' \sim_2 p_2'.$$

By the above lemma, $\mathbf{red}(p_2) \xrightarrow{a} \mathbf{red}(p_2')$ and, by definition of $\mathcal{R}$, $(\mathbf{red}(p_1'), \mathbf{red}(p_2')) \in \mathcal{R}$.

Hence $\mathcal{R}$ is a **P**-bisimulation.

(If) Consider the relation defined as follows:

$$\mathcal{R} =_{def} \{(p_1, p_2) \mid (\mathbf{red}(p_1), \mathbf{red}(p_2)) \in \sim\}.$$

$\mathcal{R}$ can be shown to be a $\mathbf{P}_{ext}$-bisimulation using the same approach as in the *only if* case.
□

This equivalence over $\mathbf{P}_{ext}$, $\sim_1 = \sim_2$, is a conservative extension of $\sim$ over **P** and, for convenience, in future we will also use $\sim$ to denote it. We will usually use its characterizations in terms of the function $\mathbf{red}(\cdot)$.

However, $\sim$ is not an adequate semantic equivalence for $\mathbf{P}_{ext}$; it turns out not to be a congruence with respect to the combinator of action-refinement, as the following example shows.

Example 3.2.3

(1) *Consider* $p = a|b$ *and* $q = (a;b) + (b;a)$. *As already noted* $p \sim q$. *However,*

$$p[a \leadsto a_s; a_f] \not\sim q[a \leadsto a_s; a_f].$$

In fact,

$$\mathbf{red}(q[a \leadsto a_s; a_f]) = ((a_s; a_f); b) + (b; (a_s; a_f)) \xrightarrow{a_s} (nil; a_f); b$$

a state in which only action a_f *is possible. No such state can be reached by* $\mathbf{red}(p[a \leadsto a_s; a_f])$ *via an action* a_s.

(2) *Consider* $q' = (a|b) + (a;b)$. *Then* $p \sim q'$ *but*

$$p[a \leadsto a_s; a_f] \not\sim q'[a \leadsto a_s; a_f].$$

In fact,

$$\mathbf{red}(q'[a \leadsto a_s; a_f]) = ((a_s; a_f)|b) + ((a_s; a_f); b) \xrightarrow{a_s} (nil; a_f); b$$

a state in which only action a_f is possible.

We have already seen how no equivalent state can be reached by $\mathbf{red}(p[a \rightsquigarrow a_s; a_f])$ *via an a_s-action.*

As $\sim$ is not a congruence with respect to action-refinement, it would not support compositional proof methods on $\mathbf{P}_{ext}$. However, bisimulation equivalence is a natural notion of observational equivalence between processes (as argued for example in [Mil89], [HM85] and [Ab87a]) and we would like to base our theory of action-refinement on a similar notion of equivalence.

We have a standard way of associating a congruence with $\sim$. It is sufficient to close $\sim$ with respect to all $\mathbf{P}_{ext}$-contexts, [Mil80]. The resulting congruence, which we denote by $\sim^c$, is known to be the largest congruence contained in $\sim$, [Mil80]. The definition of $\sim^c$ is, however, purely algebraic and does not shed much light on its behavioural significance. Moreover, it does not support useful proof techniques to show that two processes are related with respect to it. The remainder of the chapter is devoted to addressing these two issues.

3.3 Timed Equivalence

It has been pointed out in the introductory remarks that, if we want to describe the behaviour of concurrent processes to allow for a step-wise refinement of the actions they perform, the view of processes enforced by interleaving-style equivalences becomes inadequate.

Essentially this depends on the fact that an interleaving view of the behaviour of the processes under consideration strongly depends on what are regarded to be the atomic actions processes may perform.

As in this development we allow for an operation which changes the level of atomicity of actions without enforcing any mutual exclusion policy, we need a more refined behavioural description of the processes than the ones given by the interleaving based equivalences proposed in the literature [HM85], [DH84], [BHR84].

In [H88c] M. Hennessy has proposed an alternative version of bisimulation equivalence based on actions which are not necessarily instantaneous. A comprehensive description of the resulting semantic theory is given in the above quoted reference. For our purposes it will be sufficient to remind the reader that the resulting behavioural description of processes is based on the assumption that there are observers which can detect the beginning and ending of actions[1].

[1] To my knowledge, this behavioural view of processes has been studied for the first time by M. Hennessy in a pre-print entitled "On the Relationship between Time and Interleaving" (1981) which subsequently evolved to [H88c].

A standard bisimulation equivalence can be developed using the operational description of the behaviour of the processes. The result is a semantic theory of processes which distinguishes concurrency from nondeterminism and which can be completely axiomatized, [H88c].

In this section we will develop a semantic theory based on these ideas for the language **P** presented in §3.2. **P** is an extension of the language used in [H88c] as action prefixing is replaced by sequential composition of processes. The theory will be developed according to the standard two-step operational description of processes. First of all we define an operational semantics for processes based on the ideas in [H88c]. Secondly we will abstract from unwanted details by means of a semantic equivalence based on this operational semantics. This section will end with a discussion of the relevance of this notion of equivalence for processes with respect to the step-wise refinement of actions proposed in §3.2.

3.3.1 Timed Operational Semantics

We will assume that beginnings and terminations of actions are distinct subactions which may be observed. For each $a \in \mathbf{A}$ we will use $S(a)$ and $F(a)$ to denote the beginning and the termination of an a-action, respectively. We will view $S(a)$, $F(a)$ as a new class of actions and define an operational semantics in terms of new next-event relations $\xrightarrow{S(a)}_t$, $\xrightarrow{F(a)}_t$.

Notation 3.3.1 $\mathbf{A_S} =_{def} \{S(a) \mid a \in \mathbf{A}\} \cup \{F(a) \mid a \in \mathbf{A}\}$ *will be called the set of* subactions *and will be ranged over by* e, e', $e_1 \ldots$.

As pointed out in [H88c], the language for processes proposed in §3.2 is not sufficiently expressive to describe all possible states a process may reach. To overcome this problem we introduce into the language a new symbol $F(a)$ for each $a \in \mathbf{A}$. $F(a)$ will denote the state in which the atomic process a is being executed but has not yet terminated its execution.

Definition 3.3.1 (Process States) *Let* $\mathcal{S}$, *the set of* process states, *be the least set which satisfies:*

- $p \in \mathbf{P}$ implies $p \in \mathcal{S}$.
- $a \in \mathbf{A}$ implies $F(a) \in \mathcal{S}$.
- $s \in \mathcal{S}$, $p \in \mathbf{P}$ implies $s;p \in \mathcal{S}$,
- $s_1 \in \mathcal{S}$, $s_2 \in \mathcal{S}$ implies $s_1 | s_2 \in \mathcal{S}$.

$\mathcal{S}$ *will be ranged over by* s, s_1, s' ....

We can now define a standard operational semantics for states following the pattern described in §3.2. In defining the operational semantics we will use the termination predicate defined in §3.2. This predicate can be easily extended to $\mathcal{S}$ as follows:

Definition 3.3.2 $\forall s \in \mathcal{S}\ s\surd \Leftrightarrow s \in \mathbf{P}$ and $s\surd$.

Definition 3.3.3 *For each* $e \in \mathbf{A_S}$, $\xrightarrow{e}_t$ *will be the least binary relation on* $\mathcal{S}$ *which satisfies the following axioms and rules:*

(1) $a \xrightarrow{S(a)}_t F(a)$

(2) $F(a) \xrightarrow{F(a)}_t nil$

(3) $p_1 \xrightarrow{e}_t s_1'$ implies $p_1 + p_2 \xrightarrow{e}_t s_1'$, $p_2 + p_1 \xrightarrow{e}_t s_1'$

(4) $s_1 \xrightarrow{e}_t s_1'$ implies $s_1;p \xrightarrow{e}_t s_1';p$

(5) $s_1\surd$, $p \xrightarrow{e}_t s$ implies $s_1;p \xrightarrow{e}_t s$

(6) $s_1 \xrightarrow{e}_t s_1'$ implies $s_1|s_2 \xrightarrow{e}_t s_1'|s_2$, $s_2|s_1 \xrightarrow{e}_t s_2|s_1'$

(7) $p_1 \xrightarrow{e}_t s$ implies $p_1 \not| p_2 \xrightarrow{e}_t s|p_2$.

Rules (1) and (2) above are the new rules which make explicit our view of processes. They state that an atomic process a may perform the beginning of the action a and enter state $F(a)$. Moreover, when in state $F(a)$, it can only perform the termination of the action it has started.

A standard behavioural equivalence may now be defined using the above defined operational semantics and the notion of bisimulation given, for arbitrary labelled transition systems, in §3.2.

Definition 3.3.4 (Timed Equivalence) *The maximum bisimulation over the labelled transition system* $\langle \mathcal{S}, \mathbf{A_S}, \{\xrightarrow{e}_t \mid e \in \mathbf{A_S}\}\rangle$ *will be denoted by* $\sim_t$ *and will be called* timed (observational) equivalence.

Some examples of equivalent and inequivalent processes, which may be readily translated in our language, are given in [H88c].

We now concentrate on the properties of $\sim_t$ which will be relevant to the developments presented in the remainder of the chapter. First of all we show that $\sim_t$ is contained in $\sim$ for processes. To do so we need to relate the operational semantics for $\mathbf{P}$-processes given in §3.2 with the one given in the above definition.

Lemma 3.3.1 *For each* $p \in \mathbf{P}$, $a \in \mathbf{A}$, $p \stackrel{S(a)}{\longrightarrow}_t \stackrel{F(a)}{\longrightarrow}_t p'$ *iff* $p \stackrel{a}{\longrightarrow} p'$.

Proof: Both directions of the if-and-only-if can be easily shown by structural induction on p. Note that if $p \stackrel{S(a)}{\longrightarrow}_t s$ then s contains only one occurrence of $F(a)$. With this in mind the proof is straightforward and thus omitted. □

We can now show the promised theorem.

Theorem 3.3.1 $\forall p, q \in \mathbf{P}\ p \sim_t q \Rightarrow p \sim q$.

Proof: Consider the relation $\mathcal{R} =_{def} \sim_t \cap\, \mathbf{P}^2$.

We show that $\mathcal{R}$ is a bisimulation. Take $(p,q) \in \mathcal{R}$. By symmetry, it is sufficient to show that for each $a \in \mathbf{A}$

$$p \stackrel{a}{\longrightarrow} p' \text{ implies } \exists q' : q \stackrel{a}{\longrightarrow} q' \text{ and } (p',q') \in \mathcal{R}.$$

Assume $p \stackrel{a}{\longrightarrow} p'$. Then, by the above lemma, $p \stackrel{S(a)}{\longrightarrow}_t s \stackrel{F(a)}{\longrightarrow}_t p'$ for some $s \in \mathcal{S}$. As $p \sim_t q$, there exists $s' \in \mathcal{S}$ such that $s \sim_t s'$ and $q \stackrel{S(a)}{\longrightarrow}_t s' \stackrel{F(a)}{\longrightarrow}_t q'$ with $p' \sim_t q'$.

By the above lemma, $q \stackrel{a}{\longrightarrow} q'$ and, by definition of $\mathcal{R}$, $(p',q') \in \mathcal{R}$.

Hence $\mathcal{R}$ is a bisimulation. This proves the claim. □

The reverse implication does not hold as shown by the following example.

Example 3.3.1 *Consider* $p = (a;b) + (b;a)$ *and* $q = a|b$. *Then* $p \sim q$ *but* $p \not\sim_t q$.

In fact, $p \stackrel{S(a)}{\longrightarrow}_t F(a);b$ *which is a state in which* p *can only perform the termination of the action* a. *No such state can be reached by* q *via* $S(a)$. *In fact,* $q \stackrel{S(a)}{\longrightarrow}_t s$ *implies* $s = F(a)|b$ *and* s *may perform the start of action* b.

Proposition 3.3.1 $\sim_t$ *is a congruence on* $\mathbf{P}$.

Proof: Standard and thus omitted. □

It turns out that $\sim_t$ can be completely axiomatized for the language $\mathbf{P}$ presented in §3.2. The required equations are collected in Figure 3.1.

Let $=_E$ be the least **P**-congruence which satisfies the axioms in Figure 3.1.

Then we can state the following result.

Theorem 3.3.2 $\forall p, q \in \mathbf{P}\ p =_E q \Leftrightarrow p \sim_t q$.

The proof of this result is rather technical and involved; it is based on the techniques used in, e.g., [CH89], [H88c] to axiomatize non-interleaving equivalences, but the details are more delicate due to the presence of general sequential composition in the language. §3.5 is entirely devoted to the exposition of the proof. As it is independent of the rest of the chapter, we assume the theorem in the remainder of §3.3 and in §3.4.

It is interesting to note that the axiomatization of $\sim_t$ over **P** differs from that of $\sim$ only for the absence of Milner's *expansion theorem*, [Mil80,89]. In the presence of the left-merge operator, the difference between the two axiomatizations may be summed up by the law

$$a; x \lfloor\!\lfloor y \quad = \quad a; (x|y).$$

This law is satisfied by $\sim$ and can be used to derive a variation on Milner's expansion theorem suitable for the language **P**. On the other hand, it is easy to see that the above equation is *not* satisfied by $\sim_t$. For example, $a; p \lfloor\!\lfloor b \not\sim_t a; (p|b)$ for all $p \in \mathbf{P}$.

3.3.2 $\sim^c$ Coincides with $\sim_t$

The remainder of this chapter is entirely devoted to proving that $\sim_t$ coincides with $\sim^c$ on the set of processes **P**. Theorem 3.3.2 then also provides a complete equational characterization of $\sim^c$. As a first step towards proving this claim, in this section we show that $\sim^c$ is contained in $\sim_t$. To show this result we will need to *translate states into the language of observers* as stated in [H88c]. This is stated formally in the following definition:

Definition 3.3.5

(i) *Let Σ^* be the signature defined as follows:*

- $\Sigma^*_0 = \{S(a)|a \in \mathbf{A}\} \cup \{F(a)|a \in \mathbf{A}\} \cup \{nil\}$
- $\Sigma^*_2 = \{+, ;, |, \lfloor\!\lfloor\}$
- $\Sigma^*_n = \emptyset\ \forall n \neq 0, 2.$

Finally, $\Sigma^* = \bigcup_{n \geq 0} \Sigma^*_n$.

(A1) $(x + y) + z = x + (y + z)$

(A2) $x + y = y + x$

(A3) $x + x = x$

(A4) $x + nil = x$

(B1) $(x + y) \mathbin{/\!\!|} z = x \mathbin{/\!\!|} z + y \mathbin{/\!\!|} z$

(B2) $(x \mathbin{/\!\!|} y) \mathbin{/\!\!|} z = x \mathbin{/\!\!|} (y | z)$

(B3) $x \mathbin{/\!\!|} nil = x$

(B4) $nil \mathbin{/\!\!|} x = nil$

(X1) $x | y = x \mathbin{/\!\!|} y + y \mathbin{/\!\!|} x$

(C1) $(x; y); z = x; (y; z)$

(C2) $x; nil = x = nil; x$

(C3) $(x + y); z = (x; z) + (y; z)$ if $x, y \notin \surd$.

Figure 3.1. Complete Axioms for $\sim_t$

(ii) *The function* **Tr**:$\mathcal{S} \rightarrow$ $\mathbf{T}_{\Sigma^*}$ *is the unique homomorphism between the two algebras such that:*

- **Tr**$(nil)=nil$
- **Tr**$(a)=S(a);F(a)$
- **Tr**$(F(a))=nil;F(a)$.

The terms in $\mathbf{T}_{\Sigma^*}$ may be endowed with an operational description of their behaviour by defining an interleaving-style operational semantics for $\mathbf{T}_{\Sigma^*}$. This is done by associating a next-event relation $\xrightarrow{e}_i$ to each $e \in \mathbf{A_S}$ following the definition of $\xrightarrow{a}$, $a \in \mathbf{A}$, given in §3.2.

The relationships between the relations $\xrightarrow{e}_t$ and $\xrightarrow{e}_i$ are expressed by the following lemma.

Lemma 3.3.2 *For each* $s \in \mathcal{S}$, $e \in \mathbf{A_S}$ *the following statements hold:*

(1) $s\surd \Leftrightarrow \mathbf{Tr}(s)\surd$;

(2) $s \xrightarrow{e}_t s' \Rightarrow \mathbf{Tr}(s) \xrightarrow{e}_i \mathbf{Tr}(s')$;

(3) $\mathbf{Tr}(s) \xrightarrow{e}_i x \Rightarrow \exists s' : s \xrightarrow{e}_t s'$ *and* $\mathbf{Tr}(s')=x$.

Proof: All the statements can be shown by structural induction on s.

We focus on the case $s = s_1;p$ for statement (2).

(2) Assume $s_1;p \xrightarrow{e}_t s'$. There are two cases to consider:

  (a) $s_1 \xrightarrow{e}_t s_1'$ and $s' = s_1';p$.
  By inductive hypothesis, $\mathbf{Tr}(s_1) \xrightarrow{e}_i \mathbf{Tr}(s_1')$.
  Hence $\mathbf{Tr}(s_1;p)=\mathbf{Tr}(s_1);\mathbf{Tr}(\mathrm{p}) \xrightarrow{e}_i \mathbf{Tr}(s_1');\mathbf{Tr}(\mathrm{p})= \mathbf{Tr}(s_1;p)$.

  (b) $s_1\surd$ and $p \xrightarrow{e}_t s'$.
  By inductive hypothesis, $\mathbf{Tr}(\mathrm{p}) \xrightarrow{e}_i \mathbf{Tr}(s')$. By statement (1) of the lemma, $\mathbf{Tr}(s_1)$ $\surd$.
  Hence $\mathbf{Tr}(s_1);\mathbf{Tr}(\mathrm{p}) \xrightarrow{e}_i \mathbf{Tr}(s')$. □

We may define a bisimulation equivalence on $\mathbf{T}_{\Sigma^*}$ in the standard way. The details are left to the reader. Let us denote $\sim$, with abuse of notation, the resulting equivalence relation. We concentrate on investigating the relationships between $\sim_t$ and $\sim$ via $\mathbf{Tr}(\cdot)$.

Theorem 3.3.3 $\forall s_1, s_2 \in \mathcal{S}$ $s_1 \sim_t s_2 \Leftrightarrow \mathbf{Tr}(s_1) \sim \mathbf{Tr}(s_2)$.

Proof:

(Only if) Assume $s_1 \sim_t s_2$. Define the following relation on $\mathbf{T}_{\Sigma^*}$:

$$\mathcal{R} = \{(\mathbf{Tr}(s), \mathbf{Tr}(s')) \mid s \sim_t s'\}.$$

We show that $\mathcal{R}$ is a bisimulation. By symmetry, it is sufficient to show that, for each $(\mathbf{Tr}(s), \mathbf{Tr}(s')) \in \mathcal{R}$, for each $e \in \mathbf{A}_S$

$$\mathbf{Tr}(s) \xrightarrow{e}_i x \Rightarrow \exists y : \mathbf{Tr}(s') \xrightarrow{e}_i y \text{ and } (x, y) \in \mathcal{R}.$$

Assume $\mathbf{Tr}(s) \xrightarrow{e}_i x$. Then, by the above lemma, there exists $\bar{s}$ such that $s \xrightarrow{e}_t \bar{s}$ and $\mathbf{Tr}(\bar{s})$=x. As $s \sim_t s'$, there exists $\bar{s}'$ such that $s' \xrightarrow{e}_t \bar{s}'$ and $\bar{s} \sim_t \bar{s}'$.

By the above lemma, $\mathbf{Tr}(s') \xrightarrow{e}_i \mathbf{Tr}(\bar{s}')$ and by definition of $\mathcal{R}$ we have that $(\mathbf{Tr}(\bar{s}), \mathbf{Tr}(\bar{s}')) \in \mathcal{R}$.

(If) Assume $\mathbf{Tr}(s_1) \sim \mathbf{Tr}(s_2)$. Define the following relation on $\mathcal{S}$:

$$\mathcal{R} = \{(s, s') \mid \mathbf{Tr}(s) \sim \mathbf{Tr}(s')\}.$$

Reasoning as above it is easy to show that $\mathcal{R}$ is a timed-bisimulation.

This shows the claim. □

Corollary 3.3.1 $\forall p, q \in \mathbf{P}\ p \sim^c q \Rightarrow p \sim_t q$.

Proof: Let ρ_t be the syntactic substitution defined, for all $a \in \mathbf{A}$, by

$$\rho_t(a) = S(a); F(a) \text{ and } \rho_t(F(a)) = nil; F(a).$$

Note that, for each $s \in \mathcal{S}$, $\mathbf{Tr}(s) = s\rho_t$. Assume now that $p \sim^c q$. Then, for each substitution ρ, $p\rho \sim q\rho$. This implies $p\rho_t \sim q\rho_t$. By the above observation, $p\rho_t = \mathbf{Tr}(p)$ and $q\rho_t = \mathbf{Tr}(q)$. By Theorem 3.3.3, $\mathbf{Tr}(p) \sim \mathbf{Tr}(q) \Leftrightarrow p \sim_t q$. Hence $p \sim^c q$ implies $p \sim_t q$. □

To complete the proof of our claim that $\sim_t$ coincides with $\sim^c$ for the language that we have introduced, we need to show that $\sim_t$ is contained in $\sim^c$. In the presence of the equational characterization of $\sim_t$ given in Theorem 3.3.2, it is possible to give an elegant, algebraic proof of this fact. Let us recall that, by Theorem 3.3.2, $\sim_t$ is the least congruence over $\mathbf{P}$ which satisfies the set of equations in Figure 3.1. Moreover, by the construction of $\sim^c$, $\sim^c$ is a congruence over $\mathbf{P}$. Hence, in order to show that $\sim_t \subseteq \sim^c$, it is sufficient to prove that $\sim^c$ satisfies all the equations which completely characterize $\sim_t$ over the language $\mathbf{P}$.

For the sake of completeness, let us recall that $\sim$ is a congruence with respect to all the combinators apart the action refinement one. Hence $\sim^c$ can simply be obtained by closing $\sim$ with respect to all the "refinement contexts". Formally, for each $p, q \in \mathbf{P}$,

$$\begin{aligned} p \sim^c q \quad & \text{iff} \quad \text{for all } \mathbf{P}_{ext}\text{-contexts } C[\cdot],\ C[p] \sim C[q] \\ & \text{iff} \quad \text{for all } a_0, \dots, a_n \in \mathbf{A},\ r_0, \dots, r_n \in \mathbf{P}_{ext},\ p\varrho \sim q\varrho, \end{aligned}$$

where $\varrho = [a_0/\mathbf{red}(r_0)] \cdots [a_n/\mathbf{red}(r_n)]$. We may now give a first, algebraic proof of the fact that $\sim_t$ coincides with $\sim^c$ over $\mathbf{P}$.

Theorem 3.3.4 (The Characterization Theorem: Algebraic Proof) *For each* $p, q \in \mathbf{P}$, $p \sim^c q$ *iff* $p \sim_t q$.

Proof: The "only if" implication follows by Corollary 3.3.1. In view of Theorem 3.3.2, to prove the "if" implication it is sufficient to show that, for each $p, q, r_0, \dots, r_n \in \mathbf{P}$ and $a_0, \dots, a_n \in \mathbf{A}$, whenever $p = q$ is an instance of an equation in Figure 3.1 then

$$p[a_0/r_0] \cdots [a_n/r_n] \quad \sim \quad q[a_0/r_0] \cdots [a_n/r_n].$$

This follows easily for each equation in Figure 3.1 because syntactic substitution is a homomorphism with respect to all the operators in Σ and by the soundness of all the equations with respect to $\sim$. For instance, let

$$(p \lfloor\!\lfloor q) \lfloor\!\lfloor t \quad = \quad p \lfloor\!\lfloor (q | t)$$

be an instance of axiom (B2) and ϱ denote the syntactic substitution $[a_0/r_0] \cdots [a_n/r_n]$. Then

$$\begin{aligned} ((p \lfloor\!\lfloor q) \lfloor\!\lfloor t)\varrho \quad & = \quad (p\varrho \lfloor\!\lfloor q\varrho) \lfloor\!\lfloor t\varrho \\ & \sim \quad p\varrho \lfloor\!\lfloor (q\varrho | t\varrho) \quad \text{as (B2) is sound with respect to } \sim \\ & = \quad (p \lfloor\!\lfloor (q | t))\varrho. \end{aligned}$$

Thus, for each $p, q \in \mathbf{P}$, $p \sim^c q$ iff $p \sim_t q$. □

However, the above-given algebraic proof of the characterization theorem for $\sim^c$ is language-dependent and relies on the equational characterization of $\sim_t$ given in Theorem 3.3.2. Moreover, for the proof to work, it is essential that only processes and not actions are referred to in the set of complete equations. As it is unlikely that such a characterization will be available over more complex algebras, for instance those including a restriction operator [Mil80], it would be useful to give an alternative, *behavioural* proof of the fact that $\sim_t$ is contained in $\sim^c$ over $\mathbf{P}$. Such a proof will allow us both to shed more light on the properties of $\sim_t$ which make sure that such an equivalence is preserved by action-refinement and to introduce a wealth of operational techniques and results which will prove to be of considerable use in extending the work presented in this chapter to richer algebras, (see Chapter 4). In order to give a behavioural proof of our claim that $\sim_t$ is contained in $\sim^c$ over $\mathbf{P}$, the main problem is to show that $\sim_t$ is preserved by the operation of action-refinement. We already know (Proposition 3.3.1) that $\sim_t$ is a congruence with respect to the other combinators of our calculus and that it is contained in $\sim$, so if we show that $\sim_t$ is preserved by $[a \rightsquigarrow q]$ the claim would follow by the fact that $\sim^c$ is the *largest congruence* contained in $\sim$. The proof of this result is much more involved and will be facilitated by the use of a version of timed equivalence, denoted by $\sim_r$ and called *refine-equivalence*, which is induced on $\mathcal{S}$ by a family of relations on a labelled version of the calculus.

3.4 Refine Equivalence

3.4.1 Motivation

In proving that $\sim_t$ is a congruence with respect to $[a \rightsquigarrow q]$, the combinator for refining actions by processes, it will be technically useful to consider a version of our language in which we may safely talk about different occurrences of the same action symbol in a term.

The need for this technical complication can be explained by means of an example.

Example 3.4.1 *Assume we have specified a system by means of the process term $p = (a|a)|a$ and, at a lower level of abstraction, we want to refine a to $q = b;c$. As parallel processes may evolve asynchronously, $p[a \rightsquigarrow q]$ may indeed reach a stage of its evolution in which the three a-processes will each be at a (possibly) different stage of their evolution. Hence it is necessary to devise a method which allows us to express the fact that different "copies" of process q are currently active and to keep track of the state each of them has reached.*

The solution we propose in this section is to consider a labelled version of our simple language. We will restrict ourselves to considering labelled process terms in which each labelled action occurs at most once. The labelled calculus will be endowed with a relation which captures a notion of bisimulation between labelled terms. Two labelled terms will be *bisimilar* if there is a kind of label-preserving correspondence between their labelled actions which is consistent with their dynamic behaviour. This notion of bisimulation on the labelled calculus will induce an equivalence relation on the unlabelled one which we denote $\sim_r$ and call *refine-equivalence*.

This new equivalence is of independent interest as it is an alternative formulation of the idea of strong bisimulation in the presence of non-instantaneous actions. We will eventually show that $\sim_r$ and $\sim_t$ coincide, but we can show fairly straightforwardly that $\sim_r$ is preserved by action-refinement.

3.4.2 The Calculus and its Operational Semantics

We now present the formal definition of the labelled calculus we will use in proving that $\sim_t$ is a congruence with respect to action-refinement.

Definition 3.4.1 (Uniquely Labelled Processes)

(i) *The set of* labelled atomic processes, **LA**, *is defined as follows:*

$$\mathbf{LA} =_{def} \{a_i \mid a \in \mathbf{A} \wedge i \in \mathbf{N}\}.$$

(ii) *The set of* uniquely labelled processes, **LP**, *is the set of terms generated by the following grammar:*

$$\pi ::= nil \mid a_i \mid \pi + \pi \mid \pi ; \pi \mid \pi | \pi \mid \pi \not| \pi,$$

where $a_i \in \mathbf{LA}$, *subject to the constraint that each index* i *occurs at most once in* π. **LP** *will be ranged over by* $\pi, \pi', \pi_1 \ldots$.

Definition 3.4.2 (Configurations) *The set of* uniquely labelled states, *or* configurations, **L$\mathcal{S}$** *is defined by simply adapting the clauses of Definition 3.3.1 to the set of uniquely labelled processes, subject to the constraint that, for each configuration* c, *each index* i *occurs at most once in* c. **L$\mathcal{S}$** *will be ranged over by* c, d, c', d' ....

The operational semantics for labelled states will be defined in terms of next-event relations $\stackrel{\lambda}{\longmapsto}$, where λ ranges over the set of *labelled subactions*

$$\mathbf{LA_S} =_{def} \{S(a_i) \mid a_i \in \mathbf{LA}\} \cup \{F(a_i) \mid a_i \in \mathbf{LA}\}.$$

The interpretation we impose on labelled subactions is a simple variant of the one we have imposed on the subactions in $\mathbf{A_S}$. For each $a_i \in \mathbf{LA}$, the ith occurrence of the action symbol a, $S(a_i)$ stands for the beginning of its execution and $F(a_i)$ stands for its termination.

The termination predicate on **L$\mathcal{S}$**, needed in the definition of the operational semantics, and the operational semantics for **L$\mathcal{S}$** itself can be defined using the standard method. We leave the details to the reader who should also check the following result.

Proposition 3.4.1 $\forall c \in \mathbf{L}\mathcal{S}, \lambda \in \mathbf{LA_S}\ c \overset{\lambda}{\longmapsto} c' \Rightarrow c' \in \mathbf{L}\mathcal{S}$.

Having developed an operational view of configurations we would like to say when two configurations c and d are *behaviourally equivalent.* The notion of equivalent behaviour between configurations will not be absolute but will be parameterized with respect to a correspondence between their component states of the form $F(a_i)$ for some $a_i \in \mathbf{LA}$. The technical definition will take the form of a parameterized timed bisimulation. The next few definitions introduce the technical machinery we will need in introducing this kind of relation.

The parameter in the definition of bisimulation for configurations will be a family of binary relations, one for each $a \in \mathbf{A}$, on $\mathbf{N}$. If ϕ is one of those families of relations and $c \sim_\phi d$, $c, d \in \mathbf{L}\mathcal{S}$, then, for each $(i, j) \in \phi_a$, $F(a_i)$ and $F(a_j)$ *play the same role* in the dynamic behaviour of the configurations.

The informal notion of *playing the same role* stated above will be formalized by the definition of bisimulation.

Definition 3.4.3

(i) *Let $\mathcal{R}$ be a binary relation over* $\mathbf{N}$. *Then $\mathcal{R}$ is (the graph of) a partial bijection on* $\mathbf{N}$ *iff it is a bijection between two subsets of* $\mathbf{N}$.

(ii) *$\mathcal{H}$ will denote the set of* $\mathbf{A}$*-indexed families of partial bijections over* $\mathbf{N}$ *and will be ranged over by* $\phi, \varphi \ldots$.

Notation 3.4.1 *In what follows $\emptyset$ will denote both the empty set and, with abuse of notation, the* $\mathbf{A}$*-indexed family of relations such that*

$$\forall a \in \mathbf{A}\ a \mapsto \emptyset.$$

We will also sometimes write $(i, j) \in \phi_a$ as $(a, i, j) \in \phi$ or $\phi_a(i) = j$, where $\phi \in \mathcal{H}$. All the usual operations and predicates on relations will be be extended pointwise to indexed families of relations in the natural way.

For example, if $\mathcal{R}$ and $\mathcal{R}'$ are $\mathcal{H}$-indexed families of binary relations

$$\mathcal{R} \subseteq \mathcal{R}' \Leftrightarrow \forall \phi \in \mathcal{H}\ \mathcal{R}_\phi \subseteq \mathcal{R}'_\phi.$$

An $\mathbf{A}$-indexed family of partial bijections $\phi \in \mathcal{H}$ will be used to record information about which actions have been started and to whom they have been matched in the bisimulation functional. We shall call such information a *history*. Intuitively, the presence of (a, i, j) in a history ϕ will mean that in one process action a_i has started, in the other action a_j has started and we are expecting the ith occurrence of a in one to be matched to the jth in the other.

Given a configuration c, any $F(a_i)$ occurring in c will be called a *place* of c. Intuitively $F(a_i)$ stands for a state in which the execution of the atomic labelled process a_i has started but has not terminated yet. The set of a-places in c is defined as follows:

Definition 3.4.4 *For each* $a \in \mathrm{A}, c \in \mathrm{L}\mathcal{S}$

$$\mathrm{Places}(a, c) =_{def} \{i \mid F(a_i) \textit{ occurs in } c\}.$$

Configurations will be equivalent or inequivalent with respect to a history ϕ. For c and d to be equivalent with respect to $\phi \in \mathcal{H}$, we shall require that:

- all actions which have started in c or d and are not yet finished must be recorded in ϕ;
- every start move of c, $c \overset{S(a_i)}{\longmapsto} c'$, must be matched by a corresponding move $d \overset{S(a_j)}{\longmapsto} d'$ of d such that c' and d' are equivalent with respect to the augmented history $\phi \cup \{(a, i, j)\}$; and
- every end move from c, $c \overset{F(a_i)}{\longmapsto} c'$, must be matched by the end move of d associated with it in the history ϕ, i.e. $d \overset{F(a_j)}{\longmapsto} d'$ with $(a, i, j) \in \phi$ and c' and d' are equivalent with respect to the diminished history $\phi - \{(a, i, j)\}$.

This the motivation underlying the following formal definition.

Definition 3.4.5 (Strong Refine Bisimulation)

(1) *Let* $\mathcal{P}(\mathrm{L}\mathcal{S}^2)^{\mathcal{H}}$ *denote the set of* $\mathcal{H}$*- indexed families of binary relations over* $\mathrm{L}\mathcal{S}$.

(2) *Define a functional* $\mathcal{G} : \mathcal{P}(\mathrm{L}\mathcal{S}^2)^{\mathcal{H}} \rightarrow \mathcal{P}(\mathrm{L}\mathcal{S}^2)^{\mathcal{H}}$ *as follows:*

given $\mathcal{R} \in \mathcal{P}(\mathrm{L}\mathcal{S}^2)^{\mathcal{H}}$, $(c_1, c_2) \in \mathcal{G}(\mathcal{R})_\phi$ *iff*

(i) $\forall a \in \mathrm{A}$ $dom(\phi_a) = \mathrm{Places}(a, c_1)$ and $range(\phi_a) = \mathrm{Places}(a, c_2)$

(ii) $\forall a_i \in \mathrm{LA}$

(a) $c_1 \overset{S(a_i)}{\longmapsto} c_1' \Rightarrow \exists j \in \mathrm{N}, c_2' \in \mathrm{L}\mathcal{S} : c_2 \overset{S(a_j)}{\longmapsto} c_2'$ and $(c_1', c_2') \in \mathcal{R}_{\phi \cup \{(a,i,j)\}}$

(b) $c_1 \overset{F(a_i)}{\longmapsto} c_1' \Rightarrow \exists c_2' \in \mathrm{L}\mathcal{S} : \phi_a(i) = j$ and $c_2 \overset{F(a_j)}{\longmapsto} c_2'$ and $(c_1', c_2') \in \mathcal{R}_{\phi - \{(a,i,j)\}}$

(iii) $\forall a_j \in \mathrm{LA}$

(a) $c_2 \overset{S(a_j)}{\longmapsto} c_2' \Rightarrow \exists i \in \mathrm{N}, c_1' \in \mathbf{L}\mathcal{S} : c_1 \overset{S(a_i)}{\longmapsto} c_1'$ and $(c_1', c_2') \in \mathcal{R}_{\phi \cup \{(a,i,j)\}}$

(b) $c_2 \overset{F(a_j)}{\longmapsto} c_2' \Rightarrow \exists c_1' : \phi_a(i) = j$ and $c_1 \overset{F(a_i)}{\longmapsto} c_1'$ and $(c_1', c_2') \in \mathcal{R}_{\phi - \{(a,i,j)\}}$

(3) $\mathcal{R} \in \mathcal{P}(\mathbf{L}\mathcal{S}^2)^{\mathcal{H}}$ *is a* strong refine bisimulation *iff* $\mathcal{R} \subseteq \mathcal{G}(\mathcal{R})$.
$\forall \phi \in \mathcal{H} \sim_\phi =_{def} \bigcup\{\mathcal{R}_\phi \mid \mathcal{R} \subseteq \mathcal{G}(\mathcal{R})\}$.

$(\mathcal{P}(\mathbf{L}\mathcal{S}^2)^{\mathcal{H}}, \subseteq)$ is a complete lattice as $\subseteq$ is induced pointwise by the relation of set inclusion. The following result is then standard.

Proposition 3.4.2

(i) $\mathcal{G}(\cdot)$ *is a monotonic endofunction over the complete lattice* $(\mathcal{P}(\mathbf{L}\mathcal{S}^2)^{\mathcal{H}}, \subseteq)$.

(ii) $\{\sim_\phi \mid \phi \in \mathcal{H}\}$ *is the maximum fixed-point of* $\mathcal{G}(\cdot)$.

3.4.3 Definition of Refine Equivalence

The maximum fixed-point of the functional $\mathcal{G}(\cdot)$, $\{\sim_\phi \mid \phi \in \mathcal{H}\}$, can be used to induce an equivalence relation on the set $\mathcal{S}$ of unlabelled states. This relation is the *refine equivalence* promised in the previous section and is a technical tool used to give a behavioural proof of the fact that $\sim_t$ is preserved by $[a \rightsquigarrow q]$. The definition is very similar to that of timed-equivalence. There is only the additional constraint that occurrences of actions should be properly matched; namely, whenever the start of the ith occurrence of an action is matched by the start of the jth occurrence then the finish of the jth occurrence must also be used to match the finish of the ith occurrence.

This definition is more in accordance with the intuitive idea underlying strong observational equivalence in the presence of non-instantaneous actions. We still want to match the ability of processes to perform actions and their resulting potential behaviour and, to ensure that complete actions are compared, we should consider not only their beginnings and endings, but also their entire duration. This is achieved by using histories.

Definition 3.4.6

(1) *The function* $\mathrm{un}(\cdot)$:$\mathbf{L}\mathcal{S} \mapsto \mathcal{S}$ *is defined as the unique homomorphism between the two algebras such that:*

- $\mathrm{un}(nil) = nil$
- $\mathrm{un}(a_i) = a$

- $\mathbf{un}(F(a_i)) = F(a)$.

(2) $\forall s_1, s_2 \in \mathcal{S}\ s_1 \sim_r s_2 \Leftrightarrow \exists c_1, c_2 \in \mathbf{L}\mathcal{S} : \mathbf{un}(c_i) = s_i,\ i = 1, 2,$ and $\exists \phi \in \mathcal{H} : c_1 \sim_\phi c_2$.

Note: In what follows we will often make use of the obvious fact that $\mathbf{un}(\cdot)$ is onto.

Notation 3.4.2 *From now on, if* $e \in \mathbf{A_S}$ *and* $i \in \mathbf{N}$ *then* e_i *will stand for* $S(a_i)$, *if* $e = S(a)$, $F(a_i)$, *if* $e = F(a)$.

The following lemma relates the derivations of a configuration c with those of its corresponding unlabelled state $\mathbf{un}(c)$.

Lemma 3.4.1 *Let* $c \in \mathbf{L}\mathcal{S}$. *Then:*

(i) $c\surd \Leftrightarrow \mathbf{un}(c)\surd$.

(ii) $\forall e_i \in \mathbf{LA_S}\ c \stackrel{e_i}{\longmapsto} c' \Rightarrow \mathbf{un}(c) \stackrel{e}{\longrightarrow}_t \mathbf{un}(c')$.

(iii) $\forall e \in \mathbf{A_S}\ \mathbf{un}(c) \stackrel{e}{\longrightarrow}_t s \Rightarrow \exists c', j :\ c \stackrel{e_j}{\longmapsto} c' \wedge \mathbf{un}(c') = s$.

Proof: By structural induction on c. Omitted. □

One simple corollary of this lemma is the fact that $\sim_r$ is contained in $\sim_t$.

Theorem 3.4.1 $(\sim_r \subseteq \sim_t)$: $\forall s_1, s_2 \in \mathcal{S}\ s_1 \sim_r s_2 \Rightarrow s_1 \sim_t s_2$.

Proof: Define the following relation $\mathcal{R} \subseteq \mathcal{S}^2$:

$$\mathcal{R} =_{def} \{(\mathbf{un}(c), \mathbf{un}(d)) \mid \exists \phi \in \mathcal{H} :\ c \sim_\phi d\}.$$

We show that $\mathcal{R}$ is a timed-bisimulation. By symmetry, it is sufficient to show that, for each $(\mathbf{un}(c), \mathbf{un}(d)) \in \mathcal{R}$, $\mathbf{un}(c) \stackrel{e}{\longrightarrow}_t s \Rightarrow \exists s' :\ \mathbf{un}(d) \stackrel{e}{\longrightarrow}_t s' \wedge (s, s') \in \mathcal{R}$.

Assume $\mathbf{un}(c) \stackrel{S(a)}{\longrightarrow}_t s$. Then, by lemma 3.4.1,

$$\exists i, c' :\ c \stackrel{S(a_i)}{\longmapsto} c' \wedge \mathbf{un}(c') = s.$$

As $c \sim_\phi d$ for some $\phi \in \mathcal{H}$,

$$\exists j, d' :\ d \stackrel{S(a_j)}{\longmapsto} d' \wedge c' \sim_{\phi \cup \{(a,i,j)\}} d'.$$

By lemma 3.4.1, $d \overset{S(a_j)}{\longmapsto} d'$ implies $\mathbf{un}(d) \overset{S(a)}{\longrightarrow}_t \mathbf{un}(d')$. Moreover, by the definition of $\mathcal{R}$, $(\mathbf{un}(c'), \mathbf{un}(d')) \in \mathcal{R}$.

Similarly one can match the $F(a)$-moves of the configuration c.

Hence $\mathcal{R}$ is a timed-bisimulation.

Let $s_1, s_2 \in \mathcal{S}$ be such that $s_1 \sim_r s_2$. Then, by the definition of $\sim_r$, $s_1 \sim_r s_2$ implies $(s_1, s_2) \in \mathcal{R}$. It follows that $s_1 \sim_t s_2$. □

We now show that $\sim_r$ is an equivalence relation on $\mathcal{S}$.

Theorem 3.4.2 *$\sim_r$ is an equivalence relation on $\mathcal{S}$.*

Proof: We will show that $\sim_r$ is reflexive and transitive leaving the proof that it is symmetric to the reader.

Reflexivity Let $1_I =_{def} \{(i,i) \mid i \in I\}$, for $I \subseteq \mathbf{N}$. A family of relations ϕ in $\mathcal{H}$ is an identity if, and only if, for each $a \in \mathbf{A}$ $\phi_a = 1_I$ for some $I \subseteq \mathbf{N}$.

Define $\mathcal{R} \in \mathcal{P}(\mathbf{L}\mathcal{S}^2)^{\mathcal{H}}$ as follows:

$$\mathcal{R}_\phi =_{def} \begin{cases} \emptyset & \text{if } \phi \text{ is not an identity} \\ \{(c,c) \mid \forall a \in \mathbf{A}\ \mathbf{Places}(a,c) = dom(\phi_a)\} & \text{otherwise} \end{cases}$$

It can be easily seen that $\mathcal{R}$ is a strong refine bisimulation.

As $\forall s \in \mathcal{S} \exists c \in \mathbf{L}\mathcal{S} :\ \mathbf{un}(c) = s$ and $(c,c) \in \mathcal{R}_\phi$, where ϕ is such that

$$\forall a \in \mathbf{A}\ \phi_a = \{(i,i) \mid i \in \mathbf{Places}(a,c)\},$$

we have that $c \sim_\phi c$ and $s \sim_r s$.

Transitivity Define the following $\mathcal{R} \in \mathcal{P}(\mathbf{L}\mathcal{S}^2)^{\mathcal{H}}$:

$$\forall \phi \in \mathcal{H}\ \mathcal{R}_\phi =_{def} \bigcup \{\sim_\psi \circ \sim_\varphi \mid \phi = \psi \circ \varphi \wedge \psi, \varphi \in \mathcal{H}\}.$$

It can be shown that $\mathcal{R}$ is a strong refine bisimulation.

Moreover, $s_1 \sim_r s_2$ and $s_2 \sim_r s_3$ imply

$$\exists c_i,\ i = 1,2,3, : \mathbf{un}(c_i) = s_i \textit{ and } \exists \psi, \varphi : c_1 \sim_\varphi c_2 \wedge c_2 \sim_\psi c_3.$$

Hence $(c_1, c_3) \in \mathcal{R}_{\psi \circ \varphi}$. This implies $c_1 \sim_{\psi \circ \varphi} c_3$ and $s_1 \sim_r s_3$. □

Note that from the definition of $\sim_\phi$ it follows easily that

$$\forall \pi_1, \pi_2 \in \mathbf{LP}\ \pi_1 \sim_\phi \pi_2 \Rightarrow \phi = \emptyset.$$

Hence, for all $p, q \in \mathbf{P}$, $p \sim_r q$ if, and only if,

$$\exists \pi_1, \pi_2 :\ \mathbf{un}(\pi_1) = p \wedge \mathbf{un}(\pi_2) = q \wedge \pi_1 \sim_\emptyset \pi_2.$$

We will now show that in the definition of $\sim_r$ we could have used a *for all* instead of $\exists c_1, c_2 \ldots$. We hope that this result adds to the intuition of the definition of strong refine bisimulation. To do so we first show that if c and d are configurations which reduce to the same unlabelled state $s \in \mathcal{S}$ then there exists a $\phi \in \mathcal{H}$ such that $c \sim_\phi d$. First of all we prove a technical lemma which will be needed in what follows.

Lemma 3.4.2 *Let* $c, d \in \mathbf{L}\mathcal{S}$. *Assume* $\mathbf{un}(c) = \mathbf{un}(d)$ and $c \overset{F(a_i)}{\longmapsto} c'$.

Then there exists a unique $j \in \mathrm{N}, d' \in \mathbf{L}\mathcal{S}$ *such that*

$$d \overset{F(a_j)}{\longmapsto} d' \text{ and } \mathbf{un}(c') = \mathbf{un}(d').$$

Proof: By induction on the structure of c.

$c = \pi$ Vacuous.

$c = F(a_i)$ Then d is of the form $F(a_j)$ for some $j \in \mathrm{N}$. The result follows trivially.

$c = c_1; \pi$ Then d is of the form $d_1; \pi_1$ with $\mathbf{un}(c_1) = \mathbf{un}(d_1)$ and $\mathbf{un}(\pi) = \mathbf{un}(\pi_1)$. Assume $c_1; \pi \overset{F(a_i)}{\longmapsto} c_1'; \pi$ because $c_1 \overset{F(a_i)}{\longmapsto} c_1'$.

By the inductive hypothesis, $\exists! j \in \mathrm{N}, \exists! d_1' : d_1 \overset{F(a_j)}{\longmapsto} d_1'$ and $\mathbf{un}(c_1') = \mathbf{un}(d_1')$. Hence $d_1'; \pi_1$ and j are unique such that

$$d_1; \pi_1 \overset{F(a_j)}{\longmapsto} d_1'; \pi_1 \text{ and } \mathbf{un}(c_1'; \pi) = \mathbf{un}(d_1'; \pi_1).$$

$c = c_1 | c_2$ Similar. □

Lemmas 3.4.1-3.4.2 will also be useful in deriving a strong refine bisimulation $\mathcal{R}(c, d)$ and $\phi(c, d) \in \mathcal{H}$ such that $(c, d) \in \mathcal{R}(c, d)_{\phi(c,d)}$. The next definition, which is rather technical, is devoted to the construction of such a strong refine bisimulation. The construction is by induction on the size of c (and therefore also of d as $\mathbf{un}(c) = \mathbf{un}(d)$).

Definition 3.4.7 *Let $c, d \in \mathbf{L}\mathcal{S}$. Assume* $\mathbf{un}(c) = \mathbf{un}(d)$.

The following inductive construction is given:

- *If $c\surd$ and $d\surd$ then $\mathcal{R}(c,d) \in \mathcal{P}(\mathbf{L}\mathcal{S}^2)^{\mathcal{H}}$ is defined as follows:*

$$\forall \phi \in \mathcal{H}\ \mathcal{R}(c,d)_\phi =_{def} \begin{cases} \{(c,d)\} & \text{if } \phi = \emptyset \\ \emptyset & \text{otherwise} \end{cases}$$

  Moreover, $\phi(c,d) = \emptyset$.

- *If c and d are not terminated then define the following sets:*

$$\mathbf{S} =_{def} \{(c',d') \mid \exists S(a_i), S(a_j) : c \overset{S(a_i)}{\longmapsto} c' \wedge d \overset{S(a_j)}{\longmapsto} d' \wedge \mathbf{un}(c') = \mathbf{un}(d')\}$$

$$\mathbf{F} =_{def} \{(c',d') \mid \exists F(a_i), F(a_j) : c \overset{F(a_i)}{\longmapsto} c' \wedge d \overset{F(a_j)}{\longmapsto} d' \wedge \mathbf{un}(c') = \mathbf{un}(d')\}$$

$$\mathbf{X} =_{def} \{(F(a_i), F(a_j)) \mid c \overset{F(a_i)}{\longmapsto} c' \wedge d \overset{F(a_j)}{\longmapsto} d' \wedge \mathbf{un}(c') = \mathbf{un}(d')\}.$$

  Note that, by Lemma 3.4.2, $\mathbf{F}$ and $\mathbf{X}$ are graphs of injective functions.

  Then $\phi(c,d) =_{def} \{(a,i,j) \mid (F(a_i), F(a_j)) \in \mathbf{X}\}$ and $\mathcal{R}(c,d)$ is defined as follows:

$$\forall \phi \in \mathcal{H}\ \ \mathcal{R}(c,d)_\phi =_{def} \begin{cases} \bigcup_{(c',d') \in \mathbf{S} \cup \mathbf{F}} \mathcal{R}(c',d')_\phi & \text{if } \phi \neq \phi(c,d) \\ \bigcup_{(c',d') \in \mathbf{S} \cup \mathbf{F}} \mathcal{R}(c',d')_\phi \cup \{(c,d)\} & \text{if } \phi = \phi(c,d) \end{cases}$$

Note that in the above given construction the sets $\mathbf{S}$, $\mathbf{F}$, $\mathbf{X}$ are always finite. Moreover, the size of c' (and therefore also of d') has decreased. Before showing that the construction is indeed correct we state the following lemma.

Lemma 3.4.3 *For each $c, d \in \mathbf{L}\mathcal{S}$ such that* $\mathbf{un}(c) = \mathbf{un}(d)$ *the following statements hold:*

(1) $c \overset{S(a_i)}{\longmapsto} c'$ and $d \overset{S(a_j)}{\longmapsto} d'$ and $\mathbf{un}(c') = \mathbf{un}(d')$ implies $\phi(c',d') = \phi(c,d) \cup \{(a,i,j)\}$

(2) $c \overset{F(a_i)}{\longmapsto} c'$ and $d \overset{F(a_j)}{\longmapsto} d'$ and $\mathbf{un}(c') = \mathbf{un}(d')$ implies $\phi(c',d') = \phi(c,d) - \{(a,i,j)\}$.

Proof: By structural induction on $\mathbf{un}(c)$. Omitted. □

Proposition 3.4.3 *For each $c, d \in \mathbf{L}\mathcal{S}$ such that* $\mathbf{un}(c) = \mathbf{un}(d)$, *$\mathcal{R}(c,d)$ is a strong refine bisimulation.*

Proof: By induction on the size of c (and therefore also of d).

The base case, $c\surd$ and $d\surd$ is obvious.

Assume c is not terminated. By inductive hypothesis, we know that for each $(c', d') \in \mathbf{S} \cup \mathbf{F}$, $\mathcal{R}(c', d')$ is a strong refine bisimulation.

It is therefore sufficient to show that $(c, d) \in \mathcal{G}(\mathcal{R}(c, d))_{\phi(c,d)}$.

Assume $c \overset{S(a_i)}{\longmapsto} c'$. Then, by lemma 3.4.1, $\mathbf{un}(c) \overset{S(a)}{\longrightarrow}_t \mathbf{un}(c')$ and this implies the existence of j, d' such that $d \overset{S(a_j)}{\longmapsto} d'$ and $\mathbf{un}(d') = \mathbf{un}(c')$.

By induction, $(c', d') \in \mathcal{R}_{\phi(c',d')}$.

Also, by the above lemma, $\phi(c', d') = \phi(c, d) \cup \{(a, i, j)\}$ and therefore $\mathbf{Places}(a, c) = dom(\phi_a(c, d))$ and $\mathbf{Places}(a, d) = range(\phi_a(c, d))$, since the corresponding property holds of $\phi(c', d')$.

The argument for the case when $c \overset{F(a_i)}{\longmapsto} c'$ is similar.

Hence, by symmetry, $\mathcal{R}(c, d)$ is a strong refine bisimulation. □

As a corollary we get the following result which shows that in the definition of $\sim_r$ we could have made do with a *for all* instead of *there exist*.

Corollary 3.4.1 *The following statements hold:*

(i) $\forall c, d \in \mathbf{L}\mathcal{S}$ $\mathbf{un}(c) = \mathbf{un}(d) \Rightarrow c \sim_{\phi(c,d)} d$.

(ii) *Let* $s_1, s_2 \in \mathcal{S}$. *Then* $s_1 \sim_r s_2$ *iff for all* $c_1, c_2 \in \mathbf{L}\mathcal{S}$ *such that* $\mathbf{un}(c_1) = s_1$ and $\mathbf{un}(c_2) = s_2$, *there exists* $\phi \in \mathcal{H}$ *such that* $c_1 \sim_\phi c_2$.

3.4.4 Refine Equivalence and Timed Equivalence Coincide

This section is entirely devoted to showing that, for our language, $\sim_r$ and $\sim_t$ coincide. However, contrary to what happens with $\sim_t$, it will be quite straightforward to give a behavioural proof of the fact that $\sim_r$ is preserved by action-refinement. We have already shown that $\sim_r$ is contained in $\sim_t$. Unfortunately the converse is rather involved. Its proof is developed in two steps. First of all we show that $\sim_r$ is a congruence with respect to $+, ;, |, /$. Then we show that $\sim_r$ satisfies all the axioms which completely characterize $\sim_t$ over our language (the axioms are collected in Figure 3.1). The completeness theorem (Theorem 3.3.2) tells us that $\sim_t$ is the *least congruence* on $\mathbf{P}$ which satisfies the axioms; hence we will have shown the desired inclusion.

Theorem 3.4.3 *$\sim_r$ is a* **P**-*congruence.*

Proof: We will show that $\sim_r$ is preserved under $|$ leaving the other cases to the reader. Assume $s_1 \sim_r s_2$. Then, by definition of $\sim_r$,

$$\exists c_1, c_2 \in \mathbf{L}\mathcal{S} : \mathbf{un}(c_i) = s_i, i = 1, 2, \text{ and } \exists \bar{\phi} \in \mathcal{H} : c_1 \sim_{\bar{\phi}} c_2.$$

Let $s \in \mathcal{S}$. Then there exists $c \in \mathbf{L}\mathcal{S}$ such that $(c_1|c, c_2|c) \in \mathbf{L}\mathcal{S}^2$ and $\mathbf{un}(c) = s$. Define $\mathcal{R} \in \mathcal{P}(\mathbf{L}\mathcal{S}^2)^{\mathcal{H}}$ as follows:

$$\forall \phi \in \mathcal{H}\ \mathcal{R}_\phi =_{def} \{(c|c', d|d') \in \mathbf{L}\mathcal{S}^2 \mid \exists \phi_1, \phi_2 \in \mathcal{H} : \phi = \phi_1 \cup \phi_2 \wedge c \sim_{\phi_1} d \wedge c' \sim_{\phi_2} d'\}.$$

We show that $\mathcal{R}$ is a strong refine bisimulation. Assume $(c|c', d|d') \in \mathcal{R}_\phi$. Then there exist $\phi_1, \phi_2 \in \mathcal{H} : \phi_1 \cup \phi_2 = \phi$ and $c \sim_{\phi_1} d$ and $c' \sim_{\phi_2} d'$. Note that, as $(c|c', d|d') \in \mathbf{L}\mathcal{S}^2$, $\phi_1 \cap \phi_2 = \emptyset$.

(i) We show that clause (i) of the definition of strong refine bisimulation holds. This is because

$$\forall a \in \mathbf{A}\ \mathbf{Places}(a, c|c') = \mathbf{Places}(a, c) \cup \mathbf{Places}(a, c') = dom(\phi_{1_a}) \cup dom(\phi_{2_a}).$$

Moreover $dom(\phi_a) = dom(\phi_{1_a}) \cup dom(\phi_{2_a})$. Similarly for the range of the relation ϕ_a.

(ii) Assume, without loss of generality, that $c|c' \overset{S(a_i)}{\longmapsto} \bar{c}|c'$ because $c \overset{S(a_i)}{\longmapsto} \bar{c}$. Then $\exists j, \bar{d} : d \overset{S(a_j)}{\longmapsto} \bar{d}$ and $(\bar{c}, \bar{d}) \in \phi_1 \cup \{(a, i, j)\}$. This implies $d|d' \overset{S(a_j)}{\longmapsto} \bar{d}|d'$. Moreover, by the property of the operational semantics, $(\bar{c}|c', \bar{d}|d') \in \mathbf{L}\mathcal{S}^2$ and $(\bar{c}|c', \bar{d}|d') \in \mathcal{R}_{\phi_1 \cup \phi_2 \cup \{(a,i,j)\}}$, by definition of $\mathcal{R}$.

(iii) Assume, wlog, that $c|c' \overset{F(a_i)}{\longmapsto} \bar{c}|c'$ because $c \overset{F(a_i)}{\longmapsto} \bar{c}$.

As $c \sim_{\phi_1} d$ by hypothesis, $\exists \bar{d} : d \overset{F(a_j)}{\longmapsto} \bar{d}$ and $\phi_{1_a}(i) = j$ and $(\bar{c}, \bar{d}) \in \sim_{\phi_1 - \{(a,i,j)\}}$. This implies $d|d' \overset{F(a_j)}{\longmapsto} \bar{d}|d'$. Moreover, as $\phi_{1_a} \cap \phi_{2_a} = \emptyset$, $(\phi_1 \cup \phi_2)_a(i) = j$.

As $(\bar{c}|c', \bar{d}|d') \in \mathbf{L}\mathcal{S}^2$ we have that, by definition of $\mathcal{R}$, $(\bar{c}|c', \bar{d}|d') \in \mathcal{R}_{(\phi_1 - \{(a,i,j)\}) \cup \phi_2}$.

Moreover, as $\phi_1 \cap \phi_2 = \emptyset$,

$$(\phi_1 - \{(a, i, j)\}) \cup \phi_2 = (\phi_1 \cup \phi_2) - \{(a, i, j)\}.$$

By symmetry $\mathcal{R}$ is a strong refine bisimulation.

As, by definition of $\mathcal{R}$, $(c_1|c, c_2|c) \in \mathcal{R}_\psi$, where

$$\forall a \in \mathbf{A}\ \psi_a =_{def} \bar{\phi}_a \cup \{(i, i) \mid i \in \mathbf{Places}(a, c)\},$$

we conclude that $c_1|c \sim_\psi c_2|c$. Hence $s_1|s \sim_r s_2|s$. □

To complete the proof of the inclusion $\sim_t \subseteq \sim_r$, we need to show that $\sim_r$ satisfies all the laws of the complete axiomatization of $\sim_t$ on the language $\mathbf{P}$. The equational characterization of $\sim_t$ is given in Figure 3.1.

Theorem 3.4.4 *Let $\mathcal{A}$ denote the set of axioms in Figure 1.*

Then $t = t' \in \mathcal{A}$ implies $t = t'$ is valid for $\sim_r$.

Proof: The proof of this result is rather technical and tedious. For instance, when proving the validity of axioms (B1) and (X1) care must be taken in making "copies" of x, y and z with disjoint labellings.

The techniques developed in §4 are very useful in this respect. We examine only the axiom (B2), which will allow us to show the general proof technique. The remaining axioms are left to the reader.

(B2): $(x \lfloor\!\lfloor y) \lfloor\!\lfloor z = x \lfloor\!\lfloor (y|z)$. Define $\mathcal{R} \in \mathcal{P}(\mathbf{L}\mathcal{S}^2)^{\mathcal{H}}$ as follows:

$$\forall \phi \in \mathcal{H}\ \mathcal{R}_\phi =_{def} \{((d_1 \lfloor\!\lfloor d_2) \lfloor\!\lfloor d_3, d_1 \lfloor\!\lfloor (d_2|d_3)), ((d_1|d_2)|d_3, d_1|(d_2|d_3)) \in \mathbf{L}\mathcal{S}^2 \mid \Gamma\}$$

where Γ is the predicate, a function of d_1, d_2, d_3, ϕ, defined as follows:

$$\forall a \in \mathbf{A}\ \phi_a = \{(i,i) \mid i \in \bigcup_{j=1}^{3} \mathbf{Places}(a, d_j)\}.$$

We show that $\mathcal{R}$ is a strong refine bisimulation.

Assume $((d_1 \lfloor\!\lfloor d_2) \lfloor\!\lfloor d_3, d_1 \lfloor\!\lfloor (d_2|d_3)) \in \mathcal{R}_\phi$ and $(d_1 \lfloor\!\lfloor d_2) \lfloor\!\lfloor d_3 \stackrel{S(a_i)}{\longmapsto} (d_1'|d_2)|d_3$.
Then $d_1 \stackrel{S(a_i)}{\longmapsto} d_1'$.

This implies $d_1 \lfloor\!\lfloor (d_2|d_3) \stackrel{S(a_i)}{\longmapsto} d_1'|(d_2|d_3)$.

By the definition of $\mathcal{R}$, $((d_1'|d_2)|d_3, d_1'|(d_2|d_3)) \in \mathcal{R}_{\phi \cup \{(a,i,i)\}}$.

Similarly for moves of the form $(d_1 \lfloor\!\lfloor d_2) \lfloor\!\lfloor d_3 \stackrel{F(a_i)}{\longmapsto} (d_1'|d_2)|d_3$.

The case $((d_1|d_2)|d_3, d_1|(d_2|d_3))$ follows a similar pattern.

Hence $\mathcal{R}$ is a strong refine bisimulation and this is sufficient to show the validity of axiom (B2) for $\sim_r$. □

Finally we may state the main result of this section, namely that $\sim_r$ and $\sim_t$ coincide.

Theorem 3.4.5 $\forall s_1, s_2 \in \mathcal{S}\ s_1 \sim_t s_2 \Leftrightarrow s_1 \sim_r s_2$.

Proof: Follows from Theorems 3.4.1, 3.4.3 and 3.4.4. □

3.4.5 Refine Equivalence is Preserved by Action Refinement

This section is devoted to giving a behavioural proof of the fact that $\sim_r$ is preserved by action-refinement. The proof is articulated in several steps. First of all we define a suitable notion of substitution for labelled terms. Then we proceed by studying the relationships between the moves of a configuration c to which a substitution ρ has been applied and those of $\mathbf{un}(c)$ and ρ.

We conclude the section by showing that $\sim_r$, and thus $\sim_t$, is a congruence with respect to substitution.

Definition 3.4.8 *A* substitution ρ *is a mapping* $\rho : \mathbf{LA} \cup \{F(a_i) | a_i \in \mathbf{LA}\} \to \mathcal{S}$ *such that:*

(i) $\forall a_i \in \mathbf{LA}\ \rho(a_i) \in \mathbf{P}$;

(ii) $\forall i, j \in \mathrm{N}\ \rho(a_i) = \rho(a_j)$.

Let **SUB** *denote the set of all the substitutions.*

Note that we require actions to be substituted with processes and not with states (this satisfies our intuition that an action stands for a task which has not started evolving yet), but, for instance, we allow actions to be mapped to terminated processes.

Example 3.4.2 *Consider* $p = (a_1; b_1) + (a_2; c_1)$. *This process will choose non-deterministically one of the two branches of the computation it stands for when presented with the action* a *(remember that* a_1 *and* a_2 *are considered as two occurrences of action* a*). Moreover, this choice is made internally by the process itself and cannot be influenced by the experimenter. This is an example of* internal non-determinism.

Now let ρ *be a substitution mapping* a_1 *and* a_2 *to nil and renaming* b_1 *to* b *and* c_1 *to* c. *Then* $p\rho = (nil; b) + (nil; c)$. *By means of the substitution we have hidden the* a*-actions and, in doing so, we have made the choice of the branch of the computation controllable by the environment. We have made the non-determinism exhibited by the process* external.

This power of process substitution makes it more difficult to relate the moves of $c\rho$ with those of c and those of ρ. To do so we need a formal method for dealing with the situations described by the above example.

We define a way of associating to each configuration c a set of pairs of the form (A, d) where A is a set of labelled actions and states of the form $F(a_i)$. The intuition captured by these couples is the following: given $\rho \in \mathbf{SUB}$, if (A, d) is associated to c and $\rho(A)\surd$ then $c\rho$ behaves like $d\rho$.

Note: For simplicity, in the following definition of the functions η and η' we will only give the cases which are needed in the formal developments presented in this section. In fact, as we will mostly prove our technical results by some form of induction, it will be sufficient to define the effect of a substitution ρ on the initial moves of a configuration.

Definition 3.4.9

(1) *The function η is defined by structural induction on π as follows:*

(i) $\eta(nil) = \{(\emptyset, nil)\}$

(ii) $\eta(a_i) = \{(\{a_i\}, nil)\}$

(iii) $\eta(\pi_1 + \pi_2) = \{(x, \pi' + \pi_2) | (x, \pi') \in \eta(\pi_1)\} \cup \{(x, \pi_1 + \pi') | (x, \pi') \in \eta(\pi_2)\}$

(iv)

$$\eta(\pi_1; \pi_2) = \begin{cases} \eta(\pi_2) & \textit{if } \pi_1\surd \\ \{(x, \pi'; \pi_2) | (x, \pi') \in \eta(\pi_1)\} & \textit{otherwise} \end{cases}$$

(v) $\eta(\pi_1 | \pi_2) = \{(x, \pi' | \pi_2) | (x, \pi') \in \eta(\pi_1)\} \cup \{(x, \pi_1 | \pi') | (x, \pi') \in \eta(\pi_2)\}$

(vi) $\eta(\pi_1 \not| \pi_2) = \{(x, \pi' \not| \pi_2) | (x, \pi') \in \eta(\pi_1)\}$.

(2) *The function η' is the extension of η to configurations. It is defined by structural induction on c as follows:*

(i) $\eta'(\pi) = \eta(\pi)$

(ii) $\eta'(F(a_i)) = \{(\{F(a_i)\}, nil)\}$

(iii)

$$\eta'(c; \pi) = \begin{cases} \eta(\pi) & \textit{if } c\surd \\ \{(x, c'; \pi) | (x, c') \in \eta'(c)\} & \textit{otherwise} \end{cases}$$

(iv) $\eta'(c_1 | c_2) = \{(x, c_1' | c_2) | (x, c_1') \in \eta'(c_1)\} \cup \{(x, c_1 | c_2') | (x, c_2') \in \eta'(c_2)\}$.

Notation 3.4.3 *Let $X \subseteq \mathbf{LA} \cup \{F(a_i) | a_i \in \mathbf{LA}\}$. Then*

$$\rho(X)\surd \Leftrightarrow \forall a_i, F(a_i) \in X \;\; \rho(a_i)\surd \wedge \rho(F(a_i))\surd.$$

The following proposition and corollary state formally that what we hoped for in giving the above definition actually holds.

In fact, if $(X, d) \in \eta'(c)$ and $\rho(X)\surd$ then $c\rho \sim_t d\rho$.

Proposition 3.4.4 $\forall \pi \in \mathbf{LP}, \rho \in \mathbf{SUB}\ (X, \pi') \in \eta(\pi) \wedge \rho(X)\surd \Rightarrow \pi\rho \sim_t \pi'\rho$.

Proof: By structural induction on π. We give just the proof of the case $\pi = \pi_1|\pi_2$ leaving the others to the reader.

$\pi = \pi_1|\pi_2$ Assume $(X, \pi') \in \eta(\pi_1|\pi_2)$. There are three cases to examine:

(a) $(X, \pi') = (X_1 \cup X_2, \pi_1'|\pi_2')$ with $(X_1, \pi_1') \in \eta(\pi_1)$ and $(X_2, \pi_2') \in \eta(\pi_2)$.
By inductive hypothesis, $\pi_1\rho \sim_t \pi_1'\rho$ and $\pi_2\rho \sim_t \pi_2'\rho$.
Since $\sim_t$ is a congruence it follows that $(\pi_1|\pi_2)\rho = \pi_1\rho|\pi_2\rho \sim_t \pi_1'\rho|\pi_2'\rho$.

(b) $(X, \pi') = (X, \pi_1'|\pi_2)$ with $(X, \pi_1') \in \eta(\pi_1)$.
By inductive hypothesis, $\pi_1\rho \sim_t \pi_1'\rho$. Again this means that $(\pi_1|\pi_2)\rho = \pi_1\rho|\pi_2\rho \sim_t \pi_1'\rho|\pi_2\rho = (\pi_1'|\pi_2)\rho$.

(c) Symmetrical. □

Corollary 3.4.2 $\forall c \in \mathbf{L}\mathcal{S}, \rho \in \mathbf{SUB}\ (X, c') \in \eta'(c) \wedge \rho(X)\surd \Rightarrow c\rho \sim_t c'\rho$.

Proof: By structural induction on c. Omitted. □

With the next theorem we start the investigation of the possible origin of a move of $c\rho$, for $c \in \mathbf{L}\mathcal{S}$ and $\rho \in \mathbf{SUB}$. We first do so for labelled processes. Before proving the relevant theorem we state a technical lemma which we will need in the proof.

Lemma 3.4.4 *Let* $c \in \mathbf{L}\mathcal{S}$ *and* $\rho \in \mathbf{SUB}$. *Then:* $c\rho\surd$ and $(X, c') \in \eta'(c)$ imply $\rho(X)\surd$ and $c'\rho\surd$.

Proof: By structural induction on c. Omitted. □

The next theorem relates the moves of $\pi\rho$, $\pi \in \mathbf{LP}$ and $\rho \in \mathbf{SUB}$, with those of π and ρ. Intuitively it states that if $\pi\rho \xrightarrow{e}_t s$ one of the following situations occur:

- π can execute the start of action a_i, for some $a_i \in \mathbf{LA}$, and the process associated to a_i by ρ can execute the action e. From that moment onwards $F(a_i)$ will be a "place-holder" for what remains to be executed of the process $\rho(a_i)$.

- There is a couple (A, π') associated to π by η such that ρ maps all the elements of A to terminated processes and $\pi'\rho \xrightarrow{e}_t s$. This is justified by Proposition 3.4.4, which states that, in this case, $\pi\rho$ behaves like $\pi'\rho$ with respect to $\sim_t$.

Theorem 3.4.6 *Let* $\pi \in \mathbf{LP}$ *and* $\rho \in \mathbf{SUB}$. *Then* $\pi\rho \xrightarrow{e}_t s$ *implies:*

(i) $\exists a_i : \pi \overset{S(a_i)}{\longmapsto} c \wedge \rho(a_i) \xrightarrow{e}_t y \wedge s = c\rho[F(a_i) \to y]$, *or*

(ii) $\exists (X, \pi') \in \eta(\pi) : \rho(X)\surd \wedge \pi'\rho \xrightarrow{e}_t s$.

Proof: By structural induction on π.

$\pi = nil$ Vacuous.

$\pi = a_i$ Clause (i) is trivially met.

$\pi = \pi_1 + \pi_2$ Assume $(\pi_1 + \pi_2)\rho \xrightarrow{e}_t s$.

Wlog, we may assume $\pi_1\rho + \pi_2\rho \xrightarrow{e}_t s$ because $\pi_1\rho \xrightarrow{e}_t s$.

By inductive hypothesis this implies:

(i) $\exists a_i : \pi_1 \overset{S(a_i)}{\longmapsto} c \wedge \rho(a_i) \xrightarrow{e}_t y \wedge s = c\rho[F(a_i) \to y]$, or

(ii) $\exists (X, \pi_1') \in \eta(\pi_1) : \rho(X)\surd \wedge \pi_1'\rho \xrightarrow{e}_t s$.

If (i) holds then

$$\pi_1 + \pi_2 \overset{S(a_i)}{\longmapsto} c \wedge \rho(a_i) \xrightarrow{e}_t y \wedge s = c\rho[F(a_i) \to y].$$

If (ii) holds then, by definition of $\eta(\cdot)$, $(X, \pi_1' + \pi_2) \in \eta(\pi_1 + \pi_2)$.

Moreover $\rho(X)\surd$ and $\pi_1'\rho + \pi_2\rho \xrightarrow{e}_t s$.

$\pi = \pi_1; \pi_2$ Assume $(\pi_1; \pi_2)\rho = \pi_1\rho; \pi_2\rho \xrightarrow{e}_t s$. There are two possibilities:

(a) $\pi_1\rho \xrightarrow{e}_t s'$ and $s = s'; \pi_2\rho$.

By inductive hypothesis $\pi_1\rho \xrightarrow{e}_t s'$ implies:

(i) $\exists a_i : \pi_1 \overset{S(a_i)}{\longmapsto} c \wedge \rho(a_i) \xrightarrow{e}_t y \wedge s' = c\rho[F(a_i) \to y]$, or

(ii) $\exists (X, \pi_1') \in \eta(\pi_1) : \rho(X)\surd$ and $\pi_1'\rho \xrightarrow{e}_t s'$.

If (i) holds then $\pi_1; \pi_2 \overset{S(a_i)}{\longmapsto} c; \pi_2$. Moreover

$$(c; \pi_2)\rho[F(a_i) \to y] = (c\rho[F(a_i) \to y]); (\pi_2\rho[F(a_i) \to y]) = s'; \pi_2\rho.$$

If (ii) holds then, by definition of $\eta(\cdot)$, $(X, \pi_1'; \pi_2) \in \eta(\pi_1; \pi_2)$ and, by the operational semantics,

$$(\pi_1'; \pi_2)\rho = \pi_1'\rho; \pi_2\rho \xrightarrow{e}_t s'; \pi_2\rho.$$

(b) $\pi_1\rho\surd$ and $\pi_2\rho \xrightarrow{e}_t s$.

By inductive hypothesis, $\pi_2\rho \xrightarrow{e}_t s$ implies:

(i) $\exists a_i : \pi_2 \overset{S(a_i)}{\longmapsto} c \wedge \rho(a_i) \overset{e}{\longrightarrow}_t y \wedge s = c\rho[F(a_i) \rightarrow y]$, or

(ii) $\exists (X, \pi_2') \in \eta(\pi_2) : \rho(X)\surd \wedge \pi_2'\rho \overset{e}{\longrightarrow}_t s$.

If (i) holds then there are two subcases:

(1) $\pi_1\surd$. Then $\pi_1; \pi_2 \overset{S(a_i)}{\longmapsto} c$ and clause (i) is met.

(2) $\pi_1 \notin \surd$. Then, by the above lemma, $\pi_1\rho\surd$ implies that for each $(X, \pi_1') \in \eta(\pi_1)$ we have that $\rho(X)\surd$ and $\pi_1'\rho\surd$. Hence, for each $(X, \pi_1'; \pi_2) \in \eta(\pi_1; \pi_2)$ we have that $\rho(X)\surd$ and $\pi_1'\rho\surd$. By the operational semantics, for each $(X, \pi_1'; \pi_2) \in \eta(\pi_1; \pi_2)$ we have that

$$(\pi_1'; \pi_2)\rho = \pi_1'\rho; \pi_2\rho \overset{e}{\longrightarrow}_t s.$$

If (ii) holds then there are two subcases:

(1) If $\pi_1\surd$ then $(X, \pi_2') \in \eta(\pi_1; \pi_2)$ and clause (ii) holds.

(2) If $\pi_1\rho\surd$ and $\pi_1 \notin \surd$ then, again by the above lemma, for each $(Y, \pi_1') \in \eta(\pi_1)$ we have that $\rho(Y)\surd$ and $\pi_1'\rho\surd$.
Hence, by definition of $\eta(\cdot)$, for each $(Y, \pi_1'; \pi_2) \in \eta(\pi_1; \pi_2)$ we have that $\rho(Y)\surd$ and $\pi_1'\rho\surd$. Thus $(Y, \pi_1'; \pi_2) \in \eta(\pi_1; \pi_2)$ implies

$$(\pi_1'; \pi_2)\rho = \pi_1'\rho; \pi_2\rho \overset{e}{\longrightarrow}_t s.$$

$\pi = \pi_1 | \pi_2$ Assume, wlog, $\pi_1\rho|\pi_2\rho \overset{e}{\longrightarrow}_t s|\pi_2\rho$ because $\pi_1\rho \overset{e}{\longrightarrow}_t s$.

By the inductive hypothesis, $\pi_1\rho \overset{e}{\longrightarrow}_t s$ implies:

(i) $\exists a_i : \pi_1 \overset{S(a_i)}{\longmapsto} c \wedge \rho(a_i) \overset{e}{\longrightarrow}_t y \wedge s = c\rho[F(a_i) \rightarrow y]$, or

(ii) $\exists (X, \pi_1') \in \eta(\pi_1) : \rho(X)\surd \wedge \pi_1'\rho \overset{e}{\longrightarrow}_t s$.

If (i) holds then $\pi_1|\pi_2 \overset{S(a_i)}{\longmapsto} c|\pi_2$. Moreover,

$$(c|\pi_2)\rho[F(a_i) \rightarrow y] = (c\rho[F(a_i) \rightarrow y])|(\pi_2\rho[F(a_i) \rightarrow y]) = s|\pi_2\rho.$$

If (ii) holds then, by definition of $\eta(\cdot)$, $(X, \pi_1'|\pi_2) \in \eta(\pi_1|\pi_2)$.

By the operational semantics, $(\pi_1'|\pi_2)\rho = \pi_1'\rho|\pi_2\rho \overset{e}{\longrightarrow}_t s|\pi_2$.

$\pi = \pi_1 \not| \pi_2$ Similar to the one above. □

To characterize the moves of $c\rho$ for $c \in \mathrm{L}S$ we need to show how a move of $c\rho$ may depend on a possible move of the state $\rho(F(a_i))$.

This is done in the following theorem.

Theorem 3.4.7 *Let* $c \in \mathbf{L}\mathcal{S}, \rho \in \mathbf{SUB}$. *Assume* $c\rho \overset{e}{\longrightarrow}_t s$. *Then:*

(i) $\exists a_i \in \mathbf{LA}: \; c \overset{S(a_i)}{\longmapsto} c' \wedge \rho(a_i) \overset{e}{\longrightarrow}_t y \wedge s = c'\rho[F(a_i) \to y]$, *or*

(ii) $\exists (X, c') \in \eta'(c): \; \rho(X)\surd \wedge c'\rho \overset{e}{\longrightarrow}_t s$, *or*

(iii) $\exists a_i \in \mathbf{LA}: \; \rho(F(a_i)) \overset{e}{\longrightarrow}_t y \wedge s = c\rho[F(a_i) \to y]$ *and* $F(a_i)$ *occurs in* c.

Proof: By structural induction on c. The details are very similar to those in the proof of the above theorem. □

So far we have shown how to relate the moves of $c\rho$, for $c \in \mathbf{L}\mathcal{S}$ and $\rho \in \mathbf{SUB}$, to moves of c and ρ. In what follows we will need to be able to go in the opposite direction. Namely, in certain circumstances, given moves of c and moves of ρ, we want to derive the corresponding move of $c\rho$. This information will be needed in proving that refine-equivalence is preserved by action-refinement and is derived by the next lemma.

Lemma 3.4.5 *Let* $c \in \mathbf{L}\mathcal{S}, \rho \in \mathbf{SUB}$.

(a) *Assume* $c \overset{S(a_i)}{\longmapsto} c'$ *and* $\rho(a_i) \overset{e}{\longrightarrow}_t y$. *Then* $c\rho \overset{e}{\longrightarrow}_t c'\rho[F(a_i) \to y]$.

(b) *Assume* $F(a_i)$ *occurs in* c *and* $\rho(F(a_i)) \overset{e}{\longrightarrow}_t y$. *Then* $c\rho \overset{e}{\longrightarrow}_t c\rho[F(a_i) \to y]$.

Proof: Both statements can be shown by structural induction on c. We will only sketch the proof of statement (a) leaving the proof of part (b) to the reader.

$c = \pi$: We show that $\pi \overset{S(a_i)}{\longmapsto} c'$ and $\rho(a_i) \overset{e}{\longrightarrow}_t y$ imply $\pi\rho \overset{e}{\longrightarrow}_t c'\rho[F(a_i) \to y]$ by induction on the proof of $\pi \overset{S(a_i)}{\longmapsto} c'$.

We will only examine the case $\pi = \pi_1|\pi_2$ and leave the others to the reader.

$\pi = \pi_1|\pi_2$ Wlog, assume $\pi_1|\pi_2 \overset{S(a_i)}{\longmapsto} c|\pi_2$ because $\pi_1 \overset{S(a_i)}{\longmapsto} c$.

By the inductive hypothesis, $\pi_1\rho \overset{e}{\longrightarrow}_t c\rho[F(a_i) \to y]$.

By the operational semantics,

$$(\pi_1|\pi_2)\rho = \pi_1\rho|\pi_2\rho \overset{e}{\longrightarrow}_t (c\rho[F(a_i) \to y])|\pi_2\rho.$$

Moreover, as $F(a_i)$ does not occur in π_2,

$$(c|\pi_2)\rho[F(a_i) \to y] = (c\rho[F(a_i) \to y])|\pi_2\rho.$$

$c = F(a_i)$: Vacuous.

The cases $c = c_1|c_2$ and $c = c_1;\pi$ are easy. □

We now have all the technical machinery we need to prove that $\sim_r$ is preserved by action-refinement. The proof is based on the tight correspondence between the places of two configurations c, d such that $c \sim_\phi d$ which is recorded by $\phi \in \mathcal{H}$. This allows us to state formally when a place $F(a_i)$ in c *plays the same role* of a place $F(a_j)$ in d with respect to action-refinement.

Definition 3.4.10

(1) *Let $\rho, \rho' \in$ **SUB** and $\phi \in \mathcal{H}$. We say that $\rho \sim_\phi \rho'$ iff for each $a \in$ **A***

(i) $\forall i \in dom(\phi_a)\ \rho(F(a_i)) \sim_t \rho'(F(a_{\phi_a(i)}))$

(ii) $\forall i, j \in \mathrm{N}\ \rho(a_i) \sim_t \rho'(a_j)$.

(2) *Define $\mathcal{R}_{sub} \subseteq \mathcal{S}^2$ as follows:*

$$\mathcal{R}_{sub} =_{def} \{(c\rho, c'\rho') | \exists \phi \in \mathcal{H} :\ c \sim_\phi c' \wedge \rho \sim_\phi \rho'\}.$$

We want to show that $\mathcal{R}_{sub}$ is a timed-bisimulation. Before showing this result we state a technical lemma which will be useful in its proof.

Lemma 3.4.6

(i) *$\mathcal{R}_{sub}$ is an equivalence relation on $\mathcal{S}$.*

(ii) $\forall s_1, s_2 \in \mathcal{S}\ s_1 \sim_t s_2 \Rightarrow (s_1, s_2) \in \mathcal{R}_{sub}$.

Proof:

(i) This statement is a simple consequence of the properties of $\sim_\phi$ on **L**$\mathcal{S}$ and of the following observations:

- $\forall \rho, \rho'\ \rho \sim_\phi \rho' \Rightarrow \rho' \sim_{\phi^{op}} \rho$.
- $\forall \rho, \rho', \rho''\ \rho \sim_{\phi_1} \rho' \wedge \rho' \sim_{\phi_2} \rho'' \Rightarrow \rho \sim_{\phi_2 \circ \phi_1} \rho''$.

(ii) Assume $s_1, s_2 \in \mathcal{S}$ and $s_1 \sim_t s_2$. Then, by theorem 3.4.5, $s_1 \sim_r s_2$. This implies that there exist $c_1, c_2 \in \mathbf{L}\mathcal{S}$ such that $\mathbf{un}(c_i) = s_i, i = 1, 2$, and $c_1 \sim_\phi c_2$ for some $\phi \in \mathcal{H}$.

Consider $\rho \in \mathbf{SUB}$ such that $\forall a_i \in \mathbf{LA}\ \rho(a_i) = a \wedge \rho(F(a_i)) = F(a)$.

Then it is easy to see that, for each $\phi \in \mathcal{H}$, $\rho \sim_\phi \rho$. By definition of $\mathcal{R}_{sub}$, $(c_1\rho, c_2\rho) \in \mathcal{R}_{sub}$. Moreover, $c_i\rho = \mathbf{un}(c_i) = s_i, i = 1, 2$. Hence $(s_1, s_2) \in \mathcal{R}_{sub}$. □

Theorem 3.4.8 *The relation $\mathcal{R}_{sub}$ is a timed-bisimulation.*

Proof: We prove by induction on the size of c that for $(c\rho, c'\rho') \in \mathcal{R}_{sub}$

$$c\rho \xrightarrow{e}_t s \Rightarrow \exists s' : c'\rho' \xrightarrow{e}_t s' \text{ and } (s, s') \in \mathcal{R}_{sub}.$$

By symmetry this will be sufficient to prove the claim.

First of all recall that $(c\rho, c'\rho') \in \mathcal{R}_{sub}$ iff for some $\phi \in \mathcal{H}$ $c \sim_\phi c'$ and $\rho \sim_\phi \rho'$. Assume $c\rho \xrightarrow{e}_t s$. By theorem 3.4.7 there are three cases to examine:

(i) $\exists a_i : c \overset{S(a_i)}{\longmapsto} d \wedge \rho(a_i) \xrightarrow{e}_t y \wedge s = d\rho[F(a_i) \rightarrow y]$.

Then, as $c \sim_\phi c'$, there exist $j \in \mathrm{N}, d' \in \mathbf{L}\mathcal{S}$ such that

$$d \overset{S(a_j)}{\longmapsto} d' \text{ and } c' \sim_{\phi \cup \{(a,i,j)\}} d'.$$

As $\rho \sim_\phi \rho'$, $\exists y' : \rho'(a_j) \xrightarrow{e}_t y'$ and $y \sim_t y'$.

By lemma 3.4.5, $d \overset{S(a_j)}{\longmapsto} d'$ and $\rho'(a_j) \xrightarrow{e}_t y'$ imply

$$d\rho' \xrightarrow{e}_t d'\rho'[F(a_j) \rightarrow y'].$$

It is easy to see that

$$\rho[F(a_i) \rightarrow y] \sim_{\phi \cup \{(a,i,j)\}} \rho'[F(a_j) \rightarrow y'].$$

Hence, by the definition of $\mathcal{R}_{sub}$,

$$(d\rho[F(a_i) \rightarrow y], d'\rho'[F(a_j) \rightarrow y']) \in \mathcal{R}_{sub}.$$

(ii) $\exists (X, d) \in \eta'(c) : d\rho \xrightarrow{e}_t s$ and $\rho(X)\surd$.

Then, by corollary 3.4.2, $c\rho \sim_t d\rho$. By the above lemma, $(c\rho, d\rho) \in \mathcal{R}_{sub}$.

As $\mathcal{R}_{sub}$ is an equivalence relation, $(d\rho, c'\rho') \in \mathcal{R}_{sub}$. As the size of d is less than that of c, we may apply induction to obtain

$$\exists s' : c'\rho' \xrightarrow{e}_t s' \text{ and } (s, s') \in \mathcal{R}_{sub}.$$

(iii) $\exists a_i : \rho(F(a_i)) \xrightarrow{e}_t s \wedge s = c\rho[F(a_i) \to y]$ and $F(a_i)$ occurs in c. As $F(a_i)$ occurs in c and $c \sim_\phi c'$, $i \in dom(\phi_a)$. By the hypothesis that $\rho \sim_\phi \rho'$, we have that

$$\exists y' : \rho'(F(a_j)) \xrightarrow{e}_t y' \text{ and } y \sim_t y', \text{ where } j = \phi_a(i).$$

As $F(a_j)$ occurs in c' we may apply lemma 3.4.5 to obtain

$$c'\rho' \xrightarrow{e}_t c'\rho'[F(a_j) \to y'].$$

As $\rho \sim_\phi \rho'$, we have that $\rho[F(a_i) \to y] \sim_\phi \rho'[F(a_j) \to y']$.

By the definition of $\mathcal{R}_{sub}$,

$$(c\rho[F(a_i) \to y], c'\rho'[F(a_j) \to y']) \in \mathcal{R}_{sub}.\square$$

This theorem allows us to give a behavioural proof of the fact that $\sim_r$ is preserved by action-refinements.

Corollary 3.4.3 (The Refinement Theorem: Behavioural Proof) $\forall p, q, r \in \mathbf{P}$ $p \sim_r q$ *implies* $p[a \rightsquigarrow r] \sim_r q[a \rightsquigarrow r]$.

Proof: We will in fact show that $p \sim_r q$ implies that $p[a \rightsquigarrow r] \sim_t q[a \rightsquigarrow r]$, which, by Theorem 3.4.5, is sufficient to prove the claim. Assume $p, q \in \mathbf{P}$ and $p \sim_r q$. Then there exist

$$\pi_1, \pi_2 \in \mathbf{LP} : \mathbf{un}(\pi_1) = p, \mathbf{un}(\pi_2) = q \text{ and } \pi_1 \sim_\emptyset \pi_2.$$

Consider the substitution ρ defined as follows:

- $\forall i \in \mathrm{N}\ \rho(a_i) = r$
- $\forall i, b \neq a\ \rho(b_i) = b$
- $\forall i\ \rho(F(a_i)) = F(a)$.

Then $\rho \sim_\emptyset \rho$ and therefore, by the definition of $\mathcal{R}_{sub}$, $(\pi_1\rho, \pi_2\rho) \in \mathcal{R}_{sub}$.

Moreover, $\pi_1\rho = p[a/r]$ and $\pi_2\rho = q[a/r]$. By the above theorem and the definition of $\sim_t$ over $\mathbf{P}_{ert}$, $p[a \rightsquigarrow r] \sim_t q[a \rightsquigarrow r]$. $\square$

3.5 Proof of the Equational Characterization

This section is entirely devoted to a detailed proof of the equational characterization of the relation $\sim_t$ over $\mathbf{P}$ (Theorem 3.3.2). The proof is based upon standard techniques used in the literature to axiomatize non-interleaving equivalences, see, e.g., [CH89] and [H88c], but the details are much more involved because of the presence of general sequential composition in the language, rather than action-prefixing.

3.5.1 Preliminaries

Let $=_E$ denote the least congruence over $\mathbf{P}$ which satisfies the set of axioms E in Figure 1. The following proposition, whose proof is standard and thus omitted, states that the axioms are indeed sound with respect to $\sim_t$.

Proposition 3.5.1 (Soundness) *For each $p, q \in \mathbf{P}$, $p =_E q$ implies $p \sim_t q$.*

Note that axiom (C3) is *not* sound without the side-condition. For example, $(nil + b); c \not\sim_t nil; c + b; c$.

In what follows we will concentrate on the much more challenging proof of completeness of the set of axioms E with respect to $\sim_t$. In order to prove the completeness theorem, it will be convenient to reduce the processes in $\mathbf{P}$ to what will be called *head reduced forms* or *hrf's*. Before giving the definition of this class of terms, let us introduce some useful notation.

Notation 3.5.1 *In what follows, for each process $p \in \mathbf{P}$,*

$$p \longrightarrow_t \text{ iff there exist } a \in \mathbf{A}, s \in \mathcal{S} : p \stackrel{S(a)}{\longrightarrow}_t s.$$

Note that $p \not\longrightarrow_t$ iff $p \sim_t nil$.

We shall now introduce a class of processes, the head reduced forms, which will be very useful in the proof of the completeness theorem. Intuitively, sequential hrf's correspond to processes which semantically have the sequential composition operator as head operator and parallel hrf's to processes which semantically have the left-merge operator as head operator. Combinations of parallel and sequential behaviour in processes will be dealt with by means of the larger class of hrf's.

Definition 3.5.1 (Head Reduced Forms) *The sets* HRF *(*head reduced forms*),* HRF_{Seq} *(*sequential head reduced forms*) and* HRF_{Par} *(*parallel head reduced forms*) are defined simultaneously as the least sets satisfying:*

- $a; p \in \mathtt{HRF}_{Seq}$ *if* $p \in \mathtt{HRF}$*;*
- $p; q \in \mathtt{HRF}_{Seq}$ *if* $p \in \mathtt{HRF}_{Par}$*,* $q \in \mathtt{HRF}$ *and* $q \not\longrightarrow_t$*;*
- $p \not| q \in \mathtt{HRF}_{Par}$ *if* $p \in \mathtt{HRF}_{Seq}$*,* $q \in \mathtt{HRF}$ *and* $q \not\longrightarrow_t$*;*
- $\sum_{i \in I} p_i \in \mathtt{HRF}$ *if* I *is a finite index set and, for each* $i \in I$*,* $p_i \in \mathtt{HRF}_{Seq}$ *or* $p_i \in \mathtt{HRF}_{Par}$*.*

By convention, if $I = \emptyset$ *then* $\sum_{i \in \emptyset} p_i \equiv nil$*.*

The next lemma states that it is possible to reduce each process in $\mathbf{P}$ to a head reduced form using the axioms in E.

Lemma 3.5.1 (Reduction Lemma) *For each* $p \in \mathbf{P}$*, there exists a head reduced form* $h(p)$ *such that* $p =_E h(p)$*.*

Proof: By induction on the depth of p. The proof proceeds by a case analysis on the structure of p and will use all of the axioms in E apart from (A3).

- $p \equiv nil$. Then p is already a *hrf*.
- $p \equiv a$. Then $a =_E a; nil$, which is a sequential *hrf*, by axiom (C2).
- $p \equiv q + r$. By the inductive hypothesis, there exist *hrf's* $h(q)$ and $h(r)$ such that $q =_E h(q)$ and $r =_E h(r)$. If $h(q) \equiv nil$ or $h(r) \equiv nil$ then apply (A2) and (A4) to obtain a *hrf*. Otherwise $q + r =_E h(q) + h(r)$ which is a head reduced form.
- $p \equiv q; r$. By the inductive hypothesis, $q =_E h(q)$. Assume, wlog, that $h(q) \equiv \sum_{i \in I} q_i$, where each q_i is either a sequential *hrf* or a parallel *hrf*.

  If $h(q) \equiv nil$ (i.e., $I = \emptyset$) then apply (C2) and the inductive hypothesis to obtain

  $$p \equiv q; r =_E nil; h(r) =_E h(r).$$

  Otherwise $I \neq \emptyset$ and $p =_E (\sum_{i \in I} q_i); r =_E \sum_{i \in I} q_i; r$, by repeated use of axiom (C3) which is applicable as each q_i is not terminated. We show that, for each $i \in I$, $q_i; r$ may be reduced to either a sequential *hrf* or a parallel *hrf*. There are two possibilities to consider:

  (1) q_i is a sequential *hrf*. We distinguish two subcases:

  (i) $q_i \equiv a; \bar{q}$ with $\bar{q} \in \mathtt{HRF}$. Then

  $$\begin{aligned} q_i; r \equiv (a; \bar{q}); r &=_E a; (\bar{q}; r) && \text{by (C1)} \\ &=_E a; h(\bar{q}; r) && \text{by the inductive hypothesis.} \end{aligned}$$

(ii) $q_i \equiv \hat{q}_1; \hat{q}_2$ with $\hat{q}_1 \in \mathtt{HRF}_{Par}$, $\hat{q}_2 \in \mathtt{HRF}$ and $\hat{q}_2 \longrightarrow_t$. Then

$$\begin{array}{rll} q_i; r \equiv (\hat{q}_1; \hat{q}_2); r & =_E \hat{q}_1; (\hat{q}_2; r) & \text{by (C1)} \\ & =_E \hat{q}_1; h(\hat{q}_2; r) & \text{by the inductive hypothesis,} \end{array}$$

which is a sequential *hrf* as $h(\hat{q}_2; r) \longrightarrow_t$.

(2) Assume $q_i \in \mathtt{HRF}_{Par}$. Then $q_i \equiv \hat{q}_1 \lfloor \hat{q}_2$ with $\hat{q}_1 \in \mathtt{HRF}_{Seq}$, $\hat{q}_2 \in \mathtt{HRF}$ and $\hat{q}_2 \longrightarrow_t$. If $h(r) \equiv nil$ then $q_i; r =_E q_i$. Otherwise, $h(r) \longrightarrow_t$ and $q_i; r \equiv (\hat{q}_1 \lfloor \hat{q}_2); h(r)$, which is a sequential *hrf*.

- $p \equiv q \lfloor r$. By the inductive hypothesis, $q =_E h(q)$. Assume, wlog, that $h(q) \equiv \sum_{i \in I} q_i$. If $h(q) \equiv nil$ then apply (B4) to obtain $p =_E nil$. Otherwise, $p =_E (\sum_{i \in I} q_i) \lfloor r =_E \sum_{i \in I} q_i \lfloor r$ by repeated use of (B1).

  We show that, for each $i \in I$, $q_i \lfloor r$ may be reduced to either a sequential *hrf* or a parallel *hrf*. There are two cases to examine:

  (1) $q_i \in \mathtt{HRF}_{Seq}$. We distinguish two subcases:

  (i) $q_i \equiv a; \bar{q}$. Then $a; \bar{q} \lfloor r =_E a; \bar{q} \lfloor h(r)$ by the induction hypothesis. If $h(r) \equiv nil$ then apply (B3) to obtain a sequential *hrf*. Otherwise $h(r) \longrightarrow_t$ and $a; \bar{q} \lfloor h(r)$ is a parallel *hrf*.

  (ii) $q_i \equiv \hat{q}_1; \hat{q}_2$ with $\hat{q}_1 \in \mathtt{HRF}_{Par}$, $\hat{q}_2 \in \mathtt{HRF}$ and $\hat{q}_2 \longrightarrow_t$. Then

$$\begin{array}{rll} q_i \lfloor r & =_E (\hat{q}_1; \hat{q}_2) \lfloor h(r) & \text{by the inductive hypothesis} \\ & =_E \begin{cases} \hat{q}_1; \hat{q}_2 \in \mathtt{HRF}_{Seq} & \text{if } h(r) \equiv nil \\ (\hat{q}_1; \hat{q}_2) \lfloor h(r) & \text{otherwise.} \end{cases} & \end{array}$$

  (2) $q_i \equiv \hat{q}_1 \lfloor \hat{q}_2$ with $\hat{q}_1 \in \mathtt{HRF}_{Seq}$, $\hat{q}_2 \in \mathtt{HRF}$ and $\hat{q}_2 \longrightarrow_t$. Then

$$\begin{array}{rll} q_i \lfloor r \equiv (\hat{q}_1 \lfloor \hat{q}_2) \lfloor r & =_E \hat{q}_1 \lfloor (\hat{q}_2 | r) & \text{by (B2)} \\ & =_E \hat{q}_1 \lfloor h(\hat{q}_2 | r) & \text{by the inductive hypothesis,} \end{array}$$

  which is a parallel *hrf* as $\hat{q}_2 \longrightarrow_t$ implies $h(\hat{q}_2 | r) \longrightarrow_t$.

- $p \equiv q | r$. Then $p =_E q \lfloor r + r \lfloor q$ by (X1). The claim now follows from the above case.

This completes the proof of the reduction lemma. □

The proof of the completeness theorem is based upon several lemmas stating important decomposition properties. As we are dealing with finite processes, it will be convenient to prove most of these properties by induction on some notion of size of a process. In what follows, the *size* of a process p, $|p|$, will be taken to be the length of the longest sequence of subactions it can perform with respect to the timed operational semantics, i.e.

$$|p| =_{def} max\{l(\sigma) \mid \sigma \in \mathbf{A_S}^\star \text{ and } p \xrightarrow{\sigma}_t s, \text{ for some } s\},$$

where $l(\sigma)$ denotes the length of the sequence σ. This notion of size is generalized in the obvious way to $s \in \mathcal{S}$. The following proposition is easily established.

Proposition 3.5.2 *For each $s_1, s_2 \in \mathcal{S}$, $s_1 \sim_t s_2$ implies $|s_1| = |s_2|$.*

Lemma 3.5.2 *Let $s_1, s_2 \in \mathcal{S}$. Then $s_1;p \sim_t s_2;p$ implies $s_1 \sim_t s_2$.*

Proof: By induction on the combined size of s_1 and s_2.

- Suppose $s_1\surd$. Assume, towards a contradiction, that there exist $e \in \mathbf{A_S}$, s_2' such that $s_2 \xrightarrow{e}_t s_2'$. Then, by the operational semantics, $s_2;p \xrightarrow{e}_t s_2';p$. As $s_1;p \sim_t s_2;p$, there exists s such that $p \xrightarrow{e}_t s$ and $s \sim_t s_2';p$. However, this is impossible as $|s| < |p| \leq |s_2';p|$. Thus $s_2\surd$ and therefore $s_1 \sim_t s_2$.
- Assume $s_1 \xrightarrow{e}_t s_1'$. By the operational semantics, this implies that $s_1;p \xrightarrow{e}_t s_1';p$. Reasoning as above we may deduce that *not* $s_2\surd$. Hence there exists s_2' such that $s_2 \xrightarrow{e}_t s_2'$ and $s_1';p \sim_t s_2';p$. By the inductive hypothesis, $s_1' \sim_t s_2'$. As this can be done for each e, s_1' such that $s_1 \xrightarrow{e}_t s_1'$, we obtain, by symmetry, that $s_1 \sim_t s_2$. □

Two special subclasses of the set of states $\mathcal{S}$ will play an important rôle in the proof of the completeness theorem. These will be called sequential configurations and parallel configurations.

Notation 3.5.2 *For each index set $I = \{i_1, \ldots, i_k\}$, $k \geq 0$, the notation $\prod_{i \in I} c_i$ stands for $c_{i_1} | \cdots | c_{i_k}$ and is justified by commutativity and associativity of $|$ modulo $\sim_t$. By convention, $\prod_{i \in \emptyset} c_i \equiv nil$.*

Definition 3.5.2 (Parallel and Sequential Configurations) *The sets C_{Seq}, the set of* sequential configurations, *and C_{Par}, the set of* parallel configurations, *are simultaneously defined as the least subsets of $\mathcal{S}$ which satisfy the following clauses:*

- *$F(a);p \in C_{Seq}$ if $p \in \mathbf{P}$;*
- *$c \in C_{Par}$, $p \longrightarrow_t$ imply $c;p \in C_{Seq}$;*

- $\{c_i \mid i \in I\} \subseteq \mathcal{C}_{Seq}$, $|I| > 0$ *and* $p \longrightarrow_t$ *imply* $(\prod_{i \in I} c_i)|p \in \mathcal{C}_{Par}$;
- $\{c_i \mid i \in I\} \subseteq \mathcal{C}_{Seq}$, $|I| > 1$ *imply* $\prod_{i \in I} c_i \in \mathcal{C}_{Par}$.

In the above clauses I is always assumed to stand for a finite *index set.*

The following immediate properties of sequential and parallel configurations will be heavily used in the proofs of the main results of this section.

Fact 3.5.1

(1) *For each $c \in \mathcal{C}_{Seq} \cup \mathcal{C}_{Par}$, there exists $a \in \mathbf{A}$ such that $c \xrightarrow{F(a)}_t$.*

(2) *For each $c \in \mathcal{C}_{Par}$,*

(a) $c \xrightarrow{S(a)}_t$, *for some $a \in \mathbf{A}$, or*

(b) $c \xrightarrow{F(a)}_t \xrightarrow{F(b)}_t$, *for some $a, b \in \mathbf{A}$.*

Hence sequential configurations are always capable of performing the end of at least one action; on the other hand, parallel configurations are either capable of performing the start of an action or can perform at least two end moves in a row. The following lemmas study the effect of start and end-moves on processes and configurations. At this stage it is useful to introduce the notation $c \in\!\sim_t X$ to mean that there exists some $x \in X$ such that $c \sim_t x$.

Proposition 3.5.3

(i) *For each $s \in \mathcal{S}$, $s \in \mathbf{P}$ or $s \in\!\sim_t \mathcal{C}_{Seq}$ or $s \in\!\sim_t \mathcal{C}_{Par}$.*

(ii) *Let $s_1, s_2 \in \mathcal{S}$. Assume that $|s_1|, |s_2| > 0$. Then $s_1|s_2 \in \mathbf{P}$ or $s_1|s_2 \in\!\sim_t \mathcal{C}_{Par}$.*

Proof:

(i) By induction on the structure of s. We only sketch the proof of the case $s = s_1|s_2$. In this case, by the inductive hypothesis, we have that, for $i = 1, 2$,

$$s_i \in \mathbf{P} \text{ or } s_i \in\!\sim_t \mathcal{C}_{Seq} \text{ or } s_i \in\!\sim_t \mathcal{C}_{Par}.$$

If $s_1, s_2 \in \mathbf{P}$ then $s_1|s_2 \in \mathbf{P}$. If $|s_1| = 0$ and $s_2 \in\!\sim_t \mathcal{C}_{Seq}$ or $s_1 \in\!\sim_t \mathcal{C}_{Seq}$ and $|s_2| = 0$ then $s_1|s_2 \in\!\sim_t \mathcal{C}_{Seq}$. In all the other possible cases we have that $s_1|s_2 \in\!\sim_t \mathcal{C}_{Par}$.

(ii) Follows from the previous analysis for (i). □

The following lemma studies the effect of start-moves on sequential and parallel head reduced forms.

Lemma 3.5.3 *Let* $p \in \mathtt{HRF}_{Seq} \cup \mathtt{HRF}_{Par}$. *Then:*

(1) $p \in \mathtt{HRF}_{Seq}$ *and* $p \xrightarrow{S(a)}_t c$ *imply* $c \in \mathcal{C}_{Seq}$.

(2) $p \in \mathtt{HRF}_{Par}$ *and* $p \xrightarrow{S(a)}_t c$ *imply* $c \in \mathcal{C}_{Par}$ *and* c *is of the form* $d|q$, *with* $d \in \mathcal{C}_{Seq}$ *and* $|q| > 0$.

Proof: Both statements are shown by simultaneous induction on the size of p. We assume, as inductive hypothesis, that (1) and (2) hold for each q such that $|q| < |p|$.

(1) Assume $p \in \mathtt{HRF}_{Seq}$ and $p \xrightarrow{S(a)}_t c$. By the definition of the set $\mathtt{HRF}_{Seq}$ there are two cases to examine.

- p is of the form $a; q$. Then, by the operational semantics, $c \equiv F(a); q \in \mathcal{C}_{Seq}$.
- p is of the form $q; r$, with $q \in \mathtt{HRF}_{Par}$, $r \in \mathtt{HRF}$ and $|r| > 0$. By the operational semantics and the fact that $q \in \mathtt{HRF}_{Par}$ implies $q \notin \surd$, $q; r \xrightarrow{S(a)}_t c$ implies $c \equiv d; r$, with $q \xrightarrow{S(a)}_t d$. As $|r| > 0$, we have that $|q| < |p|$. Thus we may apply the inductive hypothesis for (2) to obtain, among other things, that $d \in \mathcal{C}_{Par}$. Hence $d; r \in \mathcal{C}_{Seq}$.

This completes the proof of (1).

(2) Assume $p \in \mathtt{HRF}_{Par}$ and $p \xrightarrow{S(a)}_t c$. Then, by the definition of $\mathtt{HRF}_{Par}$, p is of the form $q|\!\!/ r$, with $q \in \mathtt{HRF}_{Seq}$, $r \in \mathtt{HRF}$ and $|r| > 0$. By the operational semantics, $q|\!\!/ r \xrightarrow{S(a)}_t c$ implies $c \equiv d|r$, with $q \xrightarrow{S(a)}_t d$. As $|r| > 0$, $|q| < |p|$. We may thus apply the inductive hypothesis for (1) to obtain that $d \in \mathcal{C}_{Seq}$. Hence $d|r \in \mathcal{C}_{Par}$ and is of the required form. This completes the proof of (2). □

The following result is an immediate corollary of the previous lemma.

Corollary 3.5.1 (Effect of Start-Moves on Head Reduced Forms) *Let* $p \in \mathtt{HRF}$. *Then* $p \xrightarrow{S(a)}_t c$ *implies* $c \in \mathcal{C}_{Seq} \cup \mathcal{C}_{Par}$. *Moreover, if* $c \in \mathcal{C}_{Par}$ *then* c *is of the form* $d|q$, *with* $d \in \mathcal{C}_{Seq}$ *and* $|q| > 0$.

Lemma 3.5.4 (Effect of Start-Moves on Processes) *If* $p \in \mathbf{P}$ *and* $p \xrightarrow{S(a)}_t c$ *then* $c \in\!\sim_t \mathcal{C}_{Seq}$ *or* $c \in\!\sim_t \mathcal{C}_{Par}$. *Moreover, if* $c \in\!\sim_t \mathcal{C}_{Par}$ *then* $c \sim_t d|q$, *for some* d, q *such that* $d \in \mathcal{C}_{Seq}$ *and* $|q| > 0$.

Proof: Let $p \in \mathbf{P}$ and assume that $p \xrightarrow{S(a)}_t c$. By Lemma 3.5.1, $p =_E h(p)$ for some hrf $h(p)$. By the soundness of $=_E$ with respect to $\sim_t$, we have that $p \sim_t h(p)$. Hence there exists d such that $h(p) \xrightarrow{S(a)}_t d \sim_t c$. By the previous corollary, $d \in \mathcal{C}_{Seq} \cup \mathcal{C}_{Par}$. Thus $c \in \sim_t \mathcal{C}_{Seq} \cup \mathcal{C}_{Par}$. Moreover, again by the previous corollary, $c \in \sim_t \mathcal{C}_{Par}$ implies that d is of the required form. □

Lemma 3.5.5 (Effect of Start-Moves on Configurations) *For each $c \in \mathcal{C}_{Seq} \cup \mathcal{C}_{Par}$ the following statements hold:*

(1) $c \in \mathcal{C}_{Seq}$ and $c \xrightarrow{S(a)}_t c'$ imply $c' \in \sim_t \mathcal{C}_{Seq}$;

(2) $c \in \mathcal{C}_{Par}$ and $c \xrightarrow{S(a)}_t c'$ imply $c' \in \sim_t \mathcal{C}_{Par}$.

Proof: Both statements are proved simultaneously by induction on the size of c. We assume, as inductive hypothesis, that (1) and (2) are true of all d with $|d| < |c|$.

(1) The statement is vacuously true if c is of the form $F(a); p$. Assume now c is of the form $d; p$ with $d \in \mathcal{C}_{Par}$ and $p \longrightarrow_t$. Then $d; p \xrightarrow{S(a)}_t d'; p$ if $d \xrightarrow{S(a)}_t d'$, as for no $d \in \mathcal{C}_{Par}$, $d\sqrt{}$. By the inductive hypothesis for (2) we have that $d' \in \sim_t \mathcal{C}_{Par}$. Thus $d'; p \in \sim_t \mathcal{C}_{Seq}$.

(2) Assume c is of the form $\prod_{i \in I} c_i | p$, with $|I| > 0$ and $p \longrightarrow_t$. If $\prod_{i \in I} c_i | p \xrightarrow{S(a)}_t c'$ there are two cases to examine:

- suppose, wlog, that $\prod_{i \in I} c_i | p \xrightarrow{S(a)}_t c'_1 | \prod_{i \neq 1} c_i | p$ because $c_1 \xrightarrow{S(a)}_t c'_1$. As $c_1 \in \mathcal{C}_{Seq}$ we may apply the inductive hypothesis for (1) to obtain $c'_1 \in \sim_t \mathcal{C}_{Seq}$. Hence $c'_1 | \prod_{i \neq 1} c_i | p \in \sim_t \mathcal{C}_{Par}$;
- suppose $\prod_{i \in I} c_i | p \xrightarrow{S(a)}_t \prod_{i \in I} c_i | d$ because $p \xrightarrow{S(a)}_t d$. Then, by lemma 3.5.4, $d \in \sim_t \mathcal{C}_{Seq}$ or $d \in \sim_t \mathcal{C}_{Par}$. In both cases $\prod_{i \in I} c_i | d \in \sim_t \mathcal{C}_{Par}$.

If c is of the form $\prod_{i \in I} c_i$, $|I| > 1$, then, wlog, $\prod_{i \in I} c_i \xrightarrow{S(a)}_t c'_1 | \prod_{i \neq 1} c_i$ because $c_1 \xrightarrow{S(a)}_t c'_1$. As $c_1 \in \mathcal{C}_{Seq}$ we may apply the inductive hypothesis for (1) to obtain $c'_1 \in \sim_t \mathcal{C}_{Seq}$. Thus $c'_1 | \prod_{i \neq 1} c_i \in \sim_t \mathcal{C}_{Par}$.

This completes the proof of the lemma. □

By the above lemmas, we may assume from now on that, modulo $\sim_t$, $\mathcal{C}_{Seq}$ and $\mathcal{C}_{Par}$ are closed with respect to moves of the form $\xrightarrow{S(a)}_t$, $a \in \mathbf{A}$. The following lemma studies the effect of end-moves on configurations.

Lemma 3.5.6 (Effect of End-Moves on Configurations) *For each* $c \in C_{Seq} \cup C_{Par}$, *the following statements hold:*

(1) $c \in C_{Seq}$ *and* $c \stackrel{F(a)}{\longrightarrow}_t c'$ *imply* $c' \in \mathbf{P}$ *or* $c' \in\!\sim_t C_{Seq}$;

(2) $c \in C_{Par}$ *and* $c \stackrel{F(a)}{\longrightarrow}_t c'$ *imply* $c' \in \mathbf{P}$ *or* $c' \in\!\sim_t C_{Seq}$ *or* $c' \in\!\sim_t C_{Par}$.

Proof: Both statements are proved simultaneously by induction on the size of c. We assume, as inductive hypothesis, that (1) and (2) are true for all d with $|d| < |c|$.

(1) If c is of the form $F(a);p$ then $F(a);p \stackrel{F(a)}{\longrightarrow}_t nil;p \in \mathbf{P}$. Otherwise c is of the form $d;p$ with $d \in C_{Par}$ and $p \longrightarrow_t$. Assume that $d;p \stackrel{F(a)}{\longrightarrow}_t d';p$ because $d \stackrel{F(a)}{\longrightarrow}_t d'$. Then, by the inductive hypothesis for (2), $d' \in \mathbf{P}$ or $d' \in\!\sim_t C_{Seq}$ or $d' \in\!\sim_t C_{Par}$. If $d' \in \mathbf{P}$ then $d';p \in \mathbf{P}$. If $d' \in\!\sim_t C_{Par}$ then $d';p \in\!\sim_t C_{Seq}$. Otherwise $d' \in\!\sim_t C_{Seq}$ and there are two possibilities to consider:

- $d' \sim_t F(b);q$. Then $d';p \sim_t F(b);(q;p) \in C_{Seq}$.
- $d' \sim_t d'';q$ with $d'' \in C_{Par}$ and $q \longrightarrow_t$. Then $d';p \sim_t d'';(q;p) \in C_{Seq}$.

This completes the proof of (1).

(2) Assume $c \in C_{Par}$. We distinguish two cases:

- c is of the form $\prod_{i\in I} c_i|p$, $|I| > 0$ and $p \longrightarrow_t$. Then, wlog, $\prod_{i\in I} c_i|p \stackrel{F(a)}{\longrightarrow}_t c_1'|\prod_{i\neq 1} c_i|p$ because $c_1 \stackrel{F(a)}{\longrightarrow}_t c_1'$. By the inductive hypothesis for (1), $c_1' \in \mathbf{P}$ or $c_1' \in\!\sim_t C_{Seq}$. If $c_1' \in\!\sim_t C_{Seq}$ then $c_1'|\prod_{i\neq 1} c_i|p \in\!\sim_t C_{Par}$. Otherwise $c_1' \in \mathbf{P}$. If $|I| = 1$ then $c_1'|\prod_{i\neq 1} c_i|p \in \mathbf{P}$. If $|I| > 1$ then $c_1'|\prod_{i\neq 1} c_i|p \in\!\sim_t C_{Par}$ as $c_1'|p \longrightarrow_t$.
- c is of the form $\prod_{i\in I} c_i$, $|I| > 1$. Assume, wlog, that $\prod_{i\in I} c_i \stackrel{F(a)}{\longrightarrow}_t c_1'|\prod_{i\neq 1} c_i \equiv c'$ because $c_1 \stackrel{F(a)}{\longrightarrow}_t c_1'$. By the inductive hypothesis for (1), $c_1' \in \mathbf{P}$ or $c_1' \in\!\sim_t C_{Seq}$. If $c_1' \in\!\sim_t C_{Seq}$ then $c' \in\!\sim_t C_{Par}$. Otherwise $c_1' \in \mathbf{P}$. Now, if $|I| > 2$ then $c' \in\!\sim_t C_{Par}$. If $|I| = 2$ and $c_1' \longrightarrow_t$ then $c' \in\!\sim_t C_{Par}$. If $|I| = 2$ and $c_1' \sim_t nil$ then $c' \in\!\sim_t C_{Seq}$.

This completes the proof of (2). □

The following lemma will find application in some of the results presented in the following section.

Lemma 3.5.7 *Let* $c \in C_{Par}$. *Assume that*

- $c \stackrel{S(a)}{\not\longrightarrow}_t$, *for each* $a \in \mathbf{A}$, *and*

- *for no* $a \in \mathbf{A}, d \in_{\sim_t} \mathcal{C}_{Par}$, $c \xrightarrow{F(a)}_t d$.

Then $c \sim_t F(a)|F(b)$, *for some* $a, b \in \mathbf{A}$.

Proof: First of all, note that, by the proviso of the lemma, c must be of the form $c_1|c_2$ with $c_1, c_2 \in \mathcal{C}_{Seq}$. We claim that $c_1 \sim_t F(a)$ and $c_2 \sim_t F(b)$, for some $a, b \in \mathbf{A}$. Assume in fact, for the sake of contradiction, that $c_1 \in \mathcal{C}_{Seq}$ and, for no $a \in \mathbf{A}$, $c_1 \sim_t F(a)$. Then either $c_1 \sim_t F(a);p$, for some $a \in \mathbf{A}$ and p such that $|p| > 0$, or $c_1 \sim_t d;p$, for some $d \in \mathcal{C}_{Par}$ and p such that $|p| > 0$. In both cases, it is easy to see that, by the previous lemma, $c_1|c_2 \xrightarrow{F(a)}_t c_1'|c_2 \in_{\sim_t} \mathcal{C}_{Par}$, for some a and c_1'. This contradicts the hypotheses of the lemma. Hence, by symmetry, $c_1 \sim_t F(a)$ and $c_2 \sim_t F(b)$, for some a, b. □

The previous lemmas state the basic operational material which will find extensive application in the proof of results leading to the promised completeness theorem. Our next aim is to prove two fundamental decomposition properties used in the proof of the completeness theorem: the sequential and the parallel decomposition theorems. The proofs of these results will make use of some general unique factorization results with respect to $\sim_t$, which are based upon similar results presented in [Mol89], [MM90], for $\sim$. The proofs of the factorization and decomposition results will be the subject of the following section. In what follows rather than always working modulo $\sim_t$ we will reduce the relation $\in_{\sim_t}$ to $\in$, i.e. we assume that $\in$ will always be considered *modulo* $\sim_t$.

3.5.2 Unique Factorization and Decomposition Results

This section will be entirely devoted to proving two fundamental decomposition results which will find application in the proof of the completeness theorem: the sequential decomposition and the parallel decomposition theorems. The sequential decomposition result we are after states that, for each $c, d \in \mathcal{C}_{Par}$ and $p, q \in \mathbf{P}$,

$$c;p \sim_t d;q \quad \text{implies} \quad c \sim_t d \text{ and } p \sim_t q. \tag{3.1}$$

The parallel decomposition theorem we should like to prove, which may be seen as the dual statement of (3.1), states that, for each $c, d \in \mathcal{C}_{Seq}$ and $p, q \in \mathbf{P}$,

$$c|p \sim_t d|q \quad \text{implies} \quad c \sim_t d \text{ and } p \sim_t q. \tag{3.2}$$

In the process of proving (3.1) and (3.2) we shall establish unique factorization results, modulo $\sim_t$, for states $s \in \mathcal{S}$ with respect to the operators of sequential and parallel composition, which are similar to the ones presented in [Mol89], [MM90] modulo strong bisimulation equivalence, $\sim$. Apart from their intrinsic interest, such factorization results will allow us to give elegant proofs of the above-mentioned decomposition results.

Our first aim in this section is to show the sequential decomposition result stated in (3.1). As mentioned above, the proof of (3.1) will rely on a unique sequential factorization result, which states that each state $s \in \mathcal{S}$ may be expressed uniquely, modulo $\sim_t$, as a sequential composition of "semantically parallel states". Intuitively, a state s is "semantically parallel" if it cannot be expressed, modulo $\sim_t$, as a nontrivial sequential composition of states. Formally:

Definition 3.5.3 (Seq-irreducible and seq-prime states) *Let $s \in \mathcal{S}$. Then s is said to be:*

(i) seq-irreducible *if $s \sim_t c;p$, for some $c \in \mathcal{S}$ and $p \in \mathbf{P}$, implies $c \sim_t nil$ or $p \sim_t nil$;*

(ii) seq-prime *if s is seq-irreducible and $s \not\sim_t nil$.*

For each $s \in \mathcal{S}$, a *sequential factorization* for s modulo $\sim_t$ is a sequence of seq-primes $s_1, p_2, \ldots, p_n$, with $n \geq 0$, $s_1 \in \mathcal{S}$ and, for each $2 \leq i \leq n$, $p_i \in \mathbf{P}$, such that

$$s \sim_t s_1; p_2; \ldots; p_n.$$

By convention, if $n = 0$ then $s_1; p_2; \ldots; p_n \equiv nil$. We shall now prove that each $s \in \mathcal{S}$ has a *unique* sequential factorization into seq-primes, modulo $\sim_t$.

Theorem 3.5.1 (Unique Sequential Factorization) *Let $s \in \mathcal{S}$. Then s has a unique sequential factorization into seq-primes modulo $\sim_t$.*

Proof: The proof is divided into two parts, an existence part and a uniqueness part, both of which are shown by induction on $|s|$, the size of $s \in \mathcal{S}$.

Existence. We assume, as inductive hypothesis, that each s' with $|s'| < |s|$ has a sequential factorization. The proof proceeds by a case analysis on the possible form s may take, modulo $\sim_t$. If $s \sim_t nil$ then s has an empty factorization into seq-primes. If s is seq-prime then it is its own factorization. Otherwise we may assume, wlog, that $s \sim_t s_1; p$ with $|s_1|, |p| > 0$. By the inductive hypothesis, $c_1; q_2; \ldots; q_n$ and $p_1; \ldots; p_m$ are seq-prime factorizations for s_1 and p, respectively. Thus, by substitutivity,

$$s \sim_t c_1; q_2; \ldots; q_n; p_1; \ldots; p_m$$

is a seq-prime factorization for s. This completes the proof of the existence part of the result.

Uniqueness. We assume, as inductive hypothesis, that each $s' \in \mathcal{S}$ such that $|s'| < |s|$ has a unique seq-prime factorization. Assume that

$$\begin{aligned} s &\sim_t c_1; p_2; \ldots; p_n \\ &\sim_t d_1; q_2; \ldots; q_m \end{aligned}$$

are two factorizations of s into seq-primes. We shall show that the two factorizations are identical, modulo $\sim_t$. If $n = 0$ then $m = 0$ and we are done. If $n = 1$ then s is a seq-prime and thus $m = 1$ and $c_1 \sim_t d_1$. Hence the two factorizations are identical. Otherwise, by symmetry, we may assume that $n, m \geq 2$. We distinguish two possibilities:

(i) $p_n \sim_t q_m$, and

(ii) $p_n \not\sim_t q_m$.

We examine the two possibilities in turn.

(i) Assume that $p_n \sim_t q_m$. Then, by Lemma 3.5.2 and substitutivity, we have that

$$c_1; p_2; \ldots; p_{n-1} \sim_t d_1; q_2; \ldots; q_{m-1}. \tag{3.3}$$

As $|p_n|, |q_m| > 0$, we may apply the inductive hypothesis to (3.3) to obtain that $c_1, p_2, \ldots, p_{n-1}$ and $d_1, q_2, \ldots, q_{m-1}$ are identical seq-prime factorizations. Thus we have that $n = m$, $c_1 \sim_t d_1$ and, for each $1 < i \leq n$, $p_i \sim_t q_i$.

(ii) Assume now, for the sake of contradiction, that $p_n \not\sim_t q_m$. As $|c_1| > 0$, we have that $c_1 \xrightarrow{e}_t c_1'$ for some $e \in \mathbf{A_S}$ and c_1'. By the operational semantics, $c_1; p_2; \ldots; p_n \xrightarrow{e}_t c_1'; p_2; \ldots; p_n$. Moreover, as $|d_1| > 0$ and $c_1; p_2; \ldots; p_n \sim_t d_1; q_2; \ldots; q_n$, we have that, for some d_1',

$$d_1; q_2; \ldots; q_m \xrightarrow{e}_t d_1'; q_2; \ldots; q_m \sim_t c_1'; p_2; \ldots; p_n.$$

By the inductive hypothesis,

$$\begin{aligned} \bar{s} &\sim_t c_1'; p_2; \ldots; p_m \\ &\sim_t d_1'; q_2; \ldots; q_m \end{aligned}$$

has a unique seq-prime factorization

$$c^1; \ldots; c^h; p_2; \ldots; p_m \sim_t d^1; \ldots; d^k; q_2; \ldots; q_m,$$

where $h, k \geq 0$ and $c_1' \sim_t c^1; \ldots; c^h$ and $d_1' \sim_t d^1; \ldots; d^k$ are the unique seq-prime factorizations for c_1' and d_1', respectively. As $n, m \geq 2$, we then have that $p_n \sim_t q_m$. This contradicts the hypothesis that $p_n \not\sim_t q_m$.

This completes the proof of the uniqueness part and that of the theorem. □

We shall now show how the above-given unique sequential factorization result for states, modulo $\sim_t$, may be used to give a proof of the sequential decomposition theorem. First of all, we shall show that each parallel configuration c is indeed a seq-prime state. This result will be a corollary of the following useful proposition.

Proposition 3.5.4 *Let $c \in \mathcal{C}_{Par}$, $d \in \mathcal{S}$ and $p \in \mathbf{P}$. Assume that $c \sim_t d;p$. Then $p \sim_t nil$.*

Proof: By induction on the size of c. Assume that $c \sim_t d;p$. Note that, as $c \in \mathcal{C}_{Par}$ implies that $c \xrightarrow{F(a)}_t$ for some a, this implies that $|d| > 0$. Thus each initial move from $d;p$ has to come from d. The proof proceeds by an analysis of the following three possibilities:

(a) $c \xrightarrow{S(a)}_t c'$, for some a, c';

(b) $c \xrightarrow{F(a)}_t c' \in \mathcal{C}_{Par}$, for some a, c', and

(c) none of the two previous cases applies.

We examine each possibility in turn.

(a) In this case, by Lemma 3.5.5, $c' \in \mathcal{C}_{Par}$. Moreover, as $c \sim_t d;p$, $d;p \xrightarrow{S(a)}_t d';p \sim_t c'$ for some d'. We may now apply the inductive hypothesis to obtain $p \sim_t nil$.

(b) Similar to (a).

(c) Assume now that none of the two previous cases hold. We are then in a position to apply Lemma 3.5.3 to obtain that $c \sim_t F(a)|F(b)$, for some a, b. Then, as $c \sim_t d;p$ and $c \xrightarrow{F(a)}_t\xrightarrow{F(b)}_t nil|nil$, there exists d' such that $d;p \xrightarrow{F(a)}_t\xrightarrow{F(b)}_t d';p \sim_t nil|nil$. This implies that $d' \sim_t p \sim_t nil$.

This completes the proof of the proposition. □

Corollary 3.5.2 *Each parallel configuration is seq-prime.*

Proof: Assume that $c \in \mathcal{C}_{Par}$ and that $c \sim_t d;p$. Then, by the above proposition, we have that $p \sim_t nil$. Thus c is seq-irreducible. Moreover, as $c \in \mathcal{C}_{Par}$ implies $|c| > 0$, c is seq-prime.
□

We can now prove the promised sequential decomposition theorem.

Theorem 3.5.2 (Sequential Decomposition Theorem) *Let $c,\ d \in \mathcal{C}_{Par}$. Then $c;p \sim_t d;q$ implies $c \sim_t d$ and $p \sim_t q$.*

Proof: By Theorem 3.5.1, p and q have unique factorizations into seq-primes given by

$$p \sim_t p_1; \ldots; p_n \text{ and}$$
$$q \sim_t q_1; \ldots; q_m.$$

By Corollary 3.5.2, $c, d \in \mathcal{C}_{Par}$ implies that c and d are seq-prime. Thus, by substitutivity and Theorem 3.5.1,

$$c; p_1; \ldots; p_n \sim_t d; q_1; \ldots; q_m$$

are identical factorizations into seq-primes. This implies that $c \sim_t d$ and, for each $1 \leq i \leq n = m$, $p_i \sim_t q_i$. By substitutivity, we then get that $p \sim_t q$. □

The dual statement of Theorem 3.5.2, the parallel decomposition theorem, will now be shown by following a similar strategy. First of all, we shall prove a unique parallel factorization result for states, which states that each $s \in \mathcal{S}$ may be expressed uniquely, modulo $\sim_t$, as a parallel composition of "semantically sequential states". Intuitively, a state is "semantically sequential" if it cannot be expressed, modulo $\sim_t$, as a nontrivial parallel composition of states. Formally:

Definition 3.5.4 (Irreducible and Prime States) *Let $s \in \mathcal{S}$. Then s is said to be:*

(i) irreducible *if $s \sim_t s_1|s_2$ implies $s_1 \sim_t nil$ or $s_2 \sim_t nil$;*

(ii) prime *if s is irreducible and $s \not\sim_t nil$.*

The proof of the unique parallel factorization result for states modulo $\sim_t$ uses the following simplification lemma, which is familiar from the theory of several equivalences based on the notion of bisimulation, see e.g. [CH89], [Mol89].

Lemma 3.5.8 (Simplification Lemma) *Let $c,\ d,\ f \in \mathcal{S}$. Then:*

(1) *For each $e \in \mathrm{A_S}$, $f' \in \mathcal{S}$, $f \xrightarrow{e}_t f'$ and $c|f \sim_t d|f'$ imply $d \xrightarrow{e}_t d'$, for some d' such that $c \sim_t d'$.*

(2) *$c|f \sim_t d|f$ imply $c \sim_t d$.*

Proof: Both statements are proved simultaneously by induction on the combined size of c, d and f. We assume, as an inductive hypothesis, that (1)-(2) hold for each c', d' and f' such that $|c'| + |d'| + |f'| < |c| + |d| + |f|$.

(1) Assume that $f \xrightarrow{e}_t f'$ and $c|f \sim_t d|f'$. Then, by the operational semantics, $c|f \xrightarrow{e}_t c|f'$. As $c|f \sim_t d|f'$, there is a matching e-move from $d|f'$. We distinguish two possibilities:

- $d|f' \xrightarrow{e}_t d'|f' \sim_t c|f'$ because $d \xrightarrow{e}_t d'$. Then we may apply the inductive hypothesis for (2) to obtain $c \sim_t d'$. Hence there exists d' such that $d \xrightarrow{e}_t d' \sim_t c$.
- $d|f' \xrightarrow{e}_t d|f'' \sim_t c|f'$ because $f' \xrightarrow{e}_t f''$. Then we may apply the inductive hypothesis for (1) to obtain that $d \xrightarrow{e}_t d' \sim_t c$, for some d'.

This completes the inductive step for statement (1).

(2) Assume $c|f \sim_t d|f$ and $c \xrightarrow{e}_t c'$. Then, by the operational semantics, $c|f \xrightarrow{e}_t c'|f$. Let us examine the possible matching moves from $d|f$:

- $d|f \xrightarrow{e}_t d'|f \sim_t c'|f$ because $d \xrightarrow{e}_t d'$. Then, by the inductive hypothesis for (2), $c' \sim_t d'$.
- $d|f \xrightarrow{e}_t d|f' \sim_t c'|f$ because $f \xrightarrow{e}_t f'$. Then, by the inductive hypothesis for (1), $d \xrightarrow{e}_t d' \sim_t c'$ for some d'.

Thus for every move of c we may find a matching move by d. By symmetry, $c \sim_t d$. □

For each $s \in \mathcal{S}$, a parallel factorization for s modulo $\sim_t$ is given by a set $\{c_1, \ldots, c_k\}$, $k \geq 0$, of primes such that

$$s \sim_t c_1 | \cdots | c_k.$$

We now have all the technical material which is needed to prove that each state has a *unique* parallel factorization into primes modulo $\sim_t$. The proof of the parallel factorization result follows the one of Theorem 4.2.2 (pages 77-79) of [Mol89].

Theorem 3.5.3 (Unique Parallel Factorization) *Each $s \in \mathcal{S}$ may be expressed uniquely, modulo $\sim_t$, as a parallel composition of primes.*

Proof: The proof is divided into two parts, an existence part and a uniqueness part, both of which are shown by induction on $|s|$, the size of $s \in \mathcal{S}$.

Existence. We show, first of all, that s can be expressed, up to $\sim_t$, as a parallel composition of primes. We assume, as inductive hypothesis, that the claim holds for all s' such that

$|s'| < |s|$. The proof proceeds by an analysis on the possible form s may take, modulo $\sim_t$.

If $s \sim_t nil$ then s can be expressed as an empty product of primes. If s is prime then it is its own parallel factorization. Otherwise, we may assume that $s \sim_t s_1|s_2$ for some s_1, s_2 such that $|s_1|, |s_2| > 0$. By the inductive hypothesis, s_1 and s_2 have prime factorizations given by $c_1|\cdots|c_n$ and $d_1|\cdots|d_m$, respectively. By substitutivity, we then have that

$$s \sim_t s_1|s_2 \sim_t c_1|\cdots|c_n|d_1|\cdots|d_m$$

is a parallel factorization for s. This completes the proof of the existence part of the theorem.

Uniqueness. We assume, as inductive hypothesis, that each s' such that $|s'| < |s|$ has a unique parallel factorization up to $\sim_t$. Assume now that

$$s \sim_t c_1|\cdots|c_n \sim_t d_1|\cdots|d_m$$

are two parallel factorizations of s into primes. We prove that the two factorizations are identical, modulo $\sim_t$. We proceed by analyzing the following two possibilities:

(a) the two factorizations have a common prime factor,

(b) $c_i \not\sim_t d_j$, for all i, j, i.e. there is no common prime factor in the two factorizations.

We examine the two possibilities separately.

(a) Assume, wlog, that $c_1 \sim_t d_1$. Then, by substitutivity,

$$s \sim_t c_1|\cdots|c_n \sim_t c_1|d_2|\cdots|d_m.$$

By the Simplification Lemma, this implies that $c_2|\cdots|c_n \sim_t d_2|\cdots|d_m$. By the inductive hypothesis, which is applicable as $|c_1| > 0$, we have that $\{c_2, \ldots, c_n\}$ and $\{d_2, \ldots, d_m\}$ are identical prime factorizations, modulo $\sim_t$. Hence $\{c_1, \ldots, c_n\}$ and $\{d_1, \ldots, d_m\}$ are identical prime factorizations for s.

(b) Assume that $c_i \not\sim_t d_j$, for all i, j. If $n = 0$ then it must be the case that $m = 0$ and the two prime factorizations are both empty. Assume now, for the sake of contradiction, that $n \geq 1$. We distinguish two possibilities depending on whether $n = 1$ or $n > 1$.

(b.1) Assume that $n = 1$. Then, as c_1 is prime, we have that $m = 1$ and $c_1 \sim_t d_1$. This contradicts the assumption that the two decompositions have no common prime factor.

(b.2) By symmetry, we may assume that $n, m \geq 2$. Let, wlog, c_1 be minimal with respect to size amongst all the prime factors c_i, $1 \leq i \leq n$, and d_j, $1 \leq j \leq m$, i.e.

$$|c_1| \leq |c_i| \text{ and } |c_1| \leq |d_j|, \text{ for all } i, j.$$

Then, as $|c_1| > 0$, $c_1 \xrightarrow{e}_t c_1'$ for some $e \in \mathbf{A}_S$ and $c_1' \in \mathcal{S}$. By the inductive hypothesis, c_1' has a unique parallel factorization

$$c_1' \sim_t c^1|\ldots|c^h,\ h \geq 0.$$

By the operational semantics, $c_1|\cdots|c_n \xrightarrow{e}_t c_1'|c_2|\cdots|c_n \equiv \bar{c}$ and, by the inductive hypothesis, $\bar{c}$ has a unique prime factorization given by

$$c^1|\cdots|c^h|c_2|\cdots|c_n.$$

As $c_1|\cdots|c_n \sim_t d_1|\cdots|d_m$, we have that, wlog, $d_1|\cdots|d_m \xrightarrow{e}_t d_1'|d_2|\cdots|d_m \sim_t \bar{c}$ because $d_1 \xrightarrow{e}_t d_1'$, for some d_1'. By the inductive hypothesis, d_1' has a unique prime factorization given by $d^1|\cdots|d^k$, $k \geq 0$. Then, again by the inductive hypothesis,

$$c^1|\cdots|c^h|c_2|\cdots|c_n \sim_t d^1|\cdots|d^k|d_2|\cdots|d_m$$

are identical prime factorizations for $\bar{c}$. Note that, for each $1 \leq l \leq h$ and $2 \leq j \leq m$,

$$|c^l| < |c_1| \leq |d_j|.$$

Thus, for such l, j, we have that $c^l \not\sim_t d_j$. This implies that each d_j does not appear in $c^1|\cdots|c^h|c_2|\cdots|c_n$. Hence $m < 2$. This contradicts the assumption that $m \geq 2$.

This completes the proof of the existence part.

The proof of the theorem is now complete. □

We shall now show how the unique parallel factorization result may be used to give a proof of the parallel decomposition theorem. As it may be expected, given the dual rôles played by sequential and parallel composition and configurations in our technical development, the key to the proof of the parallel decomposition theorem is to show that each sequential configuration is prime.

Proposition 3.5.5 *Each sequential configuration is prime.*

Proof: Let $c \in \mathcal{C}_{Seq}$. We shall show that c is prime by a case analysis on the possible form c may take.

- $c \equiv F(a); p$. Assume, for the sake of contradiction, that $c \sim_t c_1|c_2$ with $|c_1|, |c_2| > 0$. We proceed by analyzing the possible form of c_1 and c_2. If $c_1 \in \mathbf{P}$ then, as $|c_1| > 0$, we have that $c_1 \xrightarrow{S(a)}_t$, for some $a \in \mathbf{A}$. This implies that $c_1|c_2 \xrightarrow{S(a)}_t$, contradicting the assumption that $c \sim_t c_1|c_2$.

  Otherwise, by symmetry, we may assume that $c_1, c_2 \in \mathcal{S} - \mathbf{P}$. Then we have that $c_1 \xrightarrow{F(a)}_t$ and $c_2 \xrightarrow{F(b)}_t$, for some a, b. This implies that $c_1|c_2 \xrightarrow{F(a)}_t \xrightarrow{F(b)}_t$, again contradicting the assumption that $c \sim_t c_1|c_2$.

  Thus each sequential configuration of the form $F(a); p$ is irreducible and obviously prime.

- $c \equiv d; p$ with $d \in \mathcal{C}_{Par}$ and $|p| > 0$. Assume, for the sake of contradiction, that $c \sim_t c_1|c_2$, with $|c_1|, |c_2| > 0$. By Lemma 3.5.3(ii), this implies that $c_1|c_2 \in \mathbf{P}$ or $c_1|c_2 \in \mathcal{C}_{Par}$. If $c_1|c_2 \in \mathbf{P}$ then we obtain a contradiction to $c \sim_t c_1|c_2$ as $c \xrightarrow{F(a)}_t$ for some a.

  If $c_1|c_2 \in \mathcal{C}_{Par}$ then, by the sequential decomposition theorem, $c \equiv d; p \sim_t c_1|c_2$ implies $p \sim_t nil$. This contradicts the assumption that $|p| > 0$. Thus each sequential configuration of the form $d; p$ is prime.

As we have considered all the possible forms sequential configurations may take, we have that each $c \in \mathcal{C}_{Seq}$ is prime. □

We can now prove the promised parallel decomposition theorem[2]. In the statement of this result we will use some new notation which is introduced in the following definition.

Definition 3.5.5 *Let $\{c_i \mid i \in I\}$, $\{d_j \mid j \in J\} \subseteq \mathcal{C}_{Seq}$, with I and J finite index sets. Then we write $\{c_i \mid i \in I\} \sim_t \{d_j \mid j \in J\}$ iff there exists a bijective map $\phi : I \rightarrow J$ such that $c_i \sim_t d_{\phi(i)}$, for each $i \in I$.*

Theorem 3.5.4 (Parallel Decomposition Theorem) *Let $\{c_i \mid i \in I\}$, $\{d_j \mid j \in J\} \subseteq \mathcal{C}_{Seq}$, $p, q \in \mathbf{P}$. Then $c \equiv (\prod_{i \in I} c_i)|p \sim_t (\prod_{j \in J} d_j)|q \equiv d$ implies $\{c_i \mid i \in I\} \sim_t \{d_j \mid j \in J\}$ and $p \sim_t q$.*

Proof: By the unique parallel factorization theorem (Theorem 3.5.3), there exist unique prime factorizations for p and q given by

(i) $p \sim_t p_1 | \cdots | p_n$,
(ii) $q \sim_t q_1 | \cdots | q_m$.

[2] We thank Frits Vaandrager for suggesting the proof of this theorem which is presented below.

Note that each prime factor of p and q has to be a process. By Proposition 3.5.5, each c_i and d_j is prime as it is a sequential configuration. Thus, by substitutivity and the unique parallel factorization theorem,

$$(\prod_{i\in I} c_i)|p_1|\cdots|p_n \sim_t (\prod_{j\in J} d_j)|q_1|\cdots|q_m$$

are identical prime factorizations up to $\sim_t$. As, obviously, $c_i \not\sim_t q_k$ and $d_j \not\sim_t p_h$, for all i, j, h, k, it must be the case that

$$\{c_i \mid i \in I\} \sim_t \{d_j \mid j \in J\} \text{ and } \{p_h \mid 1 \leq h \leq n\} \sim_t \{q_k \mid 1 \leq k \leq m\}.$$

By substitutivity, (i) and (ii), we then have that $p \sim_t q$. This completes the proof of the theorem. □

Intuitively, for a process $p \in \mathbf{P}$, its unique parallel factorization modulo $\sim_t$ may be seen as the "most parallel version" of p (modulo $\sim_t$). Similarly, its unique sequential factorization may be seen as the "most sequential version" of p (modulo $\sim_t$). For each $p \in \mathbf{P}$, by the two factorization theorems we have just shown,

$$\begin{array}{rll} p & \sim_t \; p_1|\cdots|p_n & \text{a unique product of primes} \\ & \sim_t \; q_1;\ldots;q_m & \text{a unique sequential composition of seq-primes.} \end{array}$$

With respect to standard bisimulation, $\sim$, it is possible to have that both the factorizations be nontrivial, which we take to mean that $n, m \geq 2$. For instance, for $p \equiv a|a$,

$$\begin{array}{rll} p & \sim_t \; a|a & \text{(parallel factorization)} \\ & \sim_t \; a;a & \text{(sequential factorization).} \end{array}$$

We shall now show that, up to $\sim_t$, at least one of the factorizations is trivial.

Theorem 3.5.5 *Let $p \in \mathbf{P}$. Assume that $p_1|\cdots|p_n$ and $q_1;\cdots;q_m$ are a prime factorization and a seq-prime factorization for p, up to $\sim_t$, respectively. Then $n = m = 0$ or $n = 1$ or $m = 1$.*

Proof: Assume, for the sake of contradiction, that $n, m \geq 2$ and that

$$p_1|\cdots|p_n \quad \sim_t \quad q_1;\ldots;q_m. \tag{3.4}$$

As $|p_1| > 0$, there exist $a \in \mathbf{A}$ and $c \in \mathcal{S}$ such that $p_1 \xrightarrow{S(a)}_t c$. By the operational semantics, we then have that $p_1|\cdots|p_n \xrightarrow{S(a)}_t c|\bar{p}$, where $\bar{p} \equiv p_2|\cdots|p_n$ and $|\bar{p}| > 0$ as $n \geq 2$. By (3.4) and the fact that $|q_1| > 0$, we have that, for some d, $q_1;\ldots;q_m \xrightarrow{S(a)}_t d;\bar{q} \sim_t c|\bar{p}$, where $\bar{q} \equiv q_2;\ldots;q_m$ and $|\bar{q}| > 0$ as $m \geq 2$. By Lemma 3.5.4, $c, d \in \mathcal{C}_{Seq} \cup \mathcal{C}_{Par}$. Moreover, as $c|\bar{p} \sim_t d;\bar{q}$ and

$|\bar{p}| > 0$, we have that $d \in C_{Par}$ or d is of the form $\bar{d}; t$ with $\bar{d} \in C_{Par}$ and $|t| > 0$. In both cases $d; \bar{q} \in C_{Seq}$.

We proceed by considering the cases $c \in C_{Seq}$ and $c \in C_{Par}$ separately. If $c \in C_{Seq}$ then, by the parallel decomposition theorem, $c|\bar{p} \sim_t d; \bar{q}$ implies that $\bar{p} \sim_t nil$. This contradicts the fact that $|\bar{p}| > 0$.

If $c \in C_{Par}$ then, by Lemma 3.5.4, c has the form $c_1|t$ for some $c_1 \in C_{Seq}$ and t such that $|t| > 0$. Then, by substitutivity and the parallel decomposition theorem, $c|\bar{p} \sim_t c_1|t|\bar{p} \sim_t d; \bar{q}$ implies that $t|\bar{p} \sim_t nil$. This contradicts the hypothesis that $|t|, |\bar{p}| > 0$.

Thus we have shown that either $n < 2$ or $m < 2$. Obviously, if $n = 0$ then $m = 0$. Hence, by symmetry, $n = m = 0$ or $n = 1$ or $m = 1$. □

An interesting corollary of the above theorem is the following result stating that, up to $\sim_t$, each process is either *nil* or prime or seq-prime.

Corollary 3.5.3 *Let* $p \in \mathbf{P}$. *Then* $p \sim_t nil$ *or* p *is prime or* p *is seq-prime.*

3.5.3 The Completeness Theorem

In order to prove the promised completeness result, we will need some further lemmas whose proofs will highlight the usefulness of the important decomposition results shown in the previous section.

Lemma 3.5.9 *Let* c, $d \in C_{Seq} \cup C_{Par}$. *Then:*

(i) $c \in C_{Seq}$ *and* $c \sim_t d$ *imply* $d \in C_{Seq}$.

(ii) $c \in C_{Par}$ *and* $c \sim_t d$ *imply* $d \in C_{Par}$.

Proof: We just prove (i) as (ii) will then follow by symmetry. Assume that $c \in C_{Seq}$ and $c \sim_t d$. Suppose, towards a contradiction, that $d \in C_{Par}$. Then, by the definition of C_{Par}, either $d \equiv \prod_{j \in J} d_j | p$, with $|J| > 0$, $d_j \in C_{Seq}$ for each $j \in J$ and $|p| > 0$, or $d \equiv \prod_{j \in J} d_j$, with $|J| > 1$ and $d_j \in C_{Seq}$ for each $j \in J$. If $d \equiv \prod_{j \in J} d_j | p \sim_t c$ then, by theorem 3.5.4, we obtain that $p \sim_t nil$. This contradicts the assumption that $|p| > 0$. If $d \equiv \prod_{j \in J} d_j \sim_t c$ then, again by theorem 3.5.4, we obtain that $\{c\} \sim_t \{d_j \mid j \in J\}$. However, this is impossible as $|J| > 1$. Hence $d \in C_{Seq}$. □

Lemma 3.5.10 *Let $c, d \in C_{Seq} \cup C_{Par}$. Then:*

(1) *$c \equiv F(a); p \sim_t d$ implies $d \equiv F(a); q$ and $p \sim_t q$.*

(2) *$c \equiv c_1; p \sim_t d$, with $c_1 \in C_{Par}$ and $|p| > 0$, implies $d \equiv d_1; q$ with $d_1 \in C_{Par}$, $|q| > 0$, $c_1 \sim_t d_1$ and $p \sim_t q$.*

(3) *$c \equiv c_1 | p \sim_t d$, with $c_1 \in C_{Seq}$ and $|p| > 0$, implies $d \equiv d_1 | q$, with $d_1 \in C_{Seq}$, $|q| > 0$, $c_1 \sim_t d_1$ and $p \sim_t q$.*

Proof: Assume $c, d \in C_{Seq} \cup C_{Par}$. We prove each statement separately.

(1) Assume $c \equiv F(a); p \sim_t d$. Then $c \in C_{Seq}$ and, by lemma 3.5.9, $c \sim_t d$ implies $d \in C_{Seq}$ as well. We show, first of all, that d must be of the form $F(a); q$. Assume in fact, towards a contradiction, that d is of the form $d_1; q$, with $d_1 \in C_{Par}$ and $|q| > 0$. Then, as $d_1 \in C_{Par}$, either $d_1 \xrightarrow{S(b)}_t$ or $d_1 \xrightarrow{F(b)}_t \xrightarrow{F(b')}_t$, for some $b, b' \in \mathbf{A}$. By the operational semantics the same is true of d, which contradicts the hypothesis that $c \sim_t d$. Hence d must be of the form $F(a); q$ and this easily implies $p \sim_t q$.

(2) Assume $c \equiv c_1; p \sim_t d$, with $c_1 \in C_{Par}$ and $|p| > 0$. Then $c \in C_{Seq}$ and, by lemma 3.5.9, $c \sim_t d$ implies $d \in C_{Seq}$ as well. By 1) and symmetry, we have that d must be of the form $d_1; q$, with $d_1 \in C_{Par}$ and $|q| > 0$. Thus, $c \equiv c_1; p \sim_t d_1; q \equiv d$ and we may apply corollary 3.5.2 to obtain $c_1 \sim_t d_1$ and $p \sim_t q$.

(3) Assume $c \equiv c_1 | p \sim_t d$, with $c_1 \in C_{Seq}$ and $|p| > 0$. Then $c \in C_{Par}$ and, by lemma 3.5.9, $c \sim_t d$ implies $d \in C_{Par}$ as well. We show, first of all, that d must be of the form $d_1 | q$, with $d_1 \in C_{Seq}$ and $|q| > 0$. Assume in fact, towards a contradiction, that $d \equiv \prod_{j \in J} d_j$, with $|J| > 1$ and $d_j \in C_{Seq}$ for each $j \in J$. Then we may apply theorem 3.5.4 to $c_1 | p \sim_t \prod_{j \in J} d_j$ to obtain $p \sim_t nil$. This contradicts the hypothesis that $|p| > 0$.

Assume now $d \equiv \prod_{j \in J} d_j | q$, with $|J| > 0$, $|q| > 0$ and $d_j \in C_{Seq}$ for each $j \in J$. Then we may apply theorem 3.5.4 to $c_1 | p \sim_t \prod_{j \in J} d_j | q$ to obtain $\{c_1\} \sim_t \{d_j \mid j \in J\}$ and $p \sim_t q$. Hence J is a singleton set and the statement holds. □

The following result, which will play an important rôle in the proof of the completeness theorem, expresses a very strong property of $\sim_t$ when applied to sequential and parallel head reduced forms.

Proposition 3.5.6 *Let $p, q \in \mathtt{HRF}_{Seq} \cup \mathtt{HRF}_{Par}$. Assume $p \xrightarrow{S(a)}_t c$, $q \xrightarrow{S(a)}_t d$ and $c \sim_t d$. Then $p \sim_t q$.*

Proof: By induction on the combined size of p and q. We assume, as inductive hypothesis, that the claim holds for all p', $q' \in \mathtt{HRF}_{Seq} \cup \mathtt{HRF}_{Par}$ such that $|p'| + |q'| < |p| + |q|$. Assume $p \xrightarrow{S(a)}_t c$, $q \xrightarrow{S(a)}_t d$ and $c \sim_t d$. By Lemma 3.5.3, $c \in \mathcal{C}_{Seq}$ or c is of the form $c'|r$, with $c' \in \mathcal{C}_{Seq}$ and $|r| > 0$. We consider these two possibilities in turn.

- $c \in \mathcal{C}_{Seq}$. We distinguish two subcases according to the structure of c.

  (1) $c \equiv F(a); r$. Then, by lemma 3.5.10, $c \sim_t d$ implies $d \equiv F(a); t$ and $r \sim_t t$. It is now easy to see that $p \xrightarrow{S(a)}_t F(a); r$, $p \in \mathtt{HRF}_{Seq} \cup \mathtt{HRF}_{Par}$, implies p is of the form $a; r$. Similarly q is of the form $a; t$. As $r \sim_t t$ we then have, by substitutivity, that $p \sim_t q$.

  (2) $c \equiv c'; r$, with $c' \in \mathcal{C}_{Par}$ and $|r| > 0$. Then, by lemma 3.5.10, $c \sim_t d$ implies d is of the form $d'; t$, with $d' \in \mathcal{C}_{Par}$, $|t| > 0$, $c' \sim_t d'$ and $r \sim_t t$. As $p \xrightarrow{S(a)}_t c$ and $p \in \mathtt{HRF}_{Seq} \cup \mathtt{HRF}_{Par}$, p must be of the form $p_1; r$, with $p_1 \in \mathtt{HRF}_{Par}$ and $p_1 \xrightarrow{S(a)}_t c'$. Similarly q has to be of the form $q_1; t$, with $q_1 \in \mathtt{HRF}_{Par}$ and $q_1 \xrightarrow{S(a)}_t d'$. As $|r|, |t| > 0$, $|q_1| + |p_1| < |p| + |q|$ and we may apply induction to obtain $p_1 \sim_t q_1$. By substitutivity, $p \sim_t q$.

- $c \equiv c'|r$, with $c' \in \mathcal{C}_{Seq}$ and $|r| > 0$. By lemma 3.5.10, $c \sim_t d$ implies d is of the form $d'|t$, with $d' \in \mathcal{C}_{Seq}$, $|t| > 0$, $c' \sim_t d'$ and $r \sim_t t$. It is easy to see that $p \xrightarrow{S(a)}_t c'|r$ and $p \in \mathtt{HRF}_{Seq} \cup \mathtt{HRF}_{Par}$ imply $p \equiv p_1 \not| r$, with $p_1 \in \mathtt{HRF}_{Seq}$ and $p_1 \xrightarrow{S(a)}_t c'$. Similarly, $q \xrightarrow{S(a)}_t d'|t$ implies $q \equiv q_1 \not| t$, with $q_1 \in \mathtt{HRF}_{Seq}$ and $q_1 \xrightarrow{S(a)}_t d'$. As $|r|, |t| > 0$, we may apply the inductive hypothesis to obtain $p_1 \sim_t q_1$. Hence, by substitutivity, $p \sim_t q$. □

We have now developed all the technical machinery needed in the proof of the completeness theorem.

Theorem 3.5.6 (Completeness) *Let $p, q \in \mathbf{P}$. Then $p \sim_t q$ implies $p =_E q$.*

Proof: The proof is by induction on the combined size of the terms. By the reduction lemma and the soundness of $=_E$ we may assume, wlog, that $p \equiv \sum_{i \in I} p_i$ and $q \equiv \sum_{j \in J} q_j$ are head reduced forms. By symmetry, it is then sufficient to show that

$$\forall i \in I \exists j \in J : p_i =_E q_j.$$

In fact, the equality $p =_E q$ will then be provable by applications of (A1)-(A3). Let $i \in I$. The proof proceeds by an analysis of the possible structure of p_i.

- $p_i \equiv a; r$, $r \in \mathtt{HRF}$. Then $p_i \xrightarrow{S(a)}_t F(a); r$. By the operational semantics, $p_i \xrightarrow{S(a)}_t F(a); r$ implies $p \xrightarrow{S(a)}_t F(a); r$. As $p \sim_t q$, there exists q_j such that $q_j \xrightarrow{S(a)}_t c \sim_t F(a); r$. By lemma 3.5.10, $c \sim_t F(a); r$ implies $c \equiv F(a); t$ with $r \sim_t t$. By the inductive hypothesis, $r =_E t$. Moreover, $q_j \xrightarrow{S(a)}_t F(a); t$ and $q_j \in \mathtt{HRF}_{Seq} \cup \mathtt{HRF}_{Par}$ imply $q_j \equiv a; t$. Hence, by substitutivity, $p_i \equiv a; r =_E a; t \equiv q_j$.

- $p_i \equiv r_1; r_2$, with $r_1 \in \mathtt{HRF}_{Par}$, $r_2 \in \mathtt{HRF}$ and $|r_2| > 0$. Assume $p_i \xrightarrow{S(a)}_t c$. Then c must be of the form $c'; r_2$ with $r_1 \xrightarrow{S(a)}_t c'$. By lemma 3.5.3, $c' \in C_{Par}$ and thus $c \in C_{Seq}$. As $p \sim_t q$, there exists q_j such that $q_j \xrightarrow{S(a)}_t d \sim_t c'; r_2$. By lemma 3.5.10, $d \sim_t c'; r_2$ implies d has the form $d'; t_2$, with $d' \in C_{Par}$, $|t_2| > 0$, $c' \sim_t d'$ and $r_2 \sim_t t_2$. By the inductive hypothesis, $r_2 =_E t_2$.

  It is now easy to see that $q_j \xrightarrow{S(a)}_t d'; t_2$ and $q_j \in \mathtt{HRF}_{Seq} \cup \mathtt{HRF}_{Par}$ imply $q_j \equiv t_1; t_2$, with $t_1 \in \mathtt{HRF}_{Par}$ and $t_1 \xrightarrow{S(a)}_t d'$. By proposition 3.5.6, $r_1 \xrightarrow{S(a)}_t c'$, $t_1 \xrightarrow{S(a)}_t d'$ and $c' \sim_t d'$ imply $r_1 \sim_t t_1$. Hence, by the inductive hypothesis, $r_1 =_E t_1$ and, by substitutivity, $p_i \equiv r_1; r_2 =_E t_1; t_2 \equiv q_j$.

- $p_i \equiv r_1 \lfloor r_2$, with $r_1 \in \mathtt{HRF}_{Seq}$, $r_2 \in \mathtt{HRF}$ and $|r_2| > 0$. Assume $p_i \xrightarrow{S(a)}_t c$. Then c must be of the form $c_1 | r_2$, with $r_1 \xrightarrow{S(a)}_t c_1$. By lemma 3.5.3, $c_1 \in C_{Seq}$. As $p \sim_t q$, there exists q_j such that $q_j \xrightarrow{S(a)}_t d \sim_t c_1 | r_2$. By lemma 3.5.10, it must be the case that $d \equiv d_1 | t_2$, with $d_1 \in C_{Seq}$, $|t_2| > 0$, $c_1 \sim_t d_1$ and $r_2 \sim_t t_2$. By the inductive hypothesis, $r_2 =_E t_2$.

  As $q_j \xrightarrow{S(a)}_t d_1 | t_2$ and $q_j \in \mathtt{HRF}_{Seq} \cup \mathtt{HRF}_{Par}$, q_j must be of the form $t_1 \lfloor t_2$, with $t_1 \in \mathtt{HRF}_{Seq}$ and $t_1 \xrightarrow{S(a)}_t d_1$. As $c_1 \sim_t d_1$, we then have, by proposition 3.5.6, that $r_1 \sim_t t_1$. By the inductive hypothesis, $r_1 =_E t_1$. By substitutivity, $p_i \equiv r_1 \lfloor r_2 =_E t_1 \lfloor t_2 \equiv q_j$.

This completes the proof of the theorem. □

3.6 Concluding Remarks

In this chapter, we have studied the consequences of the introduction of an operator for refining an action by a process in a simple process algebra.

More specifically we have considered a process algebra which constitutes the core of many of the existing ones, added to it a new combinator for refining an action by a process and then addressed the question of an appropriate equivalence for the augmented language.

The main result of this chapter is that, at least for the simple language we have considered, an adequate equivalence relation can be defined in a very intuitive manner. In fact, it coincides with both Hennessy's timed-equivalence, [H88c], and a slight reformulation of it which we call refine-equivalence. Moreover, these equivalences can be axiomatized in much the same way as the standard behavioural equivalences, [HM85], [DH84].

The results presented in this chapter constitute just the basis for a well-developed theory of action-refinement in process algebras. For instance, the language we have considered lacks communication between concurrent agents and a facility for their recursive definition. In the following chapter, we shall develop a richer process algebra for the specification of finite concurrent, communicating processes which incorporates a refinement operator. Moreover, building on the work presented in this chapter, we shall propose a suitable semantic equivalence for the extended language. As we shall see, the presence of communication between concurrent processes and of a restriction operator in the language will make the semantic analysis much more delicate.

Finally, let us refer briefly to some related work. In [Pr86], Pratt discusses the *pomset model* of concurrent computation. One of the most interesting operators he introduces is the operator that he calls *pomset homomorphism*. Such an operator is in essence similar to the one we have introduced in this chapter. In fact, it has the effect of substituting a pomset to a vertex of another pomset.

In [Gi84], the author, working with Pratt's pomset model, shows that, regarding each symbol of a pomset as standing for a language, causal dependency between symbols as concatenation of languages and concurrency as their shuffle, two pomsets are equivalent (in the sense that they stand for the same language) if and only if they are equivalent under the interpretation in which only languages with strings of length at most 2 are considered.

In [K88], A. Kiehn introduces a call mechanism for Petri Net system, which are finite families of place/transition nets, [Rei85], and studies some closure properties of the class of languages accepted by them. None of these papers, however, are concerned with an explicit algebraic treatment of the feature of action-refinement in the different models adopted by the authors.

In [BC88], G. Boudol and I. Castellani propose a calculus of concurrent processes which allows them to regard a finite computation as a single event. They investigate this problem both at the level of *execution* of a process and at the level of *operation* (the way processes operate on data). This kind of abstraction is in essence a dual notion of the refinement operator described in this chapter and it would be interesting to have a process algebra containing both these features. Semantic theories for processes which support the refinement of actions by processes have recently been the object of extensive study in the literature. The reference [GG88] gives a good survey of the work in this area; moreover, there the authors show that history preserving bisimulation over finite prime Event Structures, [Win87], is preserved by action-refinement (in the absence of internal, invisible actions). [NEL88] presents a natural model for a process language incorporating a refinement operator which is fully abstract with respect to a trace-based notion of equivalence over processes. [DD89b] studies action-refinement over synchronization and causal trees, [DD89a], and presents refinement theorems for two versions of branching bisimulation, [GW89b]. In [Gl90], the author studies notions of ST-bisimulation, a version of bisimulation equivalence which is very close in spirit to the refine-equivalence presented in this chapter, and ST-trace equivalence over finite, prime Event Structures and proves that they are both preserved by refinement. A refinement theorem for a version of failures semantics, [BHR84], over safe Petri nets has been presented in [Vo90]. However, apart from [NEL88], all of the above-mentioned references are concerned with the study of action-refinement at the semantic level. In this chapter, on the other hand, we have tried to provide the theoretical foundations for a syntactic treatment of the feature of action-refinement in a simple algebraic setting.

Chapter 4

Action Refinement for Communicating Processes

4.1 Introductory Remarks

In the previous chapter, we presented a simple process algebra incorporating an operator for the refinement of actions by processes and studied a reasonable notion of semantic equivalence for it. However, the simple language of Chapter 3 did not allow for the description of *communicating processes*, due to the absence of features like a parallel operator with communication and an encapsulation operator. The object of this chapter is to develop a reasonable process algebra for the description of concurrent, communicating processes which incorporates a refinement operator and to suggest a suitable notion of semantic equivalence for specifications written in this process algebra.

We take as our basic language a cross of finite **CCS**, [Mil89], and **ACP**, [BK85]. This language is based on a set of actions *Act* and contains, as usual, the binary choice combinator $+$, the restriction operator $\backslash\alpha$, where α is an action, and the binary parallel combinator $|$; $p|q$ means that the processes p and q are running in parallel and they may synchronize using complementary actions. In order to support action refinement, the usual action-prefixing operator from **CCS** is replaced by sequential composition ";". The introduction of this operator has further implications. As pointed out in Chapter 2, in the presence of sequential composition and restriction it is no longer sufficient to have only one notion of terminated process as in **CCS**. So we also have in the language a constant for the successfully terminated process, *nil*, and one for the completely deadlocked process δ. The result is a very rich and expressive language which only lacks a facility for recursive definitions for it to be considered a standard process algebra.

The first question we ask is: what action refinements, i.e. substitutions of actions by processes, should be allowed in this rich setting? There are two somewhat opposing constraints. The first has to do with the issue of what action refinements are useful in practice, in the sense that any action refinement which might be used in practice should be allowed by our definition. The second is that the allowed refinements should be restricted so that a reasonable semantic theory can be developed. One constraint that we impose is that actions can *not* be refined into terminated processes. This is unlikely to constrain practical applications and, as we shall see, the presence of such refinements would make the development of an adequate, abstract semantic equivalence which is preserved by action refinements very difficult. The other constraints we impose have to do with complementation of actions and thus with the synchronization potential of processes. Recall that in **CCS** the set of actions has the structure $Act = \Lambda \cup \bar{\Lambda} \cup \{\tau\}$, where Λ is a basic set of actions, $\bar{\Lambda}$ is the set of their complements and τ is a distinguished action meant to denote internal and unobservable actions. In view of the nature of τ it is reasonable to say that it cannot be refined. Once more this is very natural from an application point of view, although from a theoretical stand-point we might have allowed the refinement of τ by processes that can only perform internal actions. Finally, we require that if action a is refined to the process p then its complement $\bar{a}$ is refined to the "complement of p", essentially obtained from p by replacing each action with its complement. At the moment it is very difficult to say if this constraint will be restrictive in practice as there is very little documented experience of refining **CCS** specifications. However, it will be very convenient in developing our semantic theory.

The second question we address is how action refinements are to be applied. For the language considered in Chapter 3 the answer is straightforward because of the absence of restriction: an action refinement is applied to a process by syntactically substituting for each action symbol its corresponding refinement. However, this is no longer adequate in this enriched setting. For example, consider the process p:

$$((\lambda; p' + \alpha; q)|\bar{\alpha}; r) \setminus \alpha,$$

where λ does not occur in p', q and r. If we now refine λ to the process $\alpha; w$ then, intuitively, we do not wish the result of the application of this refinement to p to be the process

$$(((\alpha; w); p' + \alpha; q)|\bar{\alpha}; r) \setminus \alpha.$$

In p the action α is a local action which is known only internally to the process. It is a "bound action" and semantically p should be equivalent to

$$((\lambda; p' + \beta; q)|\bar{\beta}; r) \setminus \beta,$$

at least assuming that α and β do not appear in p', q and r. In other words, restricted actions should not be allowed to "capture" actions in the refining process. In defining the application of an action refinement we shall appeal to the standard theory of α-conversion and substitution, see [Sto88a], where the restriction operator is viewed as a binder. So, for example, assuming that β does not occur in w, the effect of refining λ by $\alpha; w$ in p will be $(((\alpha; w); p' + \beta; q)|\bar{\beta}; r) \setminus \beta$, up to α-conversion.

The final problem we address is the development of an adequate notion of semantic equivalence over the language considered in this chapter. One property we require of such an equivalence is that it abstracts from the internal evolution of processes, i.e. it interprets τ-actions as being internal or unobservable. One equivalence with this property is *weak bisimulation equivalence*, $\approx$, [Mil89]. However, as shown in the previous chapter, $\approx$ is not adequate in the presence of action refinement. In fact, it is not preserved by the refinement combinator. From the work presented in Chapter 3, one might hope that a suitable semantic equivalence for the language considered in this chapter could be obtained by considering the natural version of $\sim_t$ which abstracts from internal actions, $\approx_t$. However, it turns out that $\approx_t$ is not preserved by action refinement over a language with communication and restriction operators. An example was provided by R. van Glabbeek, [Va90], in a slightly different setting; we shall discuss its formulation for our language in § 4.4. In some sense, the fact that $\approx_t$ is not preserved by action refinement over the richer language is not surprising. In fact, the proofs of the refinement theorem for $\sim_t$ given in the previous chapter are both language dependent. Moreover, the behavioural proof presented at page 107 made an essential use of a subtler, but more natural, version of $\sim_t$, called refine equivalence and denoted by $\sim_r$. The main result of this chapter is that the appropriate version of refine equivalence which takes internal moves into account, $\approx_r$, is indeed preserved by action refinement over the richer language. It is then a simple matter to characterize the largest congruence over the richer language contained in $\approx_r$, $\approx_r^c$:

$$p \approx_r^c q \text{ iff } p + a \approx_r q + a \text{ for some action } a \text{ not occurring in } p \text{ and } q.$$

Readers familiar with the theory of weak bisimulation equivalence will recognize the need for the new action a: it is necessary because, in general, weak bisimulation equivalence is not preserved by $+$.

In Chapter 3, we showed that, for the simple language $\mathbf{P}_\rho$, $\sim_t$ gave a behavioural characterization of the largest congruence over $\mathbf{P}_\rho$ contained in strong bisimulation equivalence. Namely, for each $p, q \in \mathbf{P}_\rho$, we proved that

$$p \sim_t q \text{ iff for all } \mathbf{P}_\rho\text{-contexts } C[\cdot],\ C[p] \sim C[q].$$

Unfortunately, we have been unable to prove a similar characterization result for $\approx_r^c$ in terms of weak bisimulation equivalence over the richer language considered in this chapter.

We now give a brief outline of the contents of this chapter. In section 4.2 we introduce the language which will be studied in this chapter and present several semantic equivalences for it based on variations on the notion of bisimulation. In setting up our semantic framework, we shall rely on work presented in Chapters 2 and 3 of this thesis. Section 4.2.2 is entirely devoted to the discussion of a suitable notion of action refinement for the language we consider. There we also introduce our technique for the application of action refinements to processes. The essential idea is to consider the restriction operator as a binding operator and to adapt the notion of substitution presented in [Sto88a] to our setting. A natural "weak" version of the refine equivalence introduced in Chapter 3, $\approx_r$, is then presented and analyzed in detail in §4.3. There we show that $\approx_r$ and its closure with respect to +-contexts, $\approx_r^c$, are both preserved by action refinements. Section 4.4 is devoted to a discussion of an example showing that $\approx_t$ and $\approx_r$ are different equivalences over the language considered in this chapter.

4.2 The Language

4.2.1 The Basic Language

Following the developments presented in Chapter 2, we let Λ denote a countable set of basic uninterpreted symbols ranged over by $\alpha, \beta, \gamma, \alpha' \ldots$. The set of *actions* over Λ, $Act(\Lambda)$, is defined to be $\Lambda \cup \bar{\Lambda} \cup \{\tau\}$, where $\bar{\Lambda} =_{def} \{\bar{\alpha} \mid \alpha \in \Lambda\}$ and τ is a distinguished symbol not in $\Lambda \cup \bar{\Lambda}$. For each $\alpha \in \Lambda$, $\bar{\alpha}$ will be called the *complement* of α. The complementation $\bar{\cdot}$ is extended to the whole of $Act(\Lambda)$ by $\bar{\bar{\alpha}} = \alpha$ and $\bar{\tau} = \tau$. Intuitively, Λ may be thought of as a set of channel names to be associated with communicating processes, in which case $\alpha \in \Lambda \subseteq Act(\Lambda)$ may be viewed as the action of receiving a synchronization signal from the channel α, $\bar{\alpha}$ as the action of sending a synchronization signal to α and τ as an internal or invisible action. We shall use μ to range over $Act(\Lambda)$ and a, b over $\Lambda \cup \bar{\Lambda}$, the set of *observable actions* which we sometimes denote by $V(\Lambda)$.

Given Λ, the set of *processes* over $Act(\Lambda)$, $\mathbf{P}_\Lambda^\Gamma$, is given by the following BNF definition

$$p ::= nil \mid \delta \mid \mu\ (\mu \in Act(\Lambda)) \mid p;p \mid p+p \mid p|p \mid p \setminus \alpha\ (\alpha \in \Lambda).$$

We shall use $p, q, p' \ldots$ to range over $\mathbf{P}_\Lambda^\Gamma$. The language for processes given above is a mixture of **CCS**, [Mil80,89], and **ACP**, [BK84,85]. It is based on the signature for finite processes considered in Chapter 2. The main difference between $\mathbf{P}_\Lambda^\Gamma$ and $FREC_\Sigma$ is of a syntactic nature. In $\mathbf{P}_\Lambda^\Gamma$ we have, in fact, chosen a restriction operator with respect to channel names, rather than sets of channel names as we did in Chapter 2. This choice will lead to simplifications in some of the technical definitions which will be given in what follows. The operators *nil*, +, | and $\setminus\alpha$ are taken from **CCS**, but the **CCS** action-prefixing operator is replaced by sequential

$$
\begin{array}{llll}
\textbf{(T1)} & nil\surd & & \\
\textbf{(T2)} & p\surd \text{ and } q\surd & \text{imply} & p \odot q\surd \ \ (\odot \in \{;,+,|\}) \\
\textbf{(T3)} & p\surd & \text{implies} & p \setminus \alpha\surd
\end{array}
$$

$$
\begin{array}{llll}
\textbf{(ACT)} & \mu \xrightarrow{\mu} nil & & \\
\textbf{(SUM)} & p \xrightarrow{\mu} p' & \text{implies} & p + q \xrightarrow{\mu} p' \\
 & & & q + p \xrightarrow{\mu} p' \\
\textbf{(SC1)} & p \xrightarrow{\mu} p' & \text{implies} & p; q \xrightarrow{\mu} p'; q \\
\textbf{(SC2)} & p\surd \text{ and } q \xrightarrow{\mu} q' & \text{implies} & p; q \xrightarrow{\mu} q' \\
\textbf{(PAR)} & p \xrightarrow{\mu} p' & \text{implies} & p|q \xrightarrow{\mu} p'|q \\
 & & \text{implies} & q|p \xrightarrow{\mu} q|p' \\
\textbf{(SYN)} & p \xrightarrow{a} p' \text{ and } q \xrightarrow{\bar{a}} q' & \text{imply} & p|q \xrightarrow{\tau} p'|q' \\
\textbf{(RES)} & p \xrightarrow{\mu} p' \text{ and } \alpha \ admits\, \mu & \text{imply} & p \setminus \alpha \xrightarrow{\mu} p' \setminus \alpha
\end{array}
$$

Figure 4.1. Termination predicate and transition relations for $\mathbf{P}^{\Gamma}$

composition ";". As explained in Chapter 2, in the presence of a general sequential composition operator and of restriction it is necessary to have in the language a symbol for both successful termination, for which we use *nil*, and unsuccessful termination, which we represent by δ. We shall usually abbreviate $\mathbf{P}^{\Gamma}_{\Lambda}$ to $\mathbf{P}^{\Gamma}$. With abuse of notation, we also use Σ to refer to the set of operators used in the definition of $\mathbf{P}^{\Gamma}$ and a $\mathbf{P}^{\Gamma}$-context will be a term in the language which contains one "hole" into which a subterm may be slotted, $C[\cdot]$.

The operational semantics for $\mathbf{P}^{\Gamma}$ is given in terms of a collection of next-state relations $\xrightarrow{\mu} \subseteq \mathbf{P}^{\Gamma} \times \mathbf{P}^{\Gamma}$, one for each action μ, and a successful termination predicate $\surd$. These are given, for the sake of clarity, in Figure 4.1 and are taken directly from Chapter 2. The definition of the transition relations uses the predicate *admits* defined by

$$\alpha \ admits\, \mu \text{ iff } \mu \neq \alpha, \bar{\alpha}.$$

There are numerous variations which one could apply to these definitions (see [BG87b], [GrV89] and [BV89]), but in this chapter we shall follow the approach in Chapter 2. With these definitions we have a particular instance of a labelled transition system. (See Definition 3.2.1 on page 67) A *labelled transition system with termination* is a quadruple $\langle P, \mathbf{A}, \longrightarrow, \surd \rangle$ where

(1) P is a set of processes,

(2) $\mathbf{A}$ is a set of actions of the form $V \cup \{\tau\}$, where τ is a distinguished action symbol,

(3) $\longrightarrow \subseteq P \times \mathbf{A} \times P$ is a next-state relation and

(4) $\surd \subseteq P$ is a successful termination predicate.

This is a slight extension of the usual notion of labelled transition system defined on page 67 which suits our language. In such an LTS a *strong bisimulation* is a symmetric relation $\mathcal{R} \subseteq P \times P$ which satisfies, for each $\langle p, q \rangle \in \mathcal{R}$ and $\mu \in \mathbf{A}$,

(i) if $p \xrightarrow{\mu} p'$ then there exists q' such that $q \xrightarrow{\mu} q'$ and $\langle p', q' \rangle \in \mathcal{R}$,
(ii) if $p\surd$ then $q\surd$.

The second clause of the definition of strong bisimulation over an LTS with termination explicitly requires the matching of the termination potential of two processes, as expressed by the predicate $\surd$. This is needed to capture semantically the difference between successfully terminated and deadlocked processes.

A *weak bisimulation* is defined in essentially the same way by replacing $\xrightarrow{\mu}$ and $\surd$ by their "weak" counterparts, $\stackrel{\mu}{\Longrightarrow}$ and $\surd\!\!\surd$ respectively. These are defined exactly as in Chapter 2, but, for the sake of clarity, we now recall their definitions. Formally,

$$p \stackrel{\mu}{\Longrightarrow} q \text{ iff } p \xrightarrow{\tau}{}^{*} p_1 \xrightarrow{\mu} p_2 \xrightarrow{\tau}{}^{*} q,$$

for some p_1, p_2, and $p\surd\!\!\surd$ iff for all p', $p \xrightarrow{\tau}{}^{*} p' \not\xrightarrow{\tau}$ implies $p'\surd$. The exact definition of weak bisimulation also uses the relation $\stackrel{\varepsilon}{\Longrightarrow}$, the reflexive and transitive closure of $\xrightarrow{\tau}$, $\xrightarrow{\tau}{}^{*}$, and the notation $\hat{\mu}$, where $\hat{a}$ is simply a and $\hat{\tau} = \varepsilon$. Then a symmetric relation $\mathcal{R} \subseteq P \times P$ is a weak bisimulation if, for each $\langle p, q \rangle \in \mathcal{R}$ and $\mu \in \mathbf{A}$,

(i) if $p \xrightarrow{\mu} p'$ then there exists q' such that $q \stackrel{\hat{\mu}}{\Longrightarrow} q'$ and $\langle p', q' \rangle \in \mathcal{R}$,
(ii) if $p\surd\!\!\surd$ then $q\surd\!\!\surd$.

The definition of weak bisimulation over an LTS with termination is very similar to that of the preorder $\lesssim$ given Chapter 2. (See page 30) The only difference stems from the fact that in this chapter we shall only be concerned with finite processes and therefore shall not be interested in comparing the "divergence potential" of processes. With abuse of notation, we use $\sim$, called *strong bisimulation equivalence*, to denote the largest strong bisimulation and $\approx$, called *weak bisimulation equivalence*, to denote the largest weak bisimulation. We shall be primarily interested in these equivalence relations as applied to the LTS $\langle \mathbf{P}^{\Gamma}, Act(\Lambda), \longrightarrow, \surd \rangle$. The following proposition can be proven following standard lines.

Proposition 4.2.1 (Congruence properties of $\sim$ and $\approx$) *$\sim$ is a Σ-congruence over $\mathbf{P}^\Gamma$. $\approx$ is preserved by all the operators in Σ, apart from $+$.*

Another variation on the theme of bisimulation, which has been investigated in the previous chapter for a simple language, is obtained by splitting each visible action a into two subactions $S(a)$, the start of a, and $F(a)$, the end or finish of a. In order to define an operational semantics based on these actions, we need to enlarge the set of terms in order to describe states of processes in which, for example, actions a_1 and a_2 have started but have not yet finished.

The set of *states* $\mathcal{S}^\Gamma$, or more formally $\mathcal{S}^\Gamma_\Lambda$, is the set of terms generated by the following BNF definition, where as usual p ranges over processes,

$$s ::= nil \mid \delta \mid \mu\ (\mu \in Act(\Lambda)) \mid F(a)\ (a \in V(\Lambda)) \mid s;p \mid p+p \mid s|s \mid s \setminus \alpha\ (\alpha \in \Lambda)$$

which satisfy the following constraint

(CR) $s \setminus \alpha \in \mathcal{S}^\Gamma$ implies that $F(\alpha)$ and $F(\bar{\alpha})$ do not occur in s.

The naturality of this constraint will become clear after the definition of the operational semantics for $\mathcal{S}^\Gamma$. Notice that $\mathbf{P}^\Gamma$ is a subset of $\mathcal{S}^\Gamma$; we shall use $s, s' \ldots$ to range over $\mathcal{S}^\Gamma$. Let $Act_s(\Lambda)$, the set of *subactions*, be given by

$$Act_s(\Lambda) =_{def} \{S(a), F(a) \mid a \in V(\Lambda)\} \cup \{\tau\}.$$

The next-state relation $\stackrel{\tau}{\longrightarrow}_t$ over $\mathcal{S}^\Gamma$ is obtained by simply adapting the axiom and relevant rules presented in Figure 4.1 to $\mathcal{S}^\Gamma$.

For each e in $Act_s(\Lambda)$ other than τ, the next-state relations $\stackrel{e}{\longrightarrow}_t$ over $\mathcal{S}^\Gamma$ are based on those for $\mathcal{S}$ given in Chapter 3 and are given, for the sake of clarity, in Figure 4.2, where the predicate *admits* is extended so that

$$\alpha\ admits\ e \text{ iff } e \neq S(\alpha), F(\alpha), S(\bar{\alpha}), F(\bar{\alpha}).$$

The definition of $\stackrel{e}{\longrightarrow}_t$ is very similar to that of $\stackrel{a}{\longrightarrow}$, except that we have the new clauses

$$a \stackrel{S(a)}{\longrightarrow}_t F(a)$$
$$F(a) \stackrel{F(a)}{\longrightarrow}_t nil.$$

Intuitively, the process a can start the action a and be transformed into the state $F(a)$. This is a state in which action a is active and may terminate at any time, which corresponds to performing action $F(a)$. Note that the subactions cannot synchronize and therefore it is inaccurate to view $\mathbf{P}^\Gamma$ with this operational semantics as processes in $\mathbf{P}^\Gamma_{\Lambda_s}$, where Λ_s is some collection of basic subactions.

$$
\begin{array}{lll}
(1) & nil\surd & \\
(2) & s\surd & \text{implies} \quad s \setminus \alpha\surd \\
(3) & p\surd \text{ and } q\surd & \text{imply} \quad p+q\surd \\
(4) & s\surd \text{ and } p\surd & \text{imply} \quad s;p\surd \\
(5) & s_1\surd \text{ and } s_2\surd & \text{imply} \quad s_1|s_2\surd
\end{array}
$$

$$
\begin{array}{llll}
(S1) & a \xrightarrow{S(a)}_t F(a) & & \\
 & F(a) \xrightarrow{F(a)}_t nil & & \\
(S2) & p \xrightarrow{e}_t p' & \text{implies} & p+q \xrightarrow{e}_t p' \\
 & & & q+p \xrightarrow{e}_t p' \\
(S3) & s \xrightarrow{e}_t s' & \text{implies} & s;p \xrightarrow{e}_t s';p \\
(S4) & s\surd \text{ and } p \xrightarrow{e}_t s' & \text{imply} & s;p \xrightarrow{e}_t s' \\
(S5) & s_1 \xrightarrow{e}_t s_1' & \text{implies} & s_1|s_2 \xrightarrow{e}_t s_1'|s_2 \\
 & & & s_2|s_1 \xrightarrow{e}_t s_2|s_1' \\
(S6) & s \xrightarrow{e}_t s' \text{ and } \alpha \textit{ admits } e & \text{imply} & s \setminus \alpha \xrightarrow{e}_t s' \setminus \alpha
\end{array}
$$

Figure 4.2. Termination predicate and observable next-state relations for $\mathcal{S}^\Gamma$

The termination predicate $\surd$, defined in Figure 4.1 on $\mathbf{P}^\Gamma$, is extended in the obvious way to the set of states $\mathcal{S}^\Gamma$ (see Figure 4.2).

Proposition 4.2.2 *Let $s \in \mathcal{S}^\Gamma$ and $e \in Act_s(\Lambda)$. Then $s \xrightarrow{e}_t s'$ implies $s' \in \mathcal{S}^\Gamma$.*

We have already remarked that, for each process p, $p \in \mathcal{S}^\Gamma$. Thus, by the above proposition, each state s reachable from a process p is in $\mathcal{S}^\Gamma$ and thus satisfies condition (CR). This justifies our use of condition (CR). We are, in fact, interested in states only as means of defining the operational semantics for processes using subactions and terms built using the grammar for states which do not satisfy (CR) are *not* reachable from processes using the transition relations $\xrightarrow{e}_t$.

We now have a new labelled transition system with termination $\langle \mathcal{S}^\Gamma, Act_s(\Lambda), \longrightarrow_t, \surd \rangle$. In this structure we let $\sim_t$ and $\approx_t$ denote the resulting strong and weak bisimulation equivalence, respectively. The subscript t refers to "time" as the equivalences are obtained by assuming that actions take non-zero time. They have been studied for sublanguages of $\mathbf{P}^\Gamma$ in [H88c] and similar equivalences have been called "split-equivalences" in [GV87], [Gl90,90a].

Proposition 4.2.3 *$\sim_t$ is a congruence with respect to all the combinators in $\mathbf{P}^\Gamma$. $\approx_t$ is preserved by all the combinators in Σ, apart from $+$.*

It is interesting to note that the definitions of the split-equivalences $\sim_t$ and $\approx_t$ given above do not require the matching of actions in $Act(\Lambda)$. The addition of such a requirement to their definition would give rise to *different* equivalences. This will be demonstrated in § 4.4 by means of an example.

4.2.2 Action Refinements

An action refinement may be considered to be simply a mapping from $Act(\Lambda)$ to $\mathbf{P}^\Gamma$ and the effect of applying an action refinement to a process is a new, more detailed, process obtained by substituting for each action the corresponding refining process. In this section we formalize these ideas for the language $\mathbf{P}^\Gamma$.

We first put some natural conditions on action refinements, or more prosaically substitutions. We shall use $\rho, \rho' \ldots$ to range over them. Since τ is an internal, unobservable action, it makes no sense to be able to refine it. So, in effect, action refinements are functions from the set of visible actions $V(\Lambda)$ to $\mathbf{P}^\Gamma$. One may also argue that if actions are allowed to be refined by successfully terminated processes then all occurrences of the supposedly internal and invisible action τ will have a significant effect on the behaviour of processes. For example, let p and q

denote $(a;\tau)+b$ and $a+b$, respectively. One would expect p to be equivalent to q with respect to most "reasonable" notions of equivalence which abstracts from internal transitions. However, if an equivalence were to be preserved by refinements of actions by successfully terminated processes then $p \neq q$. For let ρ denote a refinement such that $\rho(a) = nil$ and $\rho(b) = b$. Then $p\rho$ and $q\rho$ should be $(nil;\tau)+b$ and $nil+b$, respectively, which are not bisimulation equivalent, nor indeed would they be equivalent with respect to most reasonable notions of equivalence.

As a further example, let p and q be $a;\tau;b$ and $a;b$, respectively. These two processes are again considered to be equivalent with respect to most reasonable semantic equivalences. However, under the same refinement we obtain $nil;\tau;b$ and $nil;b$, respectively. Once more, these two processes are not equivalent, at least if we use equivalences which are preserved by all language contexts. For example, $c+(nil;\tau;b)$ and $c+(nil;b)$ would be distinguished by most reasonable semantic equivalences.

A further constraint we impose on refinements is that they should, in some sense, preserve the synchronization structure of processes. Let us explain this point with an example. Let p and q denote $\alpha|\bar{\alpha}+\tau$ and $\alpha|\bar{\alpha}$, respectively. Then, once more, we would consider these two processes to be semantically equivalent. However, if we apply a refinement such as $\rho(\alpha) = \beta$ and $\rho(\bar{\alpha}) = \beta'$, the resulting processes $\beta|\beta'+\tau$ and $\beta|\beta'$ would not be equivalent. The problem is that ρ has interfered with the communication potential between the complementary actions α and $\bar{\alpha}$. We shall forbid such refinements by demanding that any action refinement ρ satisfy the following requirement

(ComPres) for each $a \in V(\Lambda)$ there exists r such that $\rho(a)|\rho(\bar{a}) \stackrel{\tau}{\Longrightarrow} r$ and $r\surd$.

This means that actions and their complements must be refined to processes which can communicate indefinitely until they have successfully terminated. This restriction is satisfied by the most common form of refinement in the literature, namely relabelling of actions (see [Mil89] for details). More generally, (ComPres) may be enforced by demanding that, for each action a, $\rho(\bar{a}) = \overline{\rho(a)}$, where, for each process p, $\bar{p}$ is the process obtained from p by substituting each action by its complement. In this case, an action refinement ρ would be uniquely determined by how it behaves on the set of channel names Λ.

Definition 4.2.1 (Action Refinements) *An* action refinement *is a mapping* $\rho : V(\Lambda) \rightarrow \mathrm{P}^{\Gamma}$ *with the following properties:*

(i) *for all* a, not $\rho(a)\surd$ *and*
(ii) ρ *satisfies (ComPres).*

Example 4.2.1 *Let $\rho : V(\Lambda) \to \mathbf{P}^\Gamma$ be the substitution which maps each a to τ, i.e. $\rho(a) = \tau$ for all a. Then ρ is an action refinement. In fact, for all a, $\rho(a)$ is not successfully terminated and*

$$\rho(a)|\rho(\bar{a}) \equiv \tau|\tau \stackrel{\tau}{\Longrightarrow} nil|nil\surd.$$

Similarly, it is easy to check that **CCS**-*relabellings, [Mil80,89], are action refinements.*

The following property derived from axiom (ComPres) will be most useful in our technical analysis of the operational properties of processes of the form $p\rho$. First of all, for each $\sigma = a_1 \ldots a_n \in V(\Lambda)^*$, let us recall that $\stackrel{\sigma}{\Longrightarrow}$ denotes the transition relation given by

$$p \stackrel{\sigma}{\Longrightarrow} q \text{ iff there exist } p_0, \ldots, p_n \text{ such that } p_0 = p,\ p_n = q \text{ and, for each } i < n,\ p_i \stackrel{a_i}{\Longrightarrow} p_{i+1}.$$

We shall assume that the complementation of actions is homomorphically extended to strings in $V(\Lambda)^* \cup \{\tau\}$. Then:

Fact 4.2.1 *Let ρ be an action refinement. Then, for each $a \in V(\Lambda)$, there exist $\sigma \in V(\Lambda)^+ \cup \{\tau\}$, $r_1, r_2 \in \mathbf{P}^\Gamma$ such that $\rho(a) \stackrel{\sigma}{\Longrightarrow} r_1$, $\rho(\bar{a}) \stackrel{\bar{\sigma}}{\Longrightarrow} r_2$, $r_1\surd$ and $r_2\surd$.*

We now turn our attention to the effect of applying an action refinement ρ to a process p. The resulting process we denote by $p\rho$ and, intuitively, it should be the process which results from substituting each occurrence of a by the process $\rho(a)$ in p, for each a. However, because of the presence of restriction, care must be taken in order to preserve the intended purpose of this combinator, namely the scoping of channel names. In this setting, restriction is a "binding operator" and, in order to take this into account, it is appropriate to define $p\rho$ by adapting to our setting a theory of substitution in the presence of binders. In what follows, we shall use the theory of substitution developed in [Sto88a].

For each process p let $FC(p)$, the set of *free channel names* in p, be defined by:

$$\begin{aligned} FC(nil) &= FC(\delta) = FC(\tau) = \emptyset \\ FC(\alpha) &= FC(\bar{\alpha}) = \{\alpha\} \\ FC(p;q) &= FC(p+q) = FC(p|q) = FC(p) \cup FC(q) \\ FC(p \setminus \alpha) &= FC(p) - \{\alpha\}. \end{aligned}$$

As in [Sto88a], we define the function *new* by

$$new\,\alpha\,p\,\rho =_{def} \{\beta \mid \text{for each } \beta' \in FC(p) - \{\alpha\},\ \beta \notin FC(\rho(\beta')) \cup FC(\rho(\bar{\beta}'))\}.$$

That is $new\,\alpha\,p\,\rho$ returns the set of innocuous channel names, none of which will capture any of the free channel names which appear when ρ is being applied to p[1]. The definition

[1]Note that the definition of *new* takes into account the fact that in **CCS** the restriction operator $\backslash\alpha$ binds both α and its complement $\bar{\alpha}$.

of substitution given below uses a choice function which takes such a set and returns some element. For example, if Λ were well-ordered we could choose the least element in the set. We also use the standard notation for the modification of substitutions from Chapter 3: $\rho[a \to p]$ is the substitution which is identical to ρ except that it maps a to p. As a variation $\rho[\alpha \mapsto \beta]$ will denote the substitution identical to ρ except that α is mapped to β and $\bar{\alpha}$ to $\bar{\beta}$. We shall use ι to denote the identity substitution.

Definition 4.2.2 (Application of Action Refinements) *For each $p \in \mathbf{P}^\Gamma$ and action refinement ρ, $p\rho$ is the process defined by:*

(i) $a\rho = \rho(a)$, $\tau\rho = \tau$, $nil\rho = nil$, $\delta\rho = \delta$

(ii) $(p \odot q)\rho = p\rho \odot q\rho$ $(\odot \in \{;, +, |\})$

(iii) $(p \setminus \alpha)\rho = (p\rho[\alpha \mapsto \beta]) \setminus \beta$ *where* $\beta = choice(new\,\alpha\,p\,\rho)$.

This is exactly the definition given in [Sto88a] except that there the binding operator is the λ-abstraction of the λ-calculus and the only operator is application. However, all of the results in [Sto88a] apply equally well here and their proofs are more or less identical. For example, suppose that for two action-refinements ρ_1 and ρ_2 we define the new refinement $\rho_2 \circ \rho_1$ by $\rho_2 \circ \rho_1(a) = \rho_1(a)\rho_2$. (We shall show that $\rho_2 \circ \rho_1$ so defined is indeed a refinement at the end of this section) Then, by Theorem 3.2 of [Sto88a] (page 321), we have:

Lemma 4.2.1 (Substitution Lemma) $(p\rho_1)\rho_2 = p(\rho_2 \circ \rho_1)$.

An example of the application of an action refinement to a process is now in order.

Example 4.2.2 *Let us assume, for the purpose of this example, that $\Lambda = \{\alpha_i \mid i \in \mathrm{N}\}$ and that, for all $i, j \in \mathrm{N}$, $\alpha_i < \alpha_j$ iff $i < j$. We assume moreover that, for each subset X of Λ, choice(X) is the least element of X with respect to $<$. Consider the process $p = ((\alpha_1; \alpha_2 + \alpha_3)|\bar{\alpha}_1) \setminus \alpha_1$ and the action refinement $\rho = \iota[\alpha_3 \mapsto \alpha_1]$. Then we have that $new\,\alpha_1\,((\alpha_1; \alpha_2 + \alpha_3)|\bar{\alpha}_1)\,\rho = \{\alpha_i \mid i \geq 3\}$ and $choice(\{\alpha_i \mid i \geq 3\}) = \alpha_3$. Thus*

$$\begin{aligned} p\rho &= (((\alpha_1; \alpha_2 + \alpha_3)|\bar{\alpha}_1)\rho[\alpha_1 \mapsto \alpha_3]) \setminus \alpha_3 \\ &= ((\alpha_3; \alpha_2 + \alpha_1)|\bar{\alpha}_3) \setminus \alpha_3. \end{aligned}$$

In general, as highlighted by the above example, the application of an action refinement to a process changes the restricted channel names and, in such a setting, it is more appropriate to replace syntactic identity with the so-called "α-conversion" (or α-congruence), $=_\alpha$. This is defined to be the least Σ-congruence over $\mathbf{P}^\Gamma$ which satisfies

(α) $\beta \notin FC(p)$ and $p\iota[\alpha \mapsto \beta] =_\alpha q$ imply $p \setminus \alpha =_\alpha q \setminus \beta$.

From Corollary 3.10 of [Sto88a] (pages 322-323) we obtain a useful syntactic characterization of $=_\alpha$.

Proposition 4.2.4 (Syntactic Characterization of $=_\alpha$) *If $p =_\alpha q$ then one of the following conditions holds:*

(1) *p and q have the form $op(p_1, \ldots, p_k)$ and $op(q_1, \ldots, q_k)$, respectively, for some $op \in \Sigma$ of arity k and $p_i =_\alpha q_i$, for all i, or*

(2) *p and q have the form $p' \setminus \alpha$ and $q' \setminus \beta$, respectively, where*

$$\beta \notin FC(p') \text{ and } p'\iota[\alpha \mapsto \beta] =_\alpha q'.$$

Another important property of substitution can be obtained from Lemma 3.1.(vi) of [Sto88a] (page 320):

Lemma 4.2.2 (Identity substitution and $=_\alpha$) *For each $p \in \mathbf{P}^\Gamma$, $p =_\alpha p\iota$.*

The following basic properties of substitution and α-congruence will find considerable application in the proofs of our results.

Fact 4.2.2 *Let $\beta \in new\,\alpha\,p\,\rho$. Then the following statements hold.*

(1) *Assume that $q =_\alpha p\rho[\alpha \mapsto \beta]$. Then $q \setminus \beta =_\alpha (p \setminus \alpha)\rho$.*

(2) $(p \setminus \alpha)\rho =_\alpha (p\rho[\alpha \mapsto \beta]) \setminus \beta$.

Proof: We only prove (1). Let $\beta \in new\,\alpha\,p\,\rho$. By the definition of substitution, we have that

$$(p \setminus \alpha)\rho = (p\rho[\alpha \mapsto \beta']) \setminus \beta',$$

where $\beta' \in choice(new\,\alpha\,p\,\rho)$. We distinguish two cases:

(A) $\beta = \beta'$. The claim then follows by the substitutivity of α-congruence.

(B) $\beta \neq \beta'$. Then $\beta \notin FC(p\rho[\alpha \mapsto \beta'])$. Moreover,

$$\begin{array}{rcll} (p\rho[\alpha \mapsto \beta'])\iota[\beta' \mapsto \beta] & = & p(\iota[\beta' \mapsto \beta] \circ \rho[\alpha \mapsto \beta']) & \text{by the substitution lemma} \\ & = & p\rho[\alpha \mapsto \beta] & \\ & =_\alpha & q & \text{by the hypothesis.} \end{array}$$

The claim now follows by rule (α) of the definition of α-congruence. This completes the proof of statement (1). □

The following useful result states that the set of free channel names of a process never increases under derivations.

Fact 4.2.3 *Let* $p \in \mathbf{P}^{\Gamma}$. *Then:*

(1) $p\surd$ *implies* $FC(p) = \emptyset$;

(2) *for each* $\mu \in Act(\Lambda)$, $q \in \mathbf{P}^{\Gamma}$, $p \xrightarrow{\mu} q$ *implies that* $FC(q) \subseteq FC(p)$.

We have so far studied several useful syntactic properties of substitution and $=_{\alpha}$. However, many of the arguments we shall apply will not only be syntactic, but will also involve semantic reasoning. For this reason it will be useful to develop behavioural properties of $=_{\alpha}$. For our purposes, it will be sufficient to prove that $=_{\alpha}$ is contained in strong bisimulation, $\sim$. First of all, we establish a lemma about relabellings. A relabelling ϱ is a mapping from Λ to Λ. It is extended to a refinement by letting $\varrho(\bar{\alpha}) = \overline{\varrho(\alpha)}$. For conciseness of notation we shall assume that, for every relabelling ϱ, $\varrho(\tau) =_{def} \tau$.

Lemma 4.2.3 *For each* $p \in \mathbf{P}^{\Gamma}$, *relabelling* ϱ *and* $\mu \in Act(\Lambda)$, $p \xrightarrow{\mu} p'$ *implies* $p\varrho \xrightarrow{\varrho(\mu)} r$ *for some* r *such that* $p'\varrho =_{\alpha} r$.

Proof: By induction on the proof of the derivation $p \xrightarrow{\mu} p'$. The only non-trivial case is when $p = q \setminus \alpha \xrightarrow{\mu} q' \setminus \alpha = p'$ because $q \xrightarrow{\mu} q'$ and α *admits* μ. In this case, $p\varrho$ has the form $(q\varrho[\alpha \mapsto \beta]) \setminus \beta$ for some β such that $\beta \neq \varrho(\beta'), \varrho(\bar{\beta}')$, for each $\beta' \in FC(q) - \{\alpha\}$. This means that β *admits* $\varrho(\mu)$. We may now apply the inductive hypothesis to $q \xrightarrow{\mu} q'$ to obtain that $q\varrho[\alpha \mapsto \beta] \xrightarrow{\varrho(\mu)} r$ for some $r =_{\alpha} q'\varrho[\alpha \mapsto \beta]$. This is because $\varrho[\alpha \mapsto \beta](\mu) = \varrho(\mu)$, as α *admits* μ. By the operational semantics, we then have that

$$(q \setminus \alpha)\varrho = (q\varrho[\alpha \mapsto \beta]) \setminus \beta \xrightarrow{\varrho(\mu)} r \setminus \beta.$$

We are thus left to show that $(q' \setminus \alpha)\varrho =_{\alpha} r \setminus \beta$. However, it is easy to see that $\beta \in new\, \alpha\, q'\varrho$. Thus the claim follows by Fact 4.2.2(1). □

The following lemma studies the relationships between the termination predicate $\surd$ and $=_{\alpha}$.

Lemma 4.2.4 *For each* $p, q \in \mathbf{P}^{\Gamma}$, $p\surd$ *and* $p =_{\alpha} q$ *imply* $q\surd$.

Proof: By induction on the termination predicate $\surd$. The details are omitted. □

Proposition 4.2.5 *For each* $p, q \in \mathbf{P}^{\Gamma}$, $p =_{\alpha} q$ *implies* $p \sim q$.

Proof: It is sufficient to show that the relation of α-congruence, $=_\alpha$, is a strong bisimulation. First of all, let us note that $=_\alpha$ is symmetric by definition. Assume now that $p =_\alpha q$. Then, by the previous lemma, $p\surd$ implies $q\surd$. We are thus left to show that

$$p \xrightarrow{\mu} p' \text{ implies } q \xrightarrow{\mu} q', \text{ for some } q' \text{ such that } p' =_\alpha q'.$$

We shall prove this statement by induction on the relation of α-congruence, $=_\alpha$. The proof proceeds by a case analysis on the structure of p and the syntactic characterization of $=_\alpha$ given in Proposition 4.2.4 will be most useful in the proof. We briefly examine two of the cases of the inductive proof.

- p has the form $p_1|p_2$. By Proposition 4.2.4, q must have the form $q_1|q_2$ with $p_i =_\alpha q_i$, $i = 1, 2$. We now examine why $p \xrightarrow{\mu} p'$. There are three possibilities and we analyze only one, when $\mu = \tau$, p' has the form $p_1'|p_2'$, $p_1 \xrightarrow{a} p_1'$ and $p_2 \xrightarrow{\bar{a}} p_2'$, for some a. By the inductive hypothesis, we then have that $q_1 \xrightarrow{a} q_1'$ and $q_2 \xrightarrow{\bar{a}} q_2'$ for some q_1' and q_2' such that $p_i' =_\alpha q_i'$, $i = 1, 2$. It follows that $q_1|q_2 \xrightarrow{\tau} q_1'|q_2'$ and $p_1'|p_2' =_\alpha q_1'|q_2'$.

- p has the form $p_1 \setminus \alpha$. By Proposition 4.2.4 there are two possible forms for q:

  (1) q has the form $q_1 \setminus \alpha$ with $p_1 =_\alpha q_1$, or

  (2) q has the form $q_1 \setminus \beta$ with $\beta \notin FC(p_1)$ and $p_1\iota[\alpha \mapsto \beta] =_\alpha q_1$.

  We only examine case (2). Assume that $p_1 \setminus \alpha \xrightarrow{\mu} p_1' \setminus \alpha$. Then $p_1 \xrightarrow{\mu} p_1'$ and α *admits* μ, i.e. $\mu \neq \alpha, \bar{\alpha}$. Now, $\iota[\alpha \mapsto \beta]$ is a relabelling and thus we may apply Lemma 4.2.3 to obtain that $p_1\iota[\alpha \mapsto \beta] \xrightarrow{\mu} r$ for some r such that $r =_\alpha p_1'\iota[\alpha \mapsto \beta]$. By the inductive hypothesis, $q_1 \xrightarrow{\mu} q_1'$ for some $q_1' =_\alpha r =_\alpha p_1'\iota[\alpha \mapsto \beta]$. Since $\beta \notin FC(p_1)$ and α *admits* μ we have that $\mu \neq \beta, \bar{\beta}$. Thus β *admits* μ and, by the operational semantics, $q_1 \setminus \beta \xrightarrow{\mu} q_1' \setminus \beta$. Moreover, by $q_1' =_\alpha p_1'\iota[\alpha \mapsto \beta]$ and $\beta \notin FC(p_1) \supseteq FC(p_1')$, we have that $q_1' \setminus \beta =_\alpha p_1' \setminus \alpha$ by applying rule (α) of the definition of $=_\alpha$. □

Corollary 4.2.1 *For each $p \in \mathbf{P}^\Gamma$, $p \sim p\iota$.*

Proof: Follows by Lemma 4.2.2 and the above proposition. □

We may now prove that the composition of refinements defined previously is indeed an action refinement. Let us recall, for the sake of clarity, that the composition of two action refinements ρ_1 and ρ_2, denoted by $\rho_2 \circ \rho_1$, is the refinement given by $(\rho_2 \circ \rho_1)(a) = \rho_1(a)\rho_2$. In order to prove that $\rho_2 \circ \rho_1$ is indeed a refinement it is sufficient to show that:

(1) $\rho_2 \circ \rho_1(a)$ is not terminated and

(2) $\rho_2 \circ \rho_1$ satisfies (ComPres).

The following result is an important consequence of the fact that, for each refinement ρ, $\rho(a)$ is not terminated.

Fact 4.2.4 *For each $p \in \mathbf{P}^\Gamma$ and action refinement ρ, $p\rho\surd$ iff $p\surd$.*

In view of the above result it is then easy to see that $\rho_2 \circ \rho_1$ satisfies condition (1) above. In order to prove that $\rho_2 \circ \rho_1$ satisfies (ComPres), we shall need the following lemma.

Notation 4.2.1 *For each binary relation $\mathcal{R}$ over $\mathbf{P}^\Gamma$, we shall write $p \xrightarrow{\mu} \mathcal{R}\, q$ iff there exists p' such that $p \xrightarrow{\mu} p'$ and $p' \mathcal{R}\, q$. A similar notation will be used with respect to the weak transition relations $\overset{\sigma}{\Longrightarrow}$, $\sigma \in Act(\Lambda)^* \cup \{\tau\}$.*

Lemma 4.2.5 *Let $p \in \mathbf{P}^\Gamma$ and ρ be an action refinement. Then:*

(1) *$p \xrightarrow{a} q$ and $\rho(a) \overset{\sigma}{\Longrightarrow} x\surd$, $\sigma \in V(\Lambda)^+ \cup \{\tau\}$, imply $p\rho \overset{\sigma}{\Longrightarrow} \sim q\rho$;*

(2) *$p \overset{\tau}{\Longrightarrow} q$ implies $p\rho \overset{\tau}{\Longrightarrow} \sim q\rho$.*

Proof: Let $p \in \mathbf{P}^\Gamma$ and ρ be an action refinement. We prove the two statements separately.

(1) By induction on the proof of the derivation $p \xrightarrow{a} q$. We proceed by a case analysis on the last rule used in the proof and only examine two possibilities.

- $p = a \xrightarrow{a} nil = q$. Then $p\rho = \rho(a) \overset{\sigma}{\Longrightarrow} x \sim nil = q\rho$ because $x\surd$.
- $p = p_1 \setminus \alpha \xrightarrow{a} p_1' \setminus \alpha = q$ because $p_1 \xrightarrow{a} p_1'$ and α *admits* a. By the definition of substitution, $(p_1 \setminus \alpha)\rho = (p_1\rho[\alpha \mapsto \beta]) \setminus \beta$ with $\beta = choice(new\, \alpha\, p_1\, \rho)$. This implies that β *admits* σ.

  By the inductive hypothesis, $p_1 \xrightarrow{a} p_1'$ and $\rho[\alpha \mapsto \beta](a) = \rho(a) \overset{\sigma}{\Longrightarrow} x\surd$ imply

  $$p_1\rho[\alpha \mapsto \beta] \overset{\sigma}{\Longrightarrow} \sim p_1'\rho[\alpha \mapsto \beta].$$

  By the operational semantics and the substitutivity of $\sim$,

  $$(p_1 \setminus \alpha)\rho = (p_1\rho[\alpha \mapsto \beta]) \setminus \beta \overset{\sigma}{\Longrightarrow} \sim (p_1'\rho[\alpha \mapsto \beta]) \setminus \beta.$$

  As $FC(p_1') \subseteq FC(p_1)$, we have that $new\, \alpha\, p_1\, \rho \subseteq new\, \alpha\, p_1'\, \rho$. We then have, by Fact 4.2.2(2), that $(p_1' \setminus \alpha)\rho =_\alpha (p_1'\rho[\alpha \mapsto \beta]) \setminus \beta$. The result now follows because $=_\alpha \subseteq \sim$ and by the transitivity of $\sim$.

(2) By induction on the length of the derivation $p \stackrel{\tau}{\Longrightarrow} q$.

- Base case: $p \stackrel{\tau}{\longrightarrow} q$. The proof proceeds by a sub-induction on the length of the proof of the derivation $p \stackrel{\tau}{\longrightarrow} q$. We only examine the most interesting case, which relies on axiom (ComPres).

  Assume that $p = p_1|p_2 \stackrel{\tau}{\longrightarrow} q_1|q_2 = q$ because $p_1 \stackrel{a}{\longrightarrow} q_1$ and $p_2 \stackrel{\bar{a}}{\longrightarrow} q_2$. As ρ satisfies (ComPres), we have that $\rho(a) \stackrel{\sigma}{\Longrightarrow} x_1\surd$ and $\rho(\bar{a}) \stackrel{\bar{\sigma}}{\Longrightarrow} x_2\surd$ for some $\sigma \in V(\Lambda)^+ \cup \{\tau\}$, x_1 and x_2. By statement (1) we then get that $p_1\rho \stackrel{\sigma}{\Longrightarrow} \sim q_1\rho$ and $p_2\rho \stackrel{\bar{\sigma}}{\Longrightarrow} \sim q_2\rho$. By the operational semantics and the substitutivity of $\sim$,

$$(p_1|p_2)\rho = p_1\rho|p_2\rho \stackrel{\tau}{\Longrightarrow} \sim q_1\rho|q_2\rho = (q_1|q_2)\rho.$$

- Inductive step, $p \stackrel{\tau}{\longrightarrow} p' \stackrel{\tau}{\Longrightarrow} q$, for some p'. Immediate. □

Statement (2) of the above lemma may be seen as a formal version of the intuitive statement that action refinements should preserve the synchronization structure of processes. In particular, it reflects the idea that action refinements should not interfere with the internal evolution of processes. The fragment of its proof that we have presented highlights the fundamental rôle played by axiom (ComPres). We may now show that $\rho_2 \circ \rho_1$ is indeed an action refinement.

Fact 4.2.5 *Let ρ_2 and ρ_1 be action refinements. Then $\rho_2 \circ \rho_1$ is also an action refinement.*

Proof: We have already seen that, for each a, $(\rho_2 \circ \rho_1)(a)$ is not terminated. We are left to show that $\rho_2 \circ \rho_1$ satisfies (ComPres). Now, for any $a \in V(\Lambda)$,

$$(\rho_2 \circ \rho_1)(a)|(\rho_2 \circ \rho_1)(\bar{a}) = (\rho_1(a)\rho_2)|(\rho_1(\bar{a})\rho_2) = (\rho_1(a)|\rho_1(\bar{a}))\rho_2.$$

As ρ_1 satisfies (ComPres), we have that $\rho_1(a)|\rho_1(\bar{a}) \stackrel{\tau}{\Longrightarrow} r\surd$, for some r. By the above lemma, we then have that

$$(\rho_1(a)|\rho_1(\bar{a}))\rho_2 \stackrel{\tau}{\Longrightarrow} x \sim r\rho_2$$

for some x. Moreover, as $r\surd$, we have that $r\rho_2\surd$. This implies that $x\surd$. Hence $\rho_2 \circ \rho_1$ satisfies (ComPres). □

4.2.3 Extending the Language

Let the set of extended processes, $\mathbf{P}^{\Gamma}_{\rho}$, be the set of all those processes which are definable by adding action refinements to the basic language $\mathbf{P}^{\Gamma}$ as extra operators:

$$p ::= nil \mid \delta \mid \mu \mid p;p \mid p+p \mid p|p \mid p\setminus\alpha \mid p[\rho].$$

It is within a language such as this (of course extended at least with recursive definitions) that the development of process specifications could take place and we are interested in a semantic theory for it. However, we shall not give an operational semantics directly for it as, intuitively, the behaviour of $p[\rho]$ should be identical to that of the process which results from applying ρ as a substitution to p. So, for any process p in $\mathbf{P}^{\Gamma}_{\rho}$, we can define the basic process in $\mathbf{P}^{\Gamma}$, $\mathbf{red}(p)$, which intuitively captures the behaviour of p by:

(i) $\mathbf{red}(p[\rho]) = \mathbf{red}(p)\rho$

(ii) $\mathbf{red}(op(p_1, \ldots, p_k)) = op(\mathbf{red}(p_1), \ldots, \mathbf{red}(p_k))$ $(op \in \Sigma)$.

In this way, any semantic equivalence Eq defined over $\mathbf{P}^{\Gamma}$ is automatically extended to $\mathbf{P}^{\Gamma}_{\rho}$ by

$$\langle p, q \rangle \in Eq \text{ iff } \langle \mathbf{red}(p), \mathbf{red}(q) \rangle \in Eq.$$

In particular, this gives a definition of $\sim$, $\sim_t$, $\approx$ and $\approx_t$ over $\mathbf{P}^{\Gamma}_{\rho}$. We are interested in developing a reasonable semantic equivalence for $\mathbf{P}^{\Gamma}_{\rho}$, in particular one which abstracts from internal actions. A minimal requirement is that it should be preserved by all $\mathbf{P}^{\Gamma}_{\rho}$-contexts. This immediately rules out $\approx$ as we know from Chapter 3 that it is not preserved by action refinements. For example, $a|b \approx a; b+b; a$, but if we use a refinement ρ such that $\rho(a) = a_1; a_2$ then $(a|b)\rho \not\approx (a; b+b; a)\rho$. In the previous chapter it was shown that, for a simple subset of $\mathbf{P}^{\Gamma}_{\rho}$ without communication, internal actions and restriction, $\sim_t$ is a reasonable equivalence; namely one which may be characterized by:

$$p \sim_t q \text{ iff for every simple context } C[\cdot],\ C[p] \sim C[q].$$

In the extended language, one would hope to have the analogous result:

$$p \approx_t q \text{ iff for every context } C[\cdot],\ C[p] \approx C[q].$$

This is not true for the trivial reason that $\approx_t$, in common with most forms of weak bisimulation equivalence, is not preserved by "+ contexts": for instance, $a \approx_t \tau; a$, but $b + a \not\approx_t b + \tau; a$. However, even the restricted statement

(*) $p \approx_t q$ only if for every action refinement ρ, $p\rho \approx_t q\rho$

is *not* true. In fact, $\approx_t$ is not preserved by action refinement over the language $\mathbf{P}^{\Gamma}$. The counter-example is due to R. van Glabbeek, [Va90], who originally developed it to show that, in general, the equivalence obtained by splitting an action in two is different from that obtained by splitting it in three. This example is discussed in § 4.4.

In the next section we define a modified version of $\approx_t$, called *weak refine equivalence* and denoted by $\approx_r$, for which (*) is true. This equivalence is essentially the weak version of $\sim_r$ defined in Chapter 3.

4.3 Refine Equivalence

4.3.1 Definition

The original motivation for developing the timed equivalences $\sim_t$ and $\approx_t$ was to develop a bisimulation theory for processes where actions are no longer atomic. The intention was to "match up" actions from equivalent processes in the usual way, but, in addition, to allow them to take some time. However, by simply requiring that the subactions $S(a)$ and $F(a)$ be matched up properly there is no guarantee that the original complete (but non-atomic) actions are also properly matched. In fact, it may be that the finish of a particular action is matched to the finish of a different action with the same name which started either before or after it. The equivalence $\sim_r$ of Chapter 3 was designed to ensure that this can not occur and that therefore the complete actions are indeed properly matched. We shall now develop a weak version of $\sim_r$, which will be denoted by $\approx_r$.

The definition requires us to identify all occurrences of actions in a process and therefore technically we use labelled actions obtained from the actions in $Act(\Lambda)$. Let $LAct(\Lambda)$ denote the set of actions $\{a_i \mid a \in V(\Lambda), i \in \mathrm{N}\} \cup \{\tau\}$ (λ will be used to range over $LAct(\Lambda)$). We use $\mathbf{LP}^\Gamma$ and $\mathbf{L}\mathcal{S}^\Gamma$ to denote the set of *uniquely labelled processes* and *uniquely labelled states*, respectively obtained by using $LAct(\Lambda)$ in place of $Act(\Lambda)$ in the definition of $\mathbf{P}^\Gamma$ and $\mathcal{S}^\Gamma$, subject to the following restrictions:

(1) each index i occurs at most once in each labelled process and labelled state, and

(2) each labelled state of the form $c \setminus \alpha$ in $\mathbf{L}\mathcal{S}^\Gamma$ satisfies the constraint that, for each i, $F(\alpha_i)$ and $F(\bar{\alpha}_i)$ do not occur in c (this constraint is the natural labelled version of condition (CR)).

Note that, in this case, restriction is still with respect to a channel name from Λ and *not* a labelled channel name. We use $\pi, \pi' \ldots$ to range over labelled processes and $c, d \ldots$ to range over labelled states.

The next-state relations $\overset{S(a_i)}{\longmapsto}$ and $\overset{F(a_i)}{\longmapsto}$ and the termination predicate $\surd$ are inherited directly from the rules in Figure 4.2 provided the *admits* predicate is extended so that:

(i) α *admits* τ

(ii) α *admits* $a_i, S(a_i), F(a_i)$ iff $a \neq \alpha, \bar{\alpha}$.

However, the transitions $\overset{a_i}{\longmapsto}$ and $\overset{\tau}{\longmapsto}$ are obtained by replacing rule (SYN) in Figure 4.1 with

$(LSYN)$ $c \overset{a_i}{\longmapsto} c'$ and $d \overset{\bar{a}_j}{\longmapsto} d'$ imply $c|d \overset{\tau}{\longmapsto} c'|d'$.

The net effect of these modifications is quite natural; restriction applies to all occurrences of actions using the channel name in question and communication may occur between any occurrences of complementary actions. We say that a labelled state c is *stable* iff $c \not\xrightarrow{\tau}$. The weak termination predicate $\surd\!\!\!/$ over $\mathbf{L}\mathcal{S}^\Gamma$ is then defined as at page 137.

Notation 4.3.1 *The set of* labelled subactions, $LAct_s(\Lambda)$, *is given by*

$$LAct_s(\Lambda) = \{S(a_i), F(a_i) \mid a_i \in LAct(\Lambda)\} \cup \{\tau\}.$$

The weak transition relations $\stackrel{\sigma}{\Longmapsto}$, *for* $\sigma \in LAct(\Lambda)^\star \cup LAct_s(\Lambda)^\star$, *are defined in the standard way.*

To ensure that complete actions, consisting of the occurrence of a start subaction $S(a_i)$ and of the corresponding finish subaction $F(a_i)$, are properly matched we need to retain information about which actions have already been started and to whom they have been matched. Such information is recorded in what was called a *history* in Chapter 3. (See Definition 3.4.3 and the following remarks for more details) As in Chapter 3, we use $\mathcal{H}$ to denote the set of all histories (h, ϕ, φ will be used to range over $\mathcal{H}$). Labelled states will be equivalent or inequivalent with respect to a given history. For c and d to be equivalent with respect to h we require that:

(1) h must be a history compatible with c and d. In other words, all actions which have started in c or d and have not yet finished must be recorded in the history h. We say that h is compatible with c and d iff for all a, $\mathbf{Places}(a, c) = dom(h_a)$ and $\mathbf{Places}(a, d) = range(h_a)$, where, for each c and a, $\mathbf{Places}(a, c)$ is defined as on page 90;

(2) start moves have to be matched. For example, every start move of c, $c \stackrel{S(a_i)}{\longmapsto} c'$, must be matched by a corresponding move $d \stackrel{S(a_j)}{\Longmapsto} d'$ of d such that c' and d' are equivalent with respect to the augmented history $h \cup \{(a, i, j)\}$. This is the way histories are built up dynamically;

(3) finish moves must be matched in a way which is consistent with the history h. For example, every finish move from c, $c \stackrel{F(a_i)}{\longmapsto} c'$, must be matched by the proper finish move from d, i.e. $d \stackrel{F(a_j)}{\Longmapsto} d'$ for some d' and j such that $(i, j) \in h_a$ and c' and d' are equivalent with respect to the diminished history $h - \{(a, i, j)\}$;

(4) silent moves must be matched as usual and

(5) successful termination must be matched as usual.

This is the motivation underlying the following definition. First of all, let us say that a family of binary relations $\{\mathcal{R}_h \mid h \in \mathcal{H}\}$ over $\mathbf{L}\mathcal{S}^\Gamma$ is *symmetric* iff $\langle c, d\rangle \in \mathcal{R}_h$ implies $\langle d, c\rangle \in \mathcal{R}_{h^{op}}$, where $h^{op} = \{(a, j, i) \mid (a, i, j) \in h\}$.

Definition 4.3.1 (Weak Refine Bisimulation) *A symmetric family of binary relations* $\{\mathcal{R}_h \mid h \in \mathcal{H}\}$ *over* $\mathbf{L}\mathcal{S}^\Gamma$ *is called a* weak refine bisimulation *whenever it satisfies, for each* $\langle c, d\rangle \in \mathcal{R}_h$ *and* $a_i \in LAct(\Lambda)$,

(1) *h is compatible with* $\langle c, d\rangle$,

(2) $c \overset{S(a_i)}{\longmapsto} c'$ *implies* $d \overset{S(a_j)}{\Longmapsto} d'$ *for some j and d′ such that* $\langle c', d'\rangle \in \mathcal{R}_{h\cup\{(a,i,j)\}}$,

(3) $c \overset{F(a_i)}{\longmapsto} c'$ *implies* $d \overset{F(a_j)}{\Longmapsto} d'$ *for some j and d′ such that* $h_a(i) = j$ *and* $\langle c', d'\rangle \in \mathcal{R}_{h-\{(a,i,j)\}}$,

(4) $c \overset{\tau}{\longmapsto} c'$ *implies* $d \overset{\varepsilon}{\Longmapsto} d'$ *for some d′ such that* $\langle c', d'\rangle \in \mathcal{R}_h$,

(5) $c\surd$ *implies* $d\surd$.

For the standard reasons it turns out that there is a largest refine bisimulation, which we denote by $\approx_{\mathcal{H}} = \{\approx_h \mid h \in \mathcal{H}\}$, *and in fact the relations* $\approx_h$ *can be defined by*

$$\approx_h = \bigcup\{\mathcal{R}_h \mid \{\mathcal{R}_h \mid h \in \mathcal{H}\} \text{ is a refine bisimulation}\}.$$

Of course, we are only interested in $\approx_{\mathcal{H}}$ in so far as it induces an equivalence relation on unlabelled states and, in particular, unlabelled processes. For $c \in \mathbf{L}\mathcal{S}^\Gamma$ let $\mathbf{un}(c)$ be the unlabelled state obtained by replacing each occurrence of a_i and $F(a_i)$ by a and $F(a)$, respectively. We may then define

$$s \approx_r s' \text{ iff for some } c, d, h,\ \mathbf{un}(c) = s,\ \mathbf{un}(d) = s' \text{ and } c \approx_h d.$$

If $\mathbf{un}(c)$ is a process, i.e. is in $\mathbf{P}^\Gamma$, then c must be in $\mathbf{LP}^\Gamma$ and any history compatible with labelled processes must be empty. Thus this definition specializes to:

$p \approx_r q$ iff there exist labelled processes π and π' such that
(a) $\mathbf{un}(\pi) = p$, $\mathbf{un}(\pi') = q$ and
(b) $\pi \approx_\emptyset \pi'$.

Definition 4.3.2 *For each* $c \in \mathbf{L}\mathcal{S}^\Gamma$, *the history associated with c, id(c), is given by*

$$id(c) =_{def} \{(a, i, i) \mid a \in V(\Lambda) \text{ and } i \in \mathsf{Places}(a, c)\}.$$

Fact 4.3.1

(1) *For each* $a \in V(\Lambda)$, *the following statements hold:*

(a) $\mathsf{Places}(a, \pi) = \emptyset$, *for each* $\pi \in \mathbf{LP}^\Gamma$;

(b) **Places**$(a, c;\pi) =$ **Places**(a, c), *for each* $c;\pi \in \mathbf{L}\mathcal{S}^{\Gamma}$;

(c) **Places**$(a, c_1|c_2) =$ **Places**$(a, c_1) \cup$ **Places**(a, c_2), *for each* $c_1|c_2 \in \mathbf{L}\mathcal{S}^{\Gamma}$; *and*

(d) **Places**$(a, c \setminus \alpha) =$ **Places**(a, c), *for each* $c \setminus \alpha \in \mathbf{L}\mathcal{S}^{\Gamma}$.

(2) *The following properties of* $id(c)$, $c \in \mathbf{L}\mathcal{S}^{\Gamma}$, *hold:*

(a) $id(\pi) = \emptyset$,

(b) $id(c;\pi) = id(c)$,

(c) $id(c|d) = id(c) \cup id(d)$ *and*

(d) $id(c \setminus \alpha) = id(c)$.

Proof: The verifications of all the properties are straightforward. The only interesting point to note is that statements (1)(d) and (2)(d) depend on the natural condition (CR) that we have imposed on states and labelled states of the form $c \setminus \alpha$. This restriction ensures that, whenever $c \overset{F(a_i)}{\longmapsto}$ and $c \setminus \alpha \in \mathbf{L}\mathcal{S}^{\Gamma}$, one has that $a \neq \alpha, \bar{\alpha}$ and thus $c \setminus \alpha \overset{F(a_i)}{\longmapsto}$. □

The following properties of $\approx_{\mathcal{H}}$ will be useful in proving that $\approx_r$ is an equivalence relation over $\mathcal{S}^{\Gamma}$.

Lemma 4.3.1 *For each* $c, d, c' \in \mathbf{L}\mathcal{S}^{\Gamma}$ *the following statements hold:*

(i) $c \approx_{id(c)} c$

(ii) $c \approx_h d$ *iff* $d \approx_{h^{op}} c$,

(iii) $c \approx_h c'$ *and* $c' \approx_{h'} d$ *imply* $c \approx_{h' \circ h} d$.

Proof: We only sketch the proof of statements (i) and (iii), as (ii) is an immediate consequence of the symmetry requirement for $\approx_{\mathcal{H}}$. We examine the two statements in turn.

(i) It is sufficient to prove that the $\mathcal{H}$-indexed family of relations $\{\mathcal{R}_h \mid h \in \mathcal{H}\}$ given by

$$\mathcal{R}_h = \{\langle c, c\rangle \mid h = id(c)\}$$

is a weak refine equivalence. The straightforward verification is omitted.

(iii) It is sufficient to prove that the $\mathcal{H}$-indexed family of relations given by

$$\mathcal{R}_h = \{\langle c, d\rangle \mid \text{there exist } c', h_1 \text{ and } h_2 \text{ such that } c \approx_{h_1} c' \approx_{h_2} d \text{ and } h = h_2 \circ h_1\}$$

is a weak refine equivalence. Again, the straightforward verifications are omitted. □

The following proposition gives the equivalent statement of Theorem 3.4.2 for the richer set of states $\mathcal{S}^\Gamma$.

Proposition 4.3.1 *$\approx_r$ is an equivalence relation over $\mathcal{S}^\Gamma$.*

Proof: Reflexivity of $\approx_r$ follows by Lemma 4.3.1(i) and the fact that, for each $s \in \mathcal{S}^\Gamma$, there exists $c \in \mathbf{L}\mathcal{S}^\Gamma$ such that $s = \mathbf{un}(c)$. Symmetry and transitivity of $\approx_r$ are simple consequences of parts (ii) and (iii) of Lemma 4.3.1, respectively. □

In the definition of weak refine bisimulation we did not require the explicit matching of moves labelled by complete actions a_i. A natural question to ask is whether the addition of a clause requiring the matching of such moves in the definition of $\approx_{\mathcal{H}}$ changes the resulting notion of equivalence. We shall now prove that, as argued informally above, $\approx_{\mathcal{H}}$ does ensure that complete actions are matched. The proof of this fact makes use of two useful "commuting transitions" results which will find application in the proof of the refinement theorem. First of all, we define a version of $\approx_{\mathcal{H}}$ which explicitly requires the matching of transitions labelled by complete actions. It will then be shown that this new relation coincides with $\approx_{\mathcal{H}}$.

Definition 4.3.3 (Augmented Weak Refine Bisimulation) *A symmetric family of binary relations $\{\mathcal{R}_h \mid h \in \mathcal{H}\}$ over $\mathbf{L}\mathcal{S}^\Gamma$ is called an* augmented weak refine bisimulation *whenever it satisfies, for each $h \in \mathcal{H}$, $\langle c, d\rangle \in \mathcal{R}_h$ and $a_i \in LAct(\Lambda)$, clauses* (1)-(5) *of the definition of $\approx_{\mathcal{H}}$ and*

(6) *$c \overset{a_i}{\longmapsto} c'$ implies $d \overset{a_j}{\Longmapsto} d'$ for some j and d' such that $\langle c', d'\rangle \in \mathcal{R}_h$.*

$\approx^a_{\mathcal{H}} = \{\approx^a_h \mid h \in \mathcal{H}\}$ will denote the largest such relation.

We shall now prove the first of the "commuting transitions" results. This result formalizes the intuition that if $c \overset{S(a_i)}{\longmapsto} c' \overset{\lambda}{\longmapsto} d$, for some c' and labelled action λ, then the move $c' \overset{\lambda}{\longmapsto} d$ does *not* depend on the occurrence of action a_i. Hence it could have occurred before the start of action a_i without influencing the resulting target state d. Formally:

Proposition 4.3.2 (Commuting Start-Moves and Complete Moves) *Let $c \in \mathbf{L}\mathcal{S}^\Gamma$ and $\lambda \in LAct(\Lambda)$. Assume that $c \overset{S(a_i)}{\longmapsto} c' \overset{\lambda}{\longmapsto} d$. Then there exists $\bar{c} \in \mathbf{L}\mathcal{S}^\Gamma$ such that $c \overset{\lambda}{\longmapsto} \bar{c} \overset{S(a_i)}{\longmapsto} d$.*

Proof: By induction on the structure of c. □

The following result is an immediate corollary of the above proposition. In its statement we shall use the notation $l(c \overset{\sigma}{\Longmapsto} d)$ to denote the *length* of the derivation $c \overset{\sigma}{\Longmapsto} d$.

Corollary 4.3.1 *Let $c \in \mathbf{L}\mathcal{S}^\Gamma$ and $\sigma \in LAct(\Lambda)^\star$. Assume that $c \overset{S(a_i)}{\longmapsto} c' \overset{\sigma}{\Longmapsto} d$. Then there exists $\bar{c} \in \mathbf{L}\mathcal{S}^\Gamma$ such that $c \overset{\sigma}{\Longmapsto} \bar{c} \overset{S(a_i)}{\longmapsto} d$ and $l(c' \overset{\sigma}{\Longmapsto} d) = l(c \overset{\sigma}{\Longmapsto} \bar{c})$.*

The following proposition and associated corollary state dual results to Proposition 4.3.2 and Corollary 4.3.1, respectively, for end-transitions. Intuitively, if $c \overset{\lambda}{\longmapsto} c' \overset{F(a_i)}{\longmapsto} d$ then the move $c' \overset{F(a_i)}{\longmapsto} d$ does not depend on the occurrence of action λ. Hence the finish action $F(a_i)$ might have occurred before the action λ without influencing the resulting target state.

Proposition 4.3.3 (Commuting End-Moves and Complete Moves) *Let $c \in \mathbf{L}\mathcal{S}^\Gamma$ and $\lambda \in LAct(\Lambda)$. Assume that $c \overset{\lambda}{\longmapsto} c' \overset{F(a_i)}{\longmapsto} d$. Then there exists $\bar{c} \in \mathbf{L}\mathcal{S}^\Gamma$ such that $c \overset{F(a_i)}{\longmapsto} \bar{c} \overset{\lambda}{\longmapsto} d$.*

Proof: By structural induction on c. □

Corollary 4.3.2 *Let $c \in \mathbf{L}\mathcal{S}^\Gamma$ and $\sigma \in LAct(\Lambda)^\star$. Assume that $c \overset{\sigma}{\Longmapsto} c' \overset{F(a_i)}{\longmapsto} d$. Then there exists $\bar{c}$ such that $c \overset{F(a_i)}{\longmapsto} \bar{c} \overset{\sigma}{\Longmapsto} d$ and $l(c \overset{\sigma}{\Longmapsto} c') = l(\bar{c} \overset{\sigma}{\Longmapsto} d)$.*

The following lemma relates the transition relation associated with a complete action a_i with those associated with its subactions $S(a_i)$ and $F(a_i)$.

Lemma 4.3.2 *Let $c, d \in \mathbf{L}\mathcal{S}^\Gamma$. Then $c \overset{a_i}{\longmapsto} d$ iff there exists c' such that $c \overset{S(a_i)}{\longmapsto} c' \overset{F(a_i)}{\longmapsto} d$.*

Proof: The "only if" implication is proven by induction on the proof of the transition $c \overset{a_i}{\longmapsto} d$. The "if" implication can be shown by structural induction on c. □

We can now prove that $\approx_{\mathcal{H}}$ and $\approx^a_{\mathcal{H}}$ coincide over $\mathbf{L}\mathcal{S}^\Gamma$.

Proposition 4.3.4 *For each $c, d \in \mathbf{L}\mathcal{S}^\Gamma$ and $h \in \mathcal{H}$, $c \approx_h d$ iff $c \approx^a_h d$.*

Proof: The "if" implication follows immediately by the definitions of the two relations. For the "only if" implication it is sufficient to show that $\approx_{\mathcal{H}}$ is an augmented weak refine bisimulation. The only interesting thing to check is that, for $c \approx_h d$,

$$c \overset{a_i}{\longmapsto} c' \text{ implies } d \overset{a_j}{\Longmapsto} d' \text{ for some } d' \text{ and } j \text{ such that } c' \approx_h d'.$$

Assume then that $c \approx_h d$ and $c \overset{a_i}{\longmapsto} c'$. By the above lemma, there exists $\bar{c}$ such that $c \overset{S(a_i)}{\longmapsto} \bar{c} \overset{F(a_i)}{\longmapsto} c'$. As $c \approx_h d$, there exist j and $\bar{d}$ such that

$$d \overset{S(a_j)}{\Longmapsto} \bar{d} \text{ and } \bar{c} \approx_{h\cup\{(a,i,j)\}} \bar{d}.$$

Again, as $\bar{c} \overset{F(a_i)}{\longmapsto} c'$, there exists d' such that

$$\bar{d} \overset{F(a_j)}{\Longmapsto} d' \text{ and } c' \approx_h d'.$$

Thus we have that $d \overset{S(a_j)}{\Longmapsto} \bar{d} \overset{F(a_j)}{\Longmapsto} d'$ for some d' such that $c' \approx_h d'$. We are now left to prove that $d \overset{a_j}{\Longmapsto} d'$. By Corollaries 4.3.1 and 4.3.2, there exist d_1 and d_2 such that

$$d \overset{\varepsilon}{\Longmapsto} d_1 \overset{S(a_j)}{\longmapsto} \bar{d} \overset{F(a_j)}{\longmapsto} d_2 \overset{\varepsilon}{\Longmapsto} d'.$$

By the above lemma, $d_1 \overset{S(a_j)}{\longmapsto} \bar{d} \overset{F(a_j)}{\longmapsto} d_2$ implies $d_1 \overset{a_j}{\longmapsto} d_2$. Thus we have that $d \overset{a_j}{\Longmapsto} d'$.

We have thus shown that $\approx_{\mathcal{H}}$ is an augmented weak refine bisimulation. □

This finishes our examination of the definition of $\approx_r$. Let us now turn our attention to some of its properties. We shall prove, first of all, that $\approx_r$ is preserved by all the combinators in Σ apart from the nondeterministic choice operator $+$. This statement will follow from some useful properties of $\approx_{\mathcal{H}}$ studied in the following lemma.

Lemma 4.3.3

(1) *Let $c, d, \pi_1, \pi_2 \in \mathbf{L}\mathcal{S}^\Gamma$ be such that $c \approx_h d$, $\pi_1 \approx_\emptyset \pi_2$ and $c;\pi_1, d;\pi_2 \in \mathbf{L}\mathcal{S}^\Gamma$. Then $c;\pi_1 \approx_h d;\pi_2$.*

(2) *Let $c_i, d_i \in \mathbf{L}\mathcal{S}^\Gamma$, $i = 1, 2$, be such that $c_1 \approx_h c_2$, $d_1 \approx_{h'} d_2$ and $c_1|d_1, c_2|d_2 \in \mathbf{L}\mathcal{S}^\Gamma$. Then $c_1|d_1 \approx_{h \cup h'} c_2|d_2$.*

(3) *Let $c, d \in \mathbf{L}\mathcal{S}^\Gamma$ and $\alpha \in \Lambda$ be such that $c \approx_h d$ and $c \setminus \alpha, d \setminus \alpha \in \mathbf{L}\mathcal{S}^\Gamma$. Then $c \setminus \alpha \approx_h d \setminus \alpha$.*

Proof: We only give the proof of statement (3). Assume that $c \approx_h d$ and $c \setminus \alpha, d \setminus \alpha \in \mathbf{L}\mathcal{S}^\Gamma$. In order to prove that $c \setminus \alpha \approx_h d \setminus \alpha$, it is sufficient to show that the family of relations $\{\mathcal{R}_h \mid h \in \mathcal{H}\}$ given by

$$\mathcal{R}_h = \{\langle c \setminus \alpha, d \setminus \alpha \rangle \mid c \approx_h d \text{ and } c \setminus \alpha, d \setminus \alpha \in \mathbf{L}\mathcal{S}^\Gamma\}$$

is a weak refine bisimulation. We check that the defining clauses of $\approx_{\mathcal{H}}$ are met by $\{\mathcal{R}_h \mid h \in \mathcal{H}\}$. Assume that $\langle c \setminus \alpha, d \setminus \alpha \rangle \in \mathcal{R}_h$. Then:

- h is compatible with $\langle c \setminus \alpha, d \setminus \alpha \rangle \in \mathcal{R}_h$. In fact, for each $a \in V(\Lambda)$, $\mathbf{Places}(a, c \setminus \alpha) = \mathbf{Places}(a, c) = dom(h_a)$ and $\mathbf{Places}(a, d \setminus \alpha) = \mathbf{Places}(a, d) = range(h_a)$.

- Assume that $c \setminus \alpha \overset{S(a_i)}{\longmapsto} c' \setminus \alpha$. Then $c \overset{S(a_i)}{\longmapsto} c'$ and $a \neq \alpha, \bar{\alpha}$. As $c \approx_h d$, there exist d' and j such that $d \overset{S(a_j)}{\Longmapsto} d'$ and $c' \approx_{h'} d'$, where $h' = h \cup \{(a,i,j)\}$. By the operational semantics, $d \setminus \alpha \overset{S(a_j)}{\Longmapsto} d' \setminus \alpha$. As $c \setminus \alpha, d \setminus \alpha \in \mathbf{L}\mathcal{S}^\Gamma$, we then have that $c' \setminus \alpha, d' \setminus \alpha \in \mathbf{L}\mathcal{S}^\Gamma$. Thus $\langle c' \setminus \alpha, d' \setminus \alpha \rangle \in \mathcal{R}_{h'}$.

- Clauses (3) and (4) are checked in similar fashion.

- Assume that $c \setminus \alpha \surd\!\!\!\surd$. Then it is easy to see that

$$\begin{aligned} c \setminus \alpha \surd\!\!\!\surd &\iff c \surd\!\!\!\surd \\ &\implies d \surd\!\!\!\surd \qquad \text{as } c \approx_h d \\ &\iff d \setminus \alpha \surd\!\!\!\surd. \end{aligned}$$

Hence $\{\mathcal{R}_h \mid h \in \mathcal{H}\}$ is a weak refine bisimulation. □

Proposition 4.3.5 $\approx_r$ *is preserved by the operators in* $\Sigma - \{+\}$.

Proof: The claim is an immediate consequence of the above lemma. For example, assume that $s_1 \approx_r s_2$ and $s_1 \setminus \alpha, s_2 \setminus \alpha \in \mathcal{S}^\Gamma$. By the definition of $\approx_r$, there exist $c_i \in \mathbf{L}\mathcal{S}^\Gamma$, $i = 1, 2$, and $h \in \mathcal{H}$ such that $c_1 \approx_h c_2$ and $\mathbf{un}(c_i) = s_i$, $i = 1, 2$. It is easy to see that, as $s_i \setminus \alpha \in \mathcal{S}^\Gamma$, $c_i \setminus \alpha \in \mathbf{L}\mathcal{S}^\Gamma$ $(i = 1, 2)$. Then, by the above lemma, $c_1 \setminus \alpha \approx_h c_2 \setminus \alpha$. Thus $s_1 \setminus \alpha \approx_r s_2 \setminus \alpha$. □

As it is the case with many forms of weak bisimulation relations, $\approx_r$ is not preserved by $+$. In fact, it is easy to see that $a \approx_r \tau; a$, but $a + b \not\approx_r \tau; a + b$. However, we are mainly interested in the operator of action refinement and this brings us to the main theorem of this chapter.

Theorem 4.3.1 (The Refinement Theorem) *Let* $p, q \in \mathbf{P}^\Gamma$. *Assume that* $p \approx_r q$. *Then, for any action refinement* ρ, $p\rho \approx_r q\rho$.

The proof of this theorem is quite complex and is relegated to the following section. It involves decomposing moves from labelled states of the form $c\rho$ into moves from c and the components of ρ and the converse, combining moves from c and the components of ρ to form moves of $c\rho$.

With this theorem we have that $\approx_r$, modified in the usual way to compensate for $+$, is preserved by all extended (i.e. $\mathbf{P}^\Gamma_\rho$) contexts. Let

$$p \approx_r^+ q \text{ iff for some action } a \text{ not appearing in } p \text{ and } q,\ p + a \approx_r q + a.$$

Theorem 4.3.2 *Let* $p, q \in \mathbf{P}^\Gamma$. *Then:*

(a) $p \approx_r^+ q$ *iff for every* $\mathbf{P}_\rho^\Gamma$*-context* $C[\cdot]$, $C[p] \approx_r^+ C[q]$.

(b) $p \approx_r^+ q$ *iff for every* $\mathbf{P}_\rho^\Gamma$*-context* $C[\cdot]$, $C[p] \approx_r C[q]$.

Proof: We prove the two statements separately.

(a) The "if" direction is immediate by taking the empty context. For the converse, one can easily show that $\approx_r^+$, unlike $\approx_r$, is preserved by $+$. For the remaining operators, the proofs that they are preserved by $\approx_r$ may be trivially adapted to $\approx_r^+$ since these relations only differ for initial τ moves.

(b) The "only if" direction follows from (a) because $\approx_r^+$ is contained in $\approx_r$. The converse is trivial since $\approx_r^+$ is defined using the context $[\cdot] + a$. □

4.3.2 The Refinement Theorem

This section will be entirely devoted to a detailed proof of Theorem 4.3.1 (the refinement theorem). Let us recall, for the sake of completeness, that the refinement theorem states that, for all $p, q \in \mathbf{P}^\Gamma$ and action refinement ρ,

$$p \approx_r q \quad \text{implies} \quad p\rho \approx_r q\rho. \tag{4.1}$$

Naturally enough, given the fundamental rôle played by labelled states in the definition of $\approx_r$, the proof of statement (4.1) will rely upon a detailed analysis of some operational properties of labelled states and we shall present labelled counterparts to several properties of processes and states studied in the previous sections. The proof of (4.1) will be given in several stages. First of all, following the standard theory of bisimulation for **CCS**-like languages, we shall present a bisimulation-like characterization of the relation $\approx_r^+$ introduced on page 157. A suitable notion of *labelled action refinement* will be then introduced as a labelled counterpart of the action refinements given on page 141. The application of labelled action refinements to labelled processes and states will be defined following the approach outlined in § 4.2.2 for processes. We shall then prove a series of composition/decomposition results for moves of labelled terms of the form $c\rho$, where $c \in LS^\Gamma$ and ρ is a labelled action refinement. These results will allow us to decompose moves of $c\rho$ into moves of c and of the components of ρ and, conversely, to compose moves of c and of the components of ρ to obtain moves of $c\rho$. Finally, the technical material developed in the stages outlined above will be used to prove the refinement theorem.

Our first aim in this section will be to give a more manageable characterization of the relation $\approx_r^+$. The following definition follows closely the one of the *strong congruence* given by R. Milner in [Mil88].

Definition 4.3.4 (Refine Equality) *For each $\phi \in \mathcal{H}$, $=_\phi$ is the symmetric relation given by:*

- *for each $\pi_1, \pi_2 \in \mathbf{LP}^\Gamma$, $\pi_1 =_\emptyset \pi_2$ iff for all $a_i \in LAct(\Lambda)$*

  (1) $\mathbf{Places}(a, \pi_1) = \mathbf{Places}(a, \pi_2) = \emptyset$,

  (2) $\pi_1 \overset{S(a_i)}{\longmapsto} c$ *implies* $\pi_2 \overset{S(a_j)}{\Longmapsto} d$, *for some j and d such that* $c \approx_{\{(a,i,j)\}} d$,

  (3) $\pi_1 \overset{\tau}{\longmapsto} \pi_1'$ *implies* $\pi_2 \overset{\tau}{\Longmapsto} \pi_2'$, *for some π_2' such that* $\pi_1' \approx_\emptyset \pi_2'$.

- $=_\phi = \approx_\phi$, *for all* $\phi \neq \emptyset$.

$=_\mathcal{H}$ *will be used to denote* $\{=_\phi \mid \phi \in \mathcal{H}\}$.

The definition of refine equality differs from the one of $\approx_\mathcal{H}$ only in the way τ-moves are required to be matched for labelled processes; for labelled processes π_1 and π_2, each initial action of π_1 or π_2 must be matched by *at least* one action of the other. This kind of matching is not required for labelled states in which some action has started, but has not yet terminated, because such terms cannot be placed in +-contexts. (See the definition of $\mathbf{L}\mathcal{S}^\Gamma$ on page 138)

Refine equality can be used to induce an equivalence relation over $\mathcal{S}^\Gamma$ following the definitions of $\sim_r$ and $\approx_r$ given previously. We shall now prove that this induced relation, which will be denoted by $=_r$, is indeed a congruence over $\mathcal{S}^\Gamma$. This will follow from some general properties of $=_\mathcal{H}$ studied in the following lemma, whose proof is similar to the one of the corresponding result for $\approx_\mathcal{H}$.

Lemma 4.3.4 (1) *Let $\pi_1 =_\emptyset \pi_2$ and $\pi_1 + \pi, \pi_2 + \pi \in \mathbf{LP}^\Gamma$. Then $\pi_1 + \pi =_\emptyset \pi_2 + \pi$.*

(2) *Assume that $c =_\phi d$ and $c;\pi, d;\pi \in \mathbf{L}\mathcal{S}^\Gamma$. Then $c;\pi =_\phi d;\pi$.*

(3) *Assume that $c =_\phi d$ and $c|d', d|d' \in \mathbf{L}\mathcal{S}^\Gamma$. Then $c|d' =_{\phi \cup id(d')} d|d'$.*

(4) *Assume that $c =_\phi d$ and $c \setminus \alpha, d \setminus \alpha \in \mathbf{L}\mathcal{S}^\Gamma$. Then $c \setminus \alpha =_\phi d \setminus \alpha$.*

Corollary 4.3.3 *$=_r$ is a congruence over $\mathcal{S}^\Gamma$.*

We shall now prove that $=_r$ coincides with $\approx_r^+$ over $\mathbf{P}^\Gamma$. The proof of the following proposition is similar to the one of analogous results in the theory of bisimulation equivalence for CCS-like languages. (See, e.g. [Mil88,89])

Proposition 4.3.6 *For each $p, q \in \mathbf{P}^\Gamma$, $p =_r q$ iff $p + a \approx_r q + a$, for some a not occurring in p and q.*

The following is a standard corollary of the above proposition.

Corollary 4.3.4 $\approx_r^+$ *is the largest congruence over* $\mathbf{P}^\Gamma$ *contained in* $\approx_r$.

We shall now introduce the labelled counterpart of the action refinements presented in § 4.2.2. The definition of these labelled action refinements is somewhat technical; the technicalities, however, are needed to deal with the added complexity introduced by the labelling on terms.

Definition 4.3.5 (Labelled Action Refinements) *A labelled action refinement* ρ *is a map* $\rho : LAct(\Lambda) \cup \{F(a_i) \mid a_i \in LAct(\Lambda)\} \to \mathbf{L}\mathcal{S}^\Gamma$ *which satisfies:*

(a) *for all* $a_i \in LAct(\Lambda)$, $\rho(a_i) \in \mathbf{LP}^\Gamma$ *and* $\rho(a_i) \notin \surd$ *(i.e. labelled actions are mapped to non-terminated labelled processes),*

(b) ρ *satisfies (ComPres), and*

(c) *for all* $a \in V(\Lambda)$, $i, j \in \mathrm{N}$, $\rho(a_i) =_\emptyset \rho(a_j)$ *(i.e. instances of the same action are mapped to congruent labelled processes).*

The application of labelled action refinements to configurations can now be defined following the approach outlined in the previous section. For each configuration c, let $FC(c)$, the set of *free channel names* in c, be defined by structural recursion on c as follows:

$$
\begin{aligned}
&FC(nil) = FC(\delta) = FC(\tau) = \emptyset \\
&FC(\alpha_i) = FC(\bar{\alpha}_i) = FC(F(\alpha_i)) = FC(F(\bar{\alpha}_i)) = \{\alpha\} \\
&FC(c;\pi) = FC(c) \cup FC(\pi) \\
&FC(\pi_1 + \pi_2) = FC(\pi_1) \cup FC(\pi_2) \\
&FC(c|d) = FC(c) \cup FC(d) \\
&FC(c \setminus \alpha) = FC(c) - \{\alpha\}.
\end{aligned}
$$

In order to simplify the definition of the function *new*, we now introduce some useful notation.

Notation 4.3.2 *For each* $c \in \mathbf{L}\mathcal{S}^\Gamma$, *the set of* symbols *in* c, $Sym(c)$, *is given by*

$$Sym(c) =_{def} \{a_i \in LAct(\Lambda) \mid a_i \text{ occurs in } c\} \cup \{F(a_i) \mid F(a_i) \text{ occurs in } c\}.$$

For each $e \in LAct(\Lambda) \cup \{F(a_i) \mid a_i \in LAct(\Lambda)\}$, $\nu(e) \in \Lambda$ *will denote the underlying channel name of* e.

Following [Sto88] and our development in the previous section, we define the function *new* by

$$new\, \alpha\, c\, \rho =_{def} \{\beta \mid \text{for all } \beta' \in FC(c) - \{\alpha\}, \beta \notin FC(\rho(e)) \text{ for all } e \in Sym(c) \text{ such that } \nu(e) = \beta'\}$$

Note that the set $new\,\alpha\,c\,\rho$ is always infinite because Λ is infinite and, for each c, $Sym(c)$ and $FC(c)$ are both finite sets. The result of applying a labelled action refinement ρ to a configuration c can now be defined following Definition 4.2.2. In the following definition we shall use the notational conventions about substitutions from the previous section, with the understanding that, for each labelled action refinement ρ and channel name α, $\rho[\alpha \mapsto \beta]$ will denote the substitution identical to ρ except that α_i is mapped to β_i and $\bar{\alpha}_i$ to $\bar{\beta}_i$, for all i.

Definition 4.3.6 (Application of Labelled Action Refinements) *For each* $c \in \mathbf{L}\mathcal{S}^{\Gamma}$ *and labelled action refinement* ρ, $c\rho$ *is the labelled state defined by:*

(i) $a_i\rho = \rho(a_i)$, $\tau\rho = \tau$, $nil\rho = nil$, $\delta\rho = \delta$, $F(a_i)\rho = \rho(F(a_i))$

(ii) $(c;\pi)\rho = c\rho;\pi\rho$, $(\pi_1 + \pi_2)\rho = \pi_1\rho + \pi_2\rho$, $(c_1|c_2)\rho = c_1\rho|c_2\rho$

(iii) $(c \setminus \alpha)\rho = (c\rho[\alpha \mapsto \beta]) \setminus \beta$, *where* $\beta = choice(new\,\alpha\,c\,\rho)$.

All the definitions and results regarding α-congruence and action refinements apply equally well to labelled action refinements. In particular, in what follows we shall often make use of the Substitution Lemma and Fact 4.2.2 applied to labelled action refinements. Moreover, following the lines of the proof of Proposition 4.2.5, it is possible to prove that $=_\alpha$ is a strong bisimulation over the LTS with termination $\langle \mathbf{L}\mathcal{S}^{\Gamma}, \longmapsto, LAct_s(\Lambda), \surd\rangle$. The largest strong bisimulation over that LTS will be denoted by $\sim$, with abuse of notation. For the sake of clarity, we state the following proposition.

Proposition 4.3.7 $=_\alpha$ *is a strong bisimulation over* $\langle \mathbf{L}\mathcal{S}^{\Gamma}, \longmapsto, LAct_s(\Lambda), \surd\rangle$.

The following technical definition will allow us to extend $\approx_{\mathcal{H}}$ to labelled action refinements in a point-wise fashion. For our purposes, it will be convenient to define such a pointwise extension of $\approx_{\mathcal{H}}$ "modulo a history ϕ". Intuitively, for two labelled action refinements ρ and ρ' to be "equivalent modulo ϕ", we shall require that, for each action a, they map occurrences of a to congruent processes and whenever $F(a_i)$ and $F(a_j)$ are related to each other by ϕ, $\rho(F(a_i))$ and $\rho'(F(a_j))$ are equivalent with respect to some history $\varphi(a,i,j)$. For use in the proof of the refinement theorem, we shall collect the histories $\varphi(a,i,j)$, for $(a,i,j) \in \phi$, in a *ϕ-vector* σ. Formally:

Definition 4.3.7

(1) *For each* $h \in \mathcal{H}$, *an* h-vector σ *is a map* $\sigma : h \to \mathcal{H}$. *For each* $(a,i,j) \in h$, $\sigma^{(a,i,j)}$ *will denote the history associated with it by* σ.

(2) *Let* ρ *and* ρ' *be two labelled action refinements. Then, for each* $h \in \mathcal{H}$ *and* h-*vector* σ, $\rho =_{(h,\sigma)} \rho'$ *iff for each* $a \in V(\Lambda)$,

(a) $\rho(a_i) =_\emptyset \rho'(a_j)$, *for all* i, j,

(b) *for each* $i \in dom(h_a)$, $\rho(S(a_i)) \approx_{\sigma(a,i,j)} \rho'(S(a_j))$, *where* $j = h_a(i)$.

(3) *Let* $c \in \mathbf{L}\mathcal{S}^\Gamma$ *and* ρ *be a labelled action refinement. Then* ρ *is* compatible *with* c *iff* $c\rho \in \mathbf{L}\mathcal{S}^\Gamma$.

The following fact is a key consequence of clause (a) of the definition of labelled action refinement and gives the labelled version of Fact 4.2.4 .

Fact 4.3.2 *Let* π *be a labelled process and* ρ *be a labelled action refinement. Then* $\pi\surd$ *iff* $\pi\rho\surd$.

As pointed out above, the proof of the refinement theorem will make an essential use of several composition and decomposition results which relate the moves of a labelled term of the form $c\rho$ to those of c and of the components of ρ. We shall now present a series of results which analyse the origin of the moves of labelled states of the form $c\rho$ in terms of moves of c and of the components of ρ. The following lemma and theorem will be useful in the decomposition results presented below.

Lemma 4.3.5 (From Start to Complete Moves) *Let* $c \in \mathbf{L}\mathcal{S}^\Gamma$. *Then:*

(1) $c \overset{S(a_i)}{\longmapsto} d$ *implies* $c \overset{a_i}{\longmapsto} c'$ *for some* c' *such that* $c' =_\alpha d\iota[F(a_i) \to nil]$;

(2) $c \overset{S(a_i)}{\longmapsto} c_1 \overset{S(\bar{a}_j)}{\longmapsto} d$ *implies* $c \overset{\tau}{\longmapsto} c'$ *for some* c' *such that* $c' =_\alpha d\iota[F(a_i) \to nil, F(\bar{a}_j) \to nil]$.

Proof: We prove the two statements separately.

(1) By induction on the proof of the derivation $c \overset{S(a_i)}{\longmapsto} d$. The proof proceeds by a case analysis on the last rule used in the proof of $c \overset{S(a_i)}{\longmapsto} d$ and we shall examine only one of the cases.

- $c \equiv c_1 \backslash \alpha \overset{S(a_i)}{\longmapsto} d_1 \backslash \alpha \equiv d$ because $c_1 \overset{S(a_i)}{\longmapsto} d_1$ and $a \neq \alpha, \bar{\alpha}$. In this case, we may apply the inductive hypothesis to $c_1 \overset{S(a_i)}{\longmapsto} d_1$ to obtain that $c_1 \overset{a_i}{\longmapsto} c_1' =_\alpha d_1\iota[F(a_i) \to nil]$, for some c_1'. By the operational semantics, $c \equiv c_1 \backslash \alpha \overset{a_i}{\longmapsto} c_1' \backslash \alpha$. We shall now show that

$$(d_1 \backslash \alpha)\iota[F(a_i) \to nil] \quad =_\alpha \quad c_1' \backslash \alpha \tag{4.2}$$

In order to prove (4.2), let us recall that

$$(d_1 \backslash \alpha)\iota[F(a_i) \to nil] = (d_1(\iota[F(a_i) \to nil][\alpha \mapsto \beta])) \backslash \beta$$

where $\beta = choice(\mathit{new}\,\alpha\, d_1\, \iota[F(a_i) \to nil])$. We distinguish two cases.

(a) If $\alpha = \beta$ then

$$\begin{aligned} d_1\iota[F(a_i) \to nil][\alpha \mapsto \beta] &= d_1\iota[F(a_i) \to nil] \\ &=_\alpha c_1'. \end{aligned}$$

In this case the claim follows by the substitutivity of $=_\alpha$.

(b) If $\alpha \neq \beta$ then it is easy to see that α does not occur in the set of free channels of $d_1(\iota[F(a_i) \to nil][\alpha \mapsto \beta])$. Then, by the substitution lemma and the fact that $a \neq \alpha, \bar{\alpha}$,

$$\begin{aligned} (d_1(\iota[F(a_i) \to nil][\alpha \mapsto \beta]))\iota[\beta \mapsto \alpha] &= d_1(\iota[\beta \mapsto \alpha] \circ \iota[F(a_i) \to nil, \alpha \mapsto \beta]) \\ &= d_1\iota[F(a_i) \to nil] \\ &=_\alpha c_1' \end{aligned}$$

Claim (4.2) now follows by rule (α) of the definition of $=_\alpha$.

(2) By induction on the structure of c. We examine only one possibility, namely the one corresponding to the restriction operator.

- $c = \bar{c} \setminus \alpha$. Assume that $\bar{c} \setminus \alpha \overset{S(a_i)}{\longmapsto} c_1 \overset{S(\bar{a}_j)}{\longmapsto} d$. Then, by the operational semantics, this is because $\bar{c} \overset{S(a_i)}{\longmapsto} c' \overset{S(\bar{a}_j)}{\longmapsto} d'$, $d = d' \setminus \alpha$ and $a \neq \alpha, \bar{\alpha}$.

  By the inductive hypothesis, $\bar{c} \overset{\tau}{\longmapsto} c_2$ for some c_2 such that $c_2 =_\alpha d'\iota[F(a_i) \to nil, F(\bar{a}_j) \to nil]$. By the operational semantics, $\bar{c} \setminus \alpha \overset{\tau}{\longmapsto} c_2 \setminus \alpha$. We are thus left to prove that

$$c_2 \setminus \alpha \quad =_\alpha \quad (d' \setminus \alpha)\iota[F(a_i) \to nil, F(\bar{a}_j) \to nil]. \tag{4.3}$$

  In order to prove the above statement, we note, first of all, that

$$(d' \setminus \alpha)\iota[F(a_i) \to nil, F(\bar{a}_j) \to nil] = (d'\iota[F(a_i) \to nil, F(\bar{a}_j) \to nil][\alpha \mapsto \beta]) \setminus \beta$$

  where $\beta = choice(new\,\alpha\, d'\,\iota[F(a_i) \to nil, F(\bar{a}_j) \to nil])$. We distinguish two possibilities depending on whether $\alpha = \beta$ or not.

  If $\alpha = \beta$ then

$$\begin{aligned} d'(\iota[F(a_i) \to nil, F(\bar{a}_j) \to nil][\alpha \mapsto \beta]) &= d'\iota[F(a_i) \to nil, F(\bar{a}_j \to nil] \\ &=_\alpha c_2. \end{aligned}$$

  The claim the follows by the substitutivity of $=_\alpha$. Otherwise $\alpha \neq \beta$. In this case $\alpha \notin FC(d'\iota[F(a_i) \to nil, F(\bar{a}_j) \to nil][\alpha \mapsto \beta])$. Moreover, by the substitution lemma and the fact that $a \neq \alpha, \bar{\alpha}$,

$$\begin{aligned} (d'\iota[F(a_i) \to nil, F(\bar{a}_j) \to nil, \alpha \mapsto \beta])\iota[\beta \mapsto \alpha] &= d'\iota[F(a_i) \to nil, F(\bar{a}_j) \to nil] \\ &=_\alpha c_2. \end{aligned}$$

  The claim now follows from rule (α) of the definition of $=_\alpha$. □

The following theorem from [Sto88a] (Theorem 3.5, pages 321-322) states the normalizing effect of substitutions for α-congruent terms.

Theorem 4.3.3 (Sto88a) *Let $c, d \in \mathbf{L}\mathcal{S}^\Gamma$. Assume that $c =_\alpha d$. Then, for every action refinement ρ, $c\rho = d\rho$, where we use $=$ to denote syntactic identity.*

We shall now start the analysis of the origin of moves of labelled states of the form $c\rho$, where ρ is a labelled action refinement compatible with c. (See Definition 4.3.7 on page 161) The following theorem gives a decomposition of the moves of labelled processes of the form $\pi\rho$ in terms of those of π and of the components of ρ. In what follows, with abuse of notation, we shall use e to range over the set of *labelled subactions* $\{S(a_i), F(a_i) \mid a_i \in LAct(\Lambda)\} \cup \{\tau\}$.

Theorem 4.3.4 ("Whence Theorem" for Labelled Processes) *Let π be a labelled process and ρ be a labelled action refinement compatible with π. Then $\pi\rho \stackrel{e}{\longmapsto} \bar{c}$ implies:*

(1) *there exist $a_i \in LAct(\Lambda)$ and $c, x \in \mathbf{L}\mathcal{S}^\Gamma$ such that $\pi \stackrel{S(a_i)}{\longmapsto} c$, $\rho(a_i) \stackrel{e}{\longmapsto} x$ and $\bar{c} =_\alpha c\rho[F(a_i) \to x]$, or*

(2) *$e = \tau$ and there exist $a_i, b_j \in LAct(\Lambda)$, $\alpha \in V(\Lambda)$ and $c, d, x, y \in \mathbf{L}\mathcal{S}^\Gamma$ such that $\pi \stackrel{S(a_i)}{\longmapsto} c \stackrel{S(b_j)}{\longmapsto} d$, $\rho(a_i) \stackrel{\alpha_h}{\longmapsto} x$, $\rho(b_j) \stackrel{\bar{\alpha}_k}{\longmapsto} y$ and $\bar{c} =_\alpha d\rho[F(a_i) \to x, F(b_j) \to y]$, or*

(3) *$e = \tau$ and $\pi \stackrel{\tau}{\longmapsto} \pi'$, for some π' such that $\bar{c} =_\alpha \pi'\rho$.*

Proof: By induction on the structure of π. We shall only present the details of two selected cases.

- $\pi = \pi_1|\pi_2$. Assume that $(\pi_1|\pi_2)\rho = \pi_1\rho|\pi_2\rho \stackrel{e}{\longmapsto} \bar{c}$. By the operational semantics, there are three possibilities to examine:

  (A) $\pi_1\rho|\pi_2\rho \stackrel{e}{\longmapsto} \bar{c}$ because $\pi_1\rho \stackrel{e}{\longmapsto} \bar{c}_1$ and $\bar{c} = \bar{c}_1|\pi_2\rho$, or

  (B) $\pi_1\rho|\pi_2\rho \stackrel{e}{\longmapsto} \bar{c}$ because $\pi_2\rho \stackrel{e}{\longmapsto} \bar{c}_2$ and $\bar{c} = \pi_1\rho|\bar{c}_2$, or

  (C) $\pi_1\rho|\pi_2\rho \stackrel{\tau}{\longmapsto} \bar{c}$ because $\pi_1\rho \stackrel{\alpha_h}{\longmapsto} \bar{c}_1$, $\pi_2\rho \stackrel{\bar{\alpha}_k}{\longmapsto} \bar{c}_2$ and $\bar{c} = \bar{c}_1|\bar{c}_2$.

  We examine each possibility in turn.

  (A) In this case, we may apply the inductive hypothesis to $\pi_1\rho \stackrel{e}{\longmapsto} \bar{c}_1$ to obtain that

  (A.1) there exist $a_i \in LAct(\Lambda)$ and $c, x \in \mathbf{L}\mathcal{S}^\Gamma$ such that $\pi_1 \stackrel{S(a_i)}{\longmapsto} c$, $\rho(a_i) \stackrel{e}{\longmapsto} x$ and $\bar{c}_1 =_\alpha c\rho[F(a_i) \to x]$, or

  (A.2) $e = \tau$ and there exist $a_i, b_j \in LAct(\Lambda)$, $\alpha \in V(\Lambda)$ and $c, d, x, y \in \mathbf{L}\mathcal{S}^\Gamma$ such that $\pi_1 \stackrel{S(a_i)}{\longmapsto} c \stackrel{S(b_j)}{\longmapsto} d$, $\rho(a_i) \stackrel{\alpha_h}{\longmapsto} x$, $\rho(b_j) \stackrel{\bar{\alpha}_k}{\longmapsto} y$ and $\bar{c}_1 =_\alpha d\rho[F(a_i) \to x, F(b_j) \to y]$, or

(A.3) $e = \tau$ and $\pi_1 \overset{\tau}{\longmapsto} \pi_1'$, for some π_1' such that $\bar{c}_1 =_\alpha \pi_1'\rho$.

If (A.1) holds then, by the operational semantics, $\pi_1|\pi_2 \overset{S(a_i)}{\longmapsto} c|\pi_2$. Moreover,

$$\begin{aligned}(c|\pi_2)\rho[F(a_i) \to x] &= (c\rho[F(a_i) \to x])|(\pi_2\rho[F(a_i) \to x]) \\ &=_\alpha \bar{c}_1|(\pi_2\rho[F(a_i) \to x]).\end{aligned}$$

Now, as $F(a_i)$ does not occur in π_2, it is easy to see that $\pi_2\rho[F(a_i) \to x] = \pi_2\rho$. Thus we have that $(c|\pi_2)\rho[F(a_i) \to x] =_\alpha \bar{c}_1|\pi_2\rho = \bar{c}$. Clause (1) of the statement of the theorem is then met.

If (A.2) holds then, by the operational semantics, $\pi_1|\pi_2 \overset{S(a_i)}{\longmapsto} c|\pi_2 \overset{S(b_j)}{\longmapsto} d|\pi_2$. Moreover,

$$\begin{aligned}&(d|\pi_2)\rho[F(a_i) \to x, F(b_j) \to y] &&= \\ &(d\rho[F(a_i) \to x, F(b_j) \to y])|(\pi_2\rho[F(a_i) \to x, F(b_j) \to y]) &&=_\alpha \\ &\bar{c}_1|(\pi_2\rho[F(a_i) \to x, F(b_j) \to y]).\end{aligned}$$

Now, as $F(a_i)$ and $F(b_j)$ do not occur in π_2, it follows that $\pi_2\rho[F(a_i) \to x, F(b_j) \to y] = \pi_2\rho$. Hence $(d|\pi_2)\rho[F(a_i) \to x, F(b_j) \to y] =_\alpha \bar{c}_1|\pi_2\rho = \bar{c}$. Clause (2) of the statement of the theorem is then met.

If (A.3) holds then, by the operational semantics, $\pi_1|\pi_2 \overset{\tau}{\longmapsto} \pi_1'|\pi_2$. Moreover,

$$(\pi_1'|\pi_2)\rho = (\pi_1'\rho|\pi_2\rho) =_\alpha \bar{c}_1|\pi_2\rho.$$

Clause (3) of the statement of the theorem is then met. This completes the verification of case (A).

(B) Symmetrical to the one above.

(C) By the inductive hypothesis, $\pi_1\rho \overset{\alpha_h}{\longmapsto} \bar{c}_1$ implies that $\pi_1 \overset{S(a_i)}{\longmapsto} c_1$, $\rho(a_i) \overset{\alpha_h}{\longmapsto} x$ and $\bar{c}_1 =_\alpha c_1\rho[F(a_i) \to x]$, for some a_i, c_1 and x. Again by the inductive hypothsis, $\pi_2\rho \overset{\bar{\alpha}_k}{\longmapsto} \bar{c}_2$ implies that $\pi_2 \overset{S(b_i)}{\longmapsto} c_2$, $\rho(b_j) \overset{\bar{\alpha}_k}{\longmapsto} y$ and $\bar{c}_2 =_\alpha c_2\rho[F(b_j) \to y]$, for some b_j, c_2 and y. By the operational semantics, $\pi_1|\pi_2 \overset{S(a_i)}{\longmapsto} c_1|\pi_2 \overset{S(b_j)}{\longmapsto} c_1|c_2$. Moreover,

$$\begin{aligned}(c_1|c_2)\rho[F(a_i) \to x, F(b_j) \to y] &= \\ (c_1\rho[F(a_i) \to x, F(b_j) \to y])|(c_2\rho[F(a_i) \to x, F(b_j) \to y]) &= \\ (c_1\rho[F(a_i) \to x])|(c_2\rho[F(b_j) \to y]),&\end{aligned}$$

because $F(a_i)$ does not occur in c_2 and $F(b_j)$ does not occur in c_1. By the inductive hypothesis, we then have that

$$(c_1|c_2)\rho[F(a_i) \to x, F(b_j) \to y] =_\alpha \bar{c}_1|\bar{c}_2.$$

Clause (3) of the statement of the theorem is then met.

- $\pi = \pi_1 \setminus \alpha$. Assume that $(\pi_1 \setminus \alpha)\rho \overset{e}{\longmapsto} \bar{c}$. First of all, note that

$$(\pi_1 \setminus \alpha)\rho = (\pi_1\rho[\alpha \mapsto \beta]) \setminus \beta,$$

where $\beta = \mathit{choice}(\mathit{new}\,\alpha\,\pi_1\,\rho)$. By the operational semantics, we then have that $\pi_1\rho[\alpha \mapsto \beta] \overset{e}{\longmapsto} \bar{c}_1$, $e \neq \beta, \bar{\beta}$ and $\bar{c} = \bar{c}_1 \setminus \beta$, for some $\bar{c}_1$. By the inductive hypothesis, $\pi_1\rho[\alpha \mapsto \beta] \overset{e}{\longmapsto} \bar{c}_1$ implies:

(A) there exist $a_i \in LAct(\Lambda)$ and $c, x \in \mathbf{L}S^\Gamma$ such that $\pi_1 \overset{S(a_i)}{\longmapsto} c$, $\rho[\alpha \mapsto \beta](a_i) \overset{e}{\longmapsto} x$ and $\bar{c}_1 =_\alpha c\rho[\alpha \mapsto \beta][F(a_i) \rightarrow x]$, or

(B) $e = \tau$ and there exist $a_i, b_j \in LAct(\Lambda)$, $\gamma \in V(\Lambda)$ and $c, d, x, y \in \mathbf{L}S^\Gamma$ such that $\pi_1 \overset{S(a_i)}{\longmapsto} c \overset{S(b_j)}{\longmapsto} d$, $\rho[\alpha \mapsto \beta](a_i) \overset{\gamma_h}{\longmapsto} x$, $\rho[\alpha \mapsto \beta](b_j) \overset{\bar{\gamma}_k}{\longmapsto} y$ and $\bar{c}_1 =_\alpha d\rho[\alpha \mapsto \beta][F(a_i) \rightarrow x, F(b_j) \rightarrow y]$, or

(C) $e = \tau$ and $\pi_1 \overset{\tau}{\longmapsto} \pi_1'$, for some π_1' such that $\bar{c}_1 =_\alpha \pi_1'\rho[\alpha \mapsto \beta]$.

We proceed by examining the three possibilities separately.

(A) As $e \neq \beta, \bar{\beta}$ and $\rho[\alpha \mapsto \beta](\alpha_i) = \beta_i$, for all i, it must be the case that $a \neq \alpha, \bar{\alpha}$. Thus α *admits* $S(a_i)$ and, by the operational semantics, we have that $\pi_1 \setminus \alpha \overset{S(a_i)}{\longmapsto} c \setminus \alpha$. Moreover, as $a \neq \alpha, \bar{\alpha}$,

$$\rho(a_i) = \rho[\alpha \mapsto \beta](a_i) \overset{e}{\longmapsto} x.$$

We shall now show that

$$\bar{c}_1 \setminus \beta \quad =_\alpha \quad (c \setminus \alpha)\rho[F(a_i) \rightarrow x], \tag{4.4}$$

where $\bar{c}_1 =_\alpha c\rho[\alpha \mapsto \beta][F(a_i) \rightarrow x]$. In order to prove (4.4), note that

$$(c \setminus \alpha)\rho[F(a_i) \rightarrow x] = (c\rho[F(a_i) \rightarrow x][\alpha \mapsto \beta']) \setminus \beta',$$

where $\beta' = \mathit{choice}(\mathit{new}\,\alpha\,c\,\rho[F(a_i) \rightarrow x])$. We proceed by distinguishing two cases, depending on whether $\beta = \beta'$ or not.

If $\beta = \beta'$ then

$$\begin{aligned} c\rho[F(a_i) \rightarrow x][\alpha \mapsto \beta] &= c\rho[\alpha \mapsto \beta][F(a_i) \rightarrow x] \quad \text{as } a \neq \alpha, \bar{\alpha} \\ &=_\alpha \bar{c}_1. \end{aligned}$$

The claim then follows by the substitutivity of $=_\alpha$.

Assume now that $\beta \neq \beta'$. Then it is easy to see that $\beta \in \mathit{new}\,\alpha\,c\,\rho[F(a_i) \rightarrow x]$. This implies that

$$\beta \notin FC(c\rho[F(a_i) \rightarrow x][\alpha \mapsto \beta']).$$

By the substitution lemma, we then have that

$$\begin{array}{rcl}(c\rho[F(a_i) \to x][\alpha \mapsto \beta'])\iota[\beta' \mapsto \beta] & = & c(\iota[\beta' \mapsto \beta] \circ \rho[F(a_i) \to x][\alpha \mapsto \beta']) \\ & = & c\rho[F(a_i) \to x][\alpha \mapsto \beta] \\ & \stackrel{\star}{=} & c\rho[\alpha \mapsto \beta][F(a_i) \to x] \\ & =_\alpha & \bar{c}_1 .\end{array}$$

Equality $\stackrel{\star}{=}$ is justified by the fact that $a \neq \alpha, \bar{\alpha}$. The claim then follows by rule (α) of the definition of $=_\alpha$. By $\pi_1 \setminus \alpha \stackrel{S(a_i)}{\longmapsto} c \setminus \alpha$, $\rho(a_i) \stackrel{e}{\longmapsto} x$ and $\bar{c}_1 \setminus \beta =_\alpha (c \setminus \alpha)\rho[F(a_i) \to x]$, we now have that clause (1) of the statement of the theorem is met.

(B) First of all, note that, because $\beta \in new\,\alpha\,\pi_1\,\rho$, $\rho[\alpha \mapsto \beta](a_i) \stackrel{\gamma_h}{\longmapsto} x$ and $\rho[\alpha \mapsto \beta](b_j) \stackrel{\bar{\gamma}_k}{\longmapsto} y$ imply that

(a) $a, b \neq \alpha, \bar{\alpha}$ or

(b) by symmetry, we may assume that, wlog, $a = \alpha$ and $b = \bar{\alpha}$.

We examine the two possibilities separately.

(a) Assume that $a, b \neq \alpha, \bar{\alpha}$. Then, by the operational semantics, $\pi_1 \setminus \alpha \stackrel{S(a_i)}{\longmapsto} c \setminus \alpha \stackrel{S(b_j)}{\longmapsto} d \setminus \alpha$. Moreover we have that $\rho(a_i) = \rho[\alpha \mapsto \beta](a_i) \stackrel{\gamma_h}{\longmapsto} x$ and $\rho(b_j) = \rho[\alpha \mapsto \beta](b_j) \stackrel{\bar{\gamma}_k}{\longmapsto} y$. We shall now show that

$$(d \setminus \alpha)\rho[F(a_i) \to x, F(b_j) \to y] \;=\; \bar{c}_1 \setminus \beta, \tag{4.5}$$

where $\bar{c}_1 =_\alpha d\rho[\alpha \mapsto \beta][F(a_i) \to x, F(b_j) \to y]$. By (4.5), we will then have that clause (2) of the statement of the theorem is met. First of all, note that

$$(d \setminus \alpha)\rho[F(a_i) \to x, F(b_j) \to y] = (d\rho[F(a_i) \to x, F(b_j) \to y][\alpha \mapsto \beta']) \setminus \beta',$$

where $\beta' = choice(new\,\alpha\,d\,\rho[F(a_i) \to x, F(b_j) \to y])$. We proceed by distinguishing two cases depending on whether $\beta = \beta'$ or not.

If $\beta = \beta'$ then we have that

$$\begin{array}{rcl}(d\rho[F(a_i) \to x, F(b_j) \to y][\alpha \mapsto \beta]) \setminus \beta & \stackrel{\star}{=} & \\ (d\rho[\alpha \mapsto \beta][F(a_i) \to x, F(b_j) \to y]) \setminus \beta & =_\alpha & \bar{c}_1 \setminus \beta .\end{array}$$

Equality $\stackrel{\star}{=}$ follows because $a, b \neq \alpha, \bar{\alpha}$. Assume now $\beta \neq \beta'$. Then it is easy to see that

$$\beta \notin FC(d\rho[F(a_i) \to x, F(b_j) \to y][\alpha \mapsto \beta']).$$

By the substitution lemma, we then have that

$$\begin{array}{rcl}(d\rho[F(a_i) \to x, F(b_j) \to y][\alpha \mapsto \beta'])\iota[\beta' \mapsto \beta] & = & \\ d(\iota[\beta' \mapsto \beta] \circ \rho[F(a_i) \to x, F(b_j) \to y][\alpha \mapsto \beta']) & = & \\ d\rho[F(a_i) \to x, F(b_j) \to y][\alpha \mapsto \beta] & \stackrel{\star}{=} & \\ d\rho[\alpha \mapsto \beta][F(a_i) \to x, F(b_j) \to y] & =_\alpha & \bar{c}_1 .\end{array}$$

Equality $\stackrel{\star}{=}$ follows because $a, b \neq \alpha, \bar{\alpha}$. (4.5) now follows by rule (α) of the definition of $=_\alpha$. This completes the proof for sub-case (a).

(b) Assume that $a = \alpha_i$ and $b_j = \bar{\alpha}_j$. Note that we then have that $\pi_1 \stackrel{S(\alpha_i)}{\longmapsto} c \stackrel{S(\bar{\alpha}_j)}{\longmapsto} d$, $\rho[\alpha \mapsto \beta](\alpha_i) = \beta_i \stackrel{\beta_i}{\longmapsto} nil = x$ and $\rho[\alpha \mapsto \beta](\bar{\alpha}_j) = \bar{\beta}_j \stackrel{\bar{\beta}_j}{\longmapsto} nil = y$. Moreover,

$$\bar{c}_1 =_\alpha d\rho[\alpha \mapsto \beta][F(\alpha_i) \to nil, F(\bar{\alpha}_j) \to nil].$$

We shall prove that clause (3) of the statement of the theorem is then met. First of all, note that, by Lemma 4.3.5, $\pi_1 \stackrel{S(\alpha_i)}{\longmapsto} c \stackrel{S(\bar{\alpha}_j)}{\longmapsto} d$ implies

$$\pi_1 \stackrel{\tau}{\longmapsto} \pi_1' =_\alpha d\iota[F(\alpha_i) \to nil, F(\bar{\alpha}_j) \to nil],$$

for some π_1'. By the operational semantics, $\pi_1 \stackrel{\tau}{\longmapsto} \pi_1'$ implies

$$\pi_1 \setminus \alpha \stackrel{\tau}{\longmapsto} \pi_1' \setminus \alpha =_\alpha (d\iota[F(\alpha_i) \to nil, F(\bar{\alpha}_j) \to nil]) \setminus \alpha.$$

In order to show that clause (3) of the statement of the theorem is met, it is sufficient to show that

$$\bar{c}_1 \setminus \beta \quad =_\alpha \quad (\pi_1' \setminus \alpha)\rho. \tag{4.6}$$

We shall now prove (4.6). First of all, note that, for $\beta' = \mathit{choice}(\mathit{new}\,\alpha\,\pi_1'\,\rho)$,

$$\begin{aligned}(\pi_1' \setminus \alpha)\rho &= (\pi_1'\rho[\alpha \mapsto \beta']) \setminus \beta' \\ &= ((d\iota[F(\alpha_i) \to nil, F(\bar{\alpha}_j) \to nil])\rho[\alpha \mapsto \beta']) \setminus \beta',\end{aligned}$$

as, by Theorem 4.3.3, applying $\rho[\alpha \mapsto \beta']$ to α-congruent terms yields *syntactically equal* terms. The proof of (4.6) now proceeds by an examination of two cases, depending on whether $\beta = \beta'$ or not.

If $\beta = \beta'$ then we have that

$$\begin{aligned}&((d\iota[F(\alpha_i) \to nil, F(\bar{\alpha}_j) \to nil])\rho[\alpha \mapsto \beta]) \setminus \beta &&= \\ &(d\rho[\alpha \mapsto \beta][F(\alpha_i) \to nil, F(\bar{\alpha}_j) \to nil]) \setminus \beta &&=_\alpha \bar{c}_1.\end{aligned}$$

Assume now that $\beta \neq \beta'$. Then it is easy to see that $\beta \notin FC((d\iota[F(\alpha_i) \to nil, F(\bar{\alpha}_j) \to nil])\rho[\alpha \mapsto \beta'])$. By the substitution lemma, we then have that

$$\begin{aligned}&((d\iota[F(\alpha_i) \to nil, F(\bar{\alpha}_j) \to nil])\rho[\alpha \mapsto \beta'])\iota[\beta' \mapsto \beta] &&= \\ &(d\iota[F(\alpha_i) \to nil, F(\bar{\alpha}_j) \to nil])(\iota[\beta' \mapsto \beta] \circ \rho[\alpha \mapsto \beta']) &&= \\ &(d\iota[F(\alpha_i) \to nil, F(\bar{\alpha}_j) \to nil])\rho[\alpha \mapsto \beta] &&= \\ &d\rho[\alpha \mapsto \beta][F(\alpha_i) \to nil, F(\bar{\alpha}_j) \to nil] &&=_\alpha \bar{c}_1.\end{aligned}$$

Thus (4.6) follows by rule (α) of the definition of $=_\alpha$. This implies that clause (3) of the statement of the theorem is met.

The verification of case (B) is thus complete.

(C) In this case, by the operational semantics, we have that $\pi_1 \setminus \alpha \overset{\tau}{\longmapsto} \pi'_1 \setminus \alpha$. In order to prove that clause (3) of the statement of the theorem is met, it is sufficient to show that

$$\bar{c}_1 \setminus \beta =_\alpha (\pi'_1 \setminus \alpha)\rho.$$

This verification follows the same pattern of the similar ones given above and is thus omitted. □

We shall now work towards a more general "whence theorem" for configurations of the form $c\rho$. Such a result will allow us to decompose moves of a configuration of the form $c\rho$ into moves of c and of the components of ρ. In order to deal with labelled action refinements mapping states of the form $F(a_i)$, or *places*, to terminated processes, we shall adapt the definition of the map η, defined for the simpler language in Chapter 3 (page 100), to the richer setting we are now working in. For each labelled state c, $\eta(c)$ will be a nonempty collection of pairs of the form (X, d), where $X \subseteq \{F(a_i) \mid a_i \in LAct(\Lambda)\}$. Intuitively, if $(X, d) \in \eta(c)$ then, for every labelled action refinement ρ mapping each element of X to a terminated process, $c\rho$ "behaves like" $d\rho$. The presence of condition (CR) on states of the form $c \setminus \alpha$ will allow us to give a neat compositional definition of the map η.

Definition 4.3.8 *The map η is defined by structural induction on c as follows:*

$$\begin{array}{lll}
\textbf{(i)}\ \ \eta(\pi) & = & \{(\emptyset, \pi)\} \\
\textbf{(ii)}\ \ \eta(F(a_i)) & = & \{(\{F(a_i)\}, nil)\} \\
\textbf{(iii)}\ \ \eta(c;\pi) & = & \{(X, c';\pi) \mid (X, c') \in \eta(c)\} \\
\textbf{(iv)}\ \ \eta(c|d) & = & \{(X, c'|d) \mid (X, c') \in \eta(c)\} \cup \{(Y, c|d') \mid (Y, d') \in \eta(d)\} \cup \\
 & & \{(X \cup Y, c'|d') \mid (X, c') \in \eta(c) \text{ and } (Y, d') \in \eta(d)\} \\
\textbf{(v)}\ \ \eta(c \setminus \alpha) & = & \{(X, c' \setminus \alpha) \mid (X, c') \in \eta(c)\}.
\end{array}$$

We shall now study some basic properties of η which will find application in what follows.

Lemma 4.3.6 *Let $c \in \mathbf{L}\mathcal{S}^\Gamma$. Then:*

(1) *$(X, d) \in \eta(c)$ implies $X \subseteq \{F(a_i) \mid a_i \in LAct(\Lambda)\}$;*

(2) *there exists d such that $(\emptyset, d) \in \eta(c)$ iff c is a labelled process;*

(3) *$(X, d) \in \eta(c)$ and $\emptyset \neq Y \subseteq X$ imply $(Y, d') \in \eta(c)$ for some d'.*

Condition (CR) plays an important rôle in the proof of the following lemma, which states a useful interaction between the map η and the operational semantics for labelled states.

Lemma 4.3.7 *Let $c \in \mathbf{L}\mathcal{S}^\Gamma$ and assume that $(\{F(a_i)\}, d) \in \eta(c)$. Then $c \overset{F(a_i)}{\longmapsto} d$.*

Proof: By structural induction on c. We only give the case in which condition (CR) is used.

- $c = c_1 \backslash \alpha$. Assume that $(\{F(a_i)\}, d) \in \eta(c_1 \backslash \alpha)$. Then $d = d_1 \backslash \alpha$ and $(\{F(a_i)\}, d_1) \in \eta(c_1)$. By the inductive hypothesis, $c_1 \overset{F(a_i)}{\longmapsto} d_1$. By condition (CR), $c_1 \setminus \alpha \in \mathbf{L}\mathcal{S}^\Gamma$ implies that $a \neq \alpha, \bar{\alpha}$. Hence α *admits* $F(a_i)$ and, by the operational semantics, $c_1 \setminus \alpha \overset{F(a_i)}{\longmapsto} d_1 \setminus \alpha = d$.
  □

We shall now investigate the relationships between the map η and the function *new*. More precisely, we shall prove that whenever $(X, d) \in \eta(c)$ and ρ is a labelled action refinement such that $\rho(X)\surd$ then, for each channel name α, $new\,\alpha\,c\,\rho = new\,\alpha\,d\,\rho$.

Proposition 4.3.8 *Let $c \in \mathbf{L}\mathcal{S}^\Gamma$ and ρ be a labelled action refinement. Then $(X, d) \in \eta(c)$ and $\rho(X)\surd$ imply that $new\,\alpha\,c\,\rho = new\,\alpha\,d\,\rho$, for all $\alpha \in \Lambda$.*

Proof: By induction on the structure of c. We shall only examine three of the possible forms of c and leave the remaining ones to the reader.

- $c = F(a_i)$. By the definition of η, (X, d) must be $(\{F(a_i)\}, nil)$. Moreover we have that $\rho(F(a_i))\surd$. By the definition of *new*, for all $\alpha \in \Lambda$,

$$\begin{aligned} new\,\alpha\,F(a_i)\,\rho &= \Lambda \\ &= new\,\alpha\,nil\,\rho, \end{aligned}$$

  because $\rho(F(a_i))\surd$ implies that $FC(\rho(F(a_i))) = \emptyset$.

- $c = c_1; \pi$. Assume that $(X, d) \in \eta(c_1; \pi)$ and $\rho(X)\surd$. By the definition of η, it must be the case that $(X, c_1') \in \eta(c_1)$ and $d = c_1'; \pi$, for some c_1'. Now, by the definition of *new*, it is easy to see that

$$\begin{aligned} new\,\alpha\,c_1; \pi\,\rho &= new\,\alpha\,c_1\,\rho \cap new\,\alpha\,\pi\,\rho \\ &= new\,\alpha\,c_1'\,\rho \cap new\,\alpha\,\pi\,\rho \quad \text{by induction} \\ &= new\,\alpha\,c_1'; \pi\,\rho. \end{aligned}$$

- $c = c_1 \setminus \beta$. Assume that $(X, d) \in \eta(c_1 \setminus \beta)$ and $\rho(X)\surd$. Then it must be the case that, for some c_1', $(X, c_1') \in \eta(c_1)$ and $d = c_1' \setminus \beta$. Let $\alpha \in \Lambda$. We proceed by distinguishing two possibilities, depending on whether $\alpha = \beta$ or $\alpha \neq \beta$.

Assume, first of all, that $\alpha = \beta$. Then it is easy to see that

$$\begin{aligned} new\,\beta\,(c_1 \setminus \beta)\,\rho &= new\,\beta\,c_1\,\rho \\ &= new\,\beta\,c_1'\,\rho \qquad \text{by induction} \\ &= new\,\beta\,(c_1' \setminus \beta)\,\rho. \end{aligned}$$

If $\alpha \neq \beta$ then, by the definition of *new* it is easy to see that

$$\begin{aligned} new\,\alpha\,(c_1 \setminus \beta)\,rho &= (new\,\alpha\,c_1\,\rho) - \bigcup\{FC(\rho(e)) \mid e \in Sym(c), \nu(e) = \beta\} \\ &= (new\,\alpha\,c_1'\,\rho) - \bigcup\{FC(\rho(e)) \mid e \in Sym(c_1), \nu(e) = \beta\} \quad \text{by induction} \\ &= new\,\alpha\,(c_1' \setminus \beta)\,\rho. \end{aligned}$$

□

The above proposition will be very useful in proving that, for all $c, d \in \mathbf{L}\mathcal{S}^\Gamma$, whenever $(X, d) \in \eta(c)$ and $\rho(X)\surd$, $c\rho$ "behaves like" $d\rho$. In fact, we shall now show that, under the above assumptions, there is a close syntactic relation between $c\rho$ and $d\rho$; namely, $c\rho$ and $d\rho$ are syntactically the "same process" up to renaming of successfully terminated processes.

Definition 4.3.9 $=_\surd$ *denotes the least congruence over* $\mathbf{L}\mathcal{S}^\Gamma$ *which satisfies the following axioms*

$$\begin{array}{lll} \textbf{(TER1)} & nil; nil & = nil \\ \textbf{(TER2)} & nil | nil & = nil \\ \textbf{(TER3)} & nil + nil & = nil \\ \textbf{(TER4)} & nil \setminus \alpha & = nil. \end{array}$$

The following lemma is easily established by induction on $\surd$.

Lemma 4.3.8 *For each* $c \in \mathbf{L}\mathcal{S}^\Gamma$, $c\surd$ *implies* $c =_\surd nil$.

Using $=_\surd$, we are now able to formalize the syntactic relationship between $c\rho$ and $d\rho$ whenever $(X, d) \in \eta(c)$ and $\rho(X)\surd$.

Proposition 4.3.9 *Let* $c \in \mathbf{L}\mathcal{S}^\Gamma$ *and* ρ *be a labelled action refinement compatible with* c. *Assume that* $(X, d) \in \eta(c)$ *and* $\rho(X)\surd$. *Then the following statements hold:*

(1) $id(c\rho) = id(d\rho)$ *and*

(2) $c\rho =_\surd d\rho$.

Proof: Both statements can be shown by structural induction on c. We only give the proof of the case $c = c_1 \setminus \alpha$ for statement (2).

- $c = c_1 \setminus \alpha$. Assume that $(X, d) \in \eta(c_1 \setminus \alpha)$ and $\rho(X)\surd$. Then, by the definition of η, it must be the case that $d = c_1' \setminus \alpha$ and $(X, c_1') \in \eta(c_1)$, for some c_1'. Now, letting $\beta = \mathit{choice}(\mathit{new}\,\alpha\, c_1\, \rho)$,

$$\begin{array}{rcll}(c_1 \setminus \alpha)\rho & = & (c_1\rho[\alpha \mapsto \beta]) \setminus \beta & \\ & =_{\surd} & (c_1'\rho[\alpha \mapsto \beta]) \setminus \beta & \text{by the inductive hypothesis,}\end{array}$$

which is applicable because, as for each $F(a_i) \in X$ we have that $a \neq \alpha, \bar{\alpha}$, $\rho[\alpha \mapsto \beta](X) = \rho(X)\surd$. By Proposition 4.3.8, $\mathit{new}\,\alpha\,(c_1 \setminus \alpha)\,\rho = \mathit{new}\,\alpha\,(c_1' \setminus \alpha)\,\rho$. It is now easy to see that $(c_1'\rho[\alpha \mapsto \beta]) \setminus \beta = (c_1' \setminus \alpha)\rho$. □

Fact 4.3.3 *Let $c \in \mathbf{L}\mathcal{S}^{\Gamma}$ and ρ be a labelled action refinement compatible with c. Then the following statements hold:*

(1) *$c\rho\surd$ implies that, for all $(X, d) \in \eta(c)$, $\rho(X)\surd$ and $d\rho\surd$;*

(2) *$c\rho\surd$ and $c \stackrel{F(a_i)}{\longmapsto} d$ imply that $\rho(F(a_i))\surd$.*

Proof: By structural induction on c. Omitted. □

As we have seen, for each $(X, d) \in \eta(c)$ and labelled action refinement ρ such that $\rho(X)\surd$, the relationship between $d\rho$ and $c\rho$ can be expressed syntactically in terms of $=_{\surd}$. However, in what follows, we shall often need to use to use $=_{\surd}$ in conjuction with the relation of α-congruence and a behavioural characterization of $=_{\surd}$ would be of considerable help. We already know that $=_{\alpha}$ is a strong bisimulation over the LTS $\langle \mathbf{L}\mathcal{S}^{\Gamma}, \longmapsto, LAct_s(\Lambda), \surd \rangle$. As it may be expected, $=_{\surd}$ will also turn out to be a strong bisimulation and this will allow us to exploit standard properties of $\sim$ in reasoning about the syntactic relations $=_{\alpha}$ and $=_{\surd}$.

Proposition 4.3.10 *Let $c, d \in \mathbf{L}\mathcal{S}^{\Gamma}$ be such that $c =_{\surd} d$. Let $\lambda \in LAct(\Lambda) \cup LAct_s(\Lambda)$. Then $c \stackrel{\lambda}{\longmapsto} c'$ implies $d \stackrel{\lambda}{\longmapsto} d'$ for some d' such that $c' =_{\surd} d'$.*

Proof: By induction on the length of the proof of the derivation $c \stackrel{\lambda}{\longmapsto} c'$. The straightforward details are omitted. □

The following corollary is an immediate consequence of the above proposition.

Corollary 4.3.5 *Let $c, d \in \mathbf{L}\mathcal{S}^{\Gamma}$ be such that $c =_{\surd} d$. Then, for each $c' \in \mathbf{L}\mathcal{S}^{\Gamma}$ and $\lambda \in LAct(\Lambda) \cup LAct_s(\Lambda)$, $c \stackrel{\lambda}{\Longmapsto} c'$ implies $d \stackrel{\lambda}{\Longmapsto} d'$ for some d' such that $c' =_{\surd} d'$.*

Fact 4.3.4 *Let $c, d \in \mathbf{L}\mathcal{S}^\Gamma$. Then $c\surd$ and $c =_\surd d$ imply $d\surd$.*

Proof: By induction on $=_\surd$. Omitted. □

We now have all the technical material needed to prove that $=_\surd$ is indeed a strong bisimulation. The following proposition is an immediate corollary of Proposition 4.3.10 and Fact 4.3.4.

Proposition 4.3.11 $=_\surd$ *is a strong bisimulation over the LTS $\langle \mathbf{L}\mathcal{S}^\Gamma, \longmapsto, LAct_s(\Lambda), \surd \rangle$.*

We can now prove the promised "whence theorem" for moves of configurations of the form $c\rho$. Intuitively, if $c\rho \overset{e}{\longmapsto} \bar{c}$ then one of the following situations occurs:

- c is capable of starting action a_i and the component of ρ corresponding to a_i, $\rho(a_i)$, is capable of performing e. From that moment onwards, $F(a_i)$ acts as a place-holder for what remains to be executed of the process $\rho(a_i)$; or
- the move is due to one of the components of ρ of the form $\rho(F(a_i))$, or
- there is a pair (X, d) associated to c by η with the property that ρ maps all the places in X to successfully terminated processes and $d\rho$ is capable of performing e, or
- $e = \tau$ and the move is due to an internal move of c, or
- $e = \tau$ and the move is generated by a synchronization between different components of ρ. (Clauses (3), (6) and (7) in the statement of the theorem below cater for all the possible interactions of this kind)

Formally:

Theorem 4.3.5 ("Whence Theorem" for Configurations) *Let $c \in \mathbf{L}\mathcal{S}^\Gamma$, $e \in LAct_s(\Lambda)$ and ρ be a labelled action refinement compatible with c. Then $c\rho \overset{e}{\longmapsto} \bar{c}$ implies:*

(1) *there exist $c', x \in \mathbf{L}\mathcal{S}^\Gamma$ and $a_i \in LAct(\Lambda)$ such that $c \overset{S(a_i)}{\longmapsto} c'$, $\rho(a_i) \overset{e}{\longmapsto} x$ and $\bar{c} =_\alpha c'\rho[F(a_i) \to x]$, or*

(2) *there exists $(X, c') \in \eta(c)$ such that $X \neq \emptyset$, $\rho(X)\surd$ and $c'\rho \overset{e}{\longmapsto} \bar{c}$, or*

(3) *$e = \tau$ and there exist $a_i, b_j \in LAct(\Lambda)$, $\alpha \in \Lambda$ and $c', d, x, y \in \mathbf{L}\mathcal{S}^\Gamma$ such that $a_i \neq b_j$, $c \overset{S(a_i)}{\longmapsto} c' \overset{S(b_j)}{\longmapsto} d$, $\rho(a_i) \overset{\alpha_h}{\longmapsto} x$, $\rho(b_j) \overset{\bar{\alpha}_k}{\longmapsto} y$ and $\bar{c} =_\alpha d\rho[F(a_i) \to x, F(b_j) \to y]$, or*

(4) *there exist $a_i \in LAct(\Lambda)$ and $x \in \mathbf{L}\mathcal{S}^\Gamma$ such that $F(a_i)$ occurs in c, $\rho(F(a_i)) \overset{e}{\longmapsto} x$ and $\bar{c} =_\alpha c\rho[F(a_i) \to x]$, or*

(5) $e = \tau$ *and* $c \overset{\tau}{\longmapsto} c'$ *for some* c' *such that* $\bar{c} =_\alpha c'\rho$, *or*

(6) $e = \tau$ *and there exist* $a_i, b_j \in LAct(\Lambda)$, $a' \in V(\Lambda)$ *and* $x, y \in \mathbf{L}\mathcal{S}^\Gamma$ *such that* $c \overset{S(a_i)}{\longmapsto} c'$, $F(b_j)$ *occurs in* c, $\rho(a_i) \overset{a'_h}{\longmapsto} x$, $\rho(F(b_j)) \overset{\bar{a}'_k}{\longmapsto} y$ *and* $\bar{c} =_\alpha c'\rho[F(a_i) \to x, F(b_j) \to y]$, *or*

(7) $e = \tau$ *and there exist* $a_i, b_j \in LAct(\Lambda)$, $x, y \in \mathbf{L}\mathcal{S}^\Gamma$ *such that* $a_i \neq b_j$, $F(a_i)$ *and* $F(b_j)$ *occur in* c, $\rho(F(a_i)) \overset{\alpha_h}{\longmapsto} x$, $\rho(F(b_j)) \overset{\bar{\alpha}_k}{\longmapsto} y$ *and* $\bar{c} =_\alpha c\rho[F(a_i) \to x, F(b_j) \to y]$.

Proof: By structural induction on c. The proof is very similar to the one of Theorem 4.3.4. Its only new feature is the use of clause (2) in dealing with moves of configurations of the form $c\rho; \pi\rho$ in which $c\rho\surd$, but not $c\surd$. The details of the proof are omitted. □

The "whence theorems" proven above give us a complete picture of the possible origin of moves of labelled processes and states of the form $\pi\rho$ and $c\rho$, respectively. In what follows, we shall often need to make use of some kind of converse information. More precisely, we shall need results that allow us to derive moves of $c\rho$ from those of a configuration c and of the components of a labelled action refinement ρ. This information will be provided by the following "whither lemmas". An example of such a result would be:

$$c \overset{\tau}{\Longmapsto} d \text{ implies } c\rho \overset{\tau}{\Longmapsto} d\rho.$$

We shall show that this is indeed true if we work up to strong bisimulation equivalence. In the formulation and the proof of the whither lemmas we shall make extensive use of the fact that both $=_\alpha$ and $=_\surd$ are strong bisimulations over the LTS $\langle \mathbf{L}\mathcal{S}^\Gamma, \longmapsto, LAct_s(\Lambda), \surd \rangle$.

Lemma 4.3.9 *Let* $c \in \mathbf{L}\mathcal{S}^\Gamma$ *and* ρ *be a labelled action refinement compatible with* c. *Assume that* $c \overset{a_i}{\longmapsto} d$ *and* $\rho(a_i) \overset{\sigma}{\Longmapsto} x\surd$, *for some* $\sigma \in LAct(\Lambda)^+ \cup \{\tau\}$. *Then* $c\rho \overset{\sigma}{\Longmapsto} \sim d\rho$.

Proof: By induction on the length of the proof of the derivation $c \overset{a_i}{\longmapsto} d$. We proceed by a case analysis on the last rule used in the proof of the derivation and only examine two of the possible cases of the proof, leaving the remaining ones to the reader.

- $c = a_i \overset{a_i}{\longmapsto} nil = d$. Then $c\rho = \rho(a_i) \overset{\sigma}{\Longmapsto} x \sim nil$, because $x\surd$.
- $c = c_1 \setminus \beta \overset{a_i}{\longmapsto} d_1 \setminus \beta$ because $c_1 \overset{a_i}{\longmapsto} d_1$ and $a \neq \beta, \bar{\beta}$. First of all, let us note that

$$(c_1 \setminus \beta)\rho = (c_1\rho[\beta \mapsto \alpha]) \setminus \alpha,$$

where $\alpha = choice(new\,\beta\, c_1\, \rho)$. In particular, α_h and $\bar{\alpha}_k$ do *not* appear in σ for all h, k. Note moreover that, as $a \neq \beta, \bar{\beta}$, $\rho[\beta \mapsto \alpha](a_i) = \rho(a_i) \overset{\sigma}{\Longmapsto} x\surd$. We may now apply the inductive hypothesis to $c_1 \overset{a_i}{\longmapsto} d_1$ and $\rho[\beta \mapsto \alpha](a_i) \overset{\sigma}{\Longmapsto} x\surd$ to obtain

$$c_1\rho[\beta \mapsto \alpha] \overset{\sigma}{\Longmapsto} y \sim d_1\rho[\beta \mapsto \alpha],$$

for some y. By the operational semantics and the substitutivity of $\sim$, recalling the fact that α *admits* σ, we obtain that

$$(c_1 \setminus \beta)\rho = (c_1\rho[\beta \mapsto \alpha]) \setminus \alpha \overset{\sigma}{\Longmapsto} y \setminus \alpha \sim (d_1\rho[\beta \mapsto \alpha]) \setminus \alpha.$$

As $\alpha \in new\,\beta\, c_1\, \rho$, we have that $\alpha \in new\,\beta\, d_1\, \rho$. By the labelled version of Fact 4.2.2, we then have that $(d_1 \setminus \beta)\rho =_\alpha (d_1\rho[\beta \mapsto \alpha]) \setminus \alpha$. The claim now follows by the transitivity of $\sim$. □

The above lemma, together with the fact that labelled action refinements satisfy axiom (Com-Pres), will be used to provide a converse to clause (5) of Theorem 4.3.5. This lemma may be seen as the "labelled" counterpart of Lemma 4.2.5.(2) on page 147.

Lemma 4.3.10 *Let $c \in \mathbf{L}\mathcal{S}^\Gamma$ and ρ be a labelled action refinement compatible with c. Then $c \overset{\tau}{\Longmapsto} d$ implies $c\rho \overset{\tau}{\Longmapsto} \sim d\rho$.*

Proof: By induction on the length of the derivation $c \overset{\tau}{\Longmapsto} d$. The details are very similar to the ones in Lemma 4.2.5 and the reader is referred to that proof for more information. □

The following is an immediate corollary of the previous result.

Corollary 4.3.6 *Let $c \in \mathbf{L}\mathcal{S}^\Gamma$ and ρ be a labelled action refinement compatible with c. Then $c \overset{\varepsilon}{\Longmapsto} d$ implies $c\rho \overset{\varepsilon}{\Longmapsto} \sim d\rho$.*

The following lemma provides the converse to clause (1) of Theorem 4.3.5.

Lemma 4.3.11 *Let $c \in \mathbf{L}\mathcal{S}^\Gamma$ and ρ be a labelled action refinement compatible with c. Then $c \overset{S(a_i)}{\Longmapsto} c'$ and $\rho(a_i) \overset{e}{\Longmapsto} x$ imply $c\rho \overset{e}{\Longmapsto} \sim c'\rho[F(a_i) \rightarrow x]$.*

Proof: By induction on the length of the derivation $c \overset{S(a_i)}{\Longmapsto} c'$. The straightforward details are left to the reader. □

In the proof of the following lemma, which provides a converse to clause (2) of Theorem 4.3.5, we shall make use of Corollary 4.3.1 on page 155.

Lemma 4.3.12 *Let $c \in \mathbf{L}\mathcal{S}^\Gamma$ and ρ be a labelled action refinement compatible with c. Assume that $a_i \neq b_j$, $c \overset{S(a_i)}{\Longmapsto} c' \overset{S(b_j)}{\Longmapsto} d$, $\rho(a_i) \overset{\alpha_h}{\Longmapsto} x$ and $\rho(b_j) \overset{\bar{\alpha}_k}{\Longmapsto} y$. Then $c\rho \overset{\tau}{\Longmapsto} \sim d\rho[F(a_i) \rightarrow x, F(b_j) \rightarrow y]$.*

Proof: By induction on the combined length of the derivations $c \overset{S(a_i)}{\Longmapsto} c' \overset{S(b_j)}{\Longmapsto} d$.

- Base case, $c \overset{S(a_i)}{\longmapsto} c' \overset{S(b_j)}{\longmapsto} d$. The proof of the base case proceeds by induction on the structure of c. We only give the details for two of the cases.

  (1) $c = c_1|c_2$. Assume that $c_1|c_2 \overset{S(a_i)}{\longmapsto} c' \overset{S(b_j)}{\longmapsto} d$. Then, by the operational semantics and by symmetry, we may restrict ourselves to considering the following two possibilities:
  (a) $c_1|c_2 \overset{S(a_i)}{\longmapsto} c_1'|c_2 \overset{S(b_j)}{\longmapsto} d_1|c_2 = d$ because $c_1 \overset{S(a_i)}{\longmapsto} c_1' \overset{S(b_j)}{\longmapsto} d_1$ or
  (b) $c_1|c_2 \overset{S(a_i)}{\longmapsto} d_1|c_2 \overset{S(b_j)}{\longmapsto} d_1|d_2 = d$ because $c_1 \overset{S(a_i)}{\longmapsto} d_1$ and $c_2 \overset{S(b_j)}{\longmapsto} d_2$.

  We examine the two possibilities separately.

  (a) In this case we may apply the inductive hypothesis to obtain that $c_1\rho \overset{\tau}{\Longmapsto} \sim d_1\rho[F(a_i) \to x, F(b_j) \to y]$. By the operational semantics and the substitutivity of $\sim$, we then have that

  $$(c_1|c_2)\rho = c_1\rho|c_2\rho \overset{\tau}{\Longmapsto} \sim (d_1\rho[F(a_i) \to x, F(b_j) \to y])|c_2\rho.$$

  Moreover, as $c_1|c_2 \in \mathbf{L}\mathcal{S}^\Gamma$, we have that $F(a_i)$ and $F(b_j)$ do not occur in c_2 and this implies that

  $$(d_1\rho[F(a_i) \to x, F(b_j) \to y])|c_2\rho = (d_1|c_2)\rho[F(a_i) \to x, F(b_j) \to y].$$

  This completes the verification of case (a).

  (b) In this case, by Lemma 4.3.11, we obtain that $c_1\rho \overset{\alpha_h}{\Longmapsto} \sim d_1\rho[F(a_i) \to x]$ and $c_2\rho \overset{\bar{\alpha}_k}{\Longmapsto} \sim d_2\rho[F(b_j) \to y]$. By the operational semantics and the substitutivity of $\sim$, we have that

  $$(c_1|c_2)\rho = c_1\rho|c_2\rho \overset{\tau}{\Longmapsto} \sim (d_1\rho[F(a_i) \to x])|(d_2\rho[F(b_j) \to y]).$$

  Moreover, as $d_1|d_2 \in \mathbf{L}\mathcal{S}^\Gamma$, we have that $F(a_i)$ does not occur in d_2 and $F(b_j)$ does not occur in d_1. This implies that

  $$(d_1\rho[F(a_i) \to x])|(d_2\rho[F(b_j) \to y]) = (d_1|d_2)\rho[F(a_i) \to x, F(b_j) \to y].$$

  This completes the verification of case (b).

  (2) $c = c_1 \setminus \beta$. Assume that $c_1 \setminus \beta \overset{S(a_i)}{\longmapsto} c_1' \setminus \beta \overset{S(b_j)}{\longmapsto} d_1 \setminus \beta$ because $c_1 \overset{S(a_i)}{\longmapsto} c_1' \overset{S(b_j)}{\longmapsto} d_1$ and $a, b \neq \beta, \bar{\beta}$. Note now that

  $$(c_1 \setminus \beta)\rho = (c_1\rho[\beta \mapsto \alpha]) \setminus \alpha,$$

  where $\alpha = choice(new\beta\, c_1\, \rho)$. As $a, b \neq \beta, \bar{\beta}$, we have that $\rho[\beta \mapsto \alpha](a_i) = \rho(a_i) \overset{\alpha_h}{\Longmapsto} x$ and $\rho[\beta \mapsto \alpha](b_j) = \rho(b_j) \overset{\bar{\alpha}_k}{\Longmapsto} y$. We may now apply the inductive hypothesis to obtain

  $$c_1\rho[\beta \mapsto \alpha] \overset{\tau}{\Longmapsto} \sim d_1\rho[\beta \mapsto \alpha][F(a_i) \to x, F(b_j) \to y].$$

By the operational semantics and the substitutivity of $\sim$,

$$(c_1 \setminus \beta)\rho = (c_1\rho[\beta \mapsto \alpha]) \setminus \alpha \overset{\tau}{\Longmapsto} \sim (d_1\rho[\beta \mapsto \alpha][F(a_i) \to x, F(b_j) \to y]) \setminus \alpha.$$

It is now easy to see that $\alpha \in new\beta\, d_1\, \rho[F(a_i) \to x, F(b_j) \to y]$ and that

$$(d_1 \setminus \beta)\rho[F(a_i) \to x, F(b_j) \to y] =_\alpha d_1\rho[\beta \mapsto \alpha][F(a_i) \to x, F(b_j) \to y].$$

- Inductive step, $c \overset{S(a_i)}{\Longmapsto} c' \overset{S(b_j)}{\Longmapsto} d$ and the combined length of the derivations is greater than 2. We proceed by analyzing the possible structure of the two derivations $c \overset{S(a_i)}{\Longmapsto} c' \overset{S(b_j)}{\Longmapsto} d$. There are four possibilities to examine:

(1) $c \overset{S(a_i)}{\longmapsto} c' \overset{S(b_j)}{\longmapsto} \hat{c} \overset{\tau}{\Longmapsto} d$, for some $\hat{c}$, or

(2) $c \overset{S(a_i)}{\longmapsto} c' \overset{\tau}{\longmapsto} \hat{c} \overset{S(b_j)}{\Longmapsto} d$, for some $\hat{c}$, or

(3) $c \overset{S(a_i)}{\longmapsto} \hat{c} \overset{\tau}{\Longmapsto} c' \overset{S(b_j)}{\Longmapsto} d$, or

(4) $c \overset{\tau}{\longmapsto} \hat{c} \overset{S(a_i)}{\Longmapsto} c' \overset{S(b_j)}{\Longmapsto} d$.

We only examine cases (1) and (3), as the remaining ones are dealt with in similar fashion.

(1) In this case, by the inductive hypothesis, we have that $c\rho \overset{\tau}{\Longmapsto} \sim \hat{c}\rho[F(a_i) \to x, F(b_j) \to y]$. By Lemma 4.3.10, $\hat{c} \overset{\tau}{\Longmapsto} d$ implies that $\hat{c}\rho[F(a_i) \to x, F(b_j) \to y] \overset{\tau}{\Longmapsto} \sim d\rho[F(a_i) \to x, F(b_j) \to y]$. It is easy to see that this implies

$$c\rho \overset{\tau}{\Longmapsto} \sim d\rho[F(a_i) \to x, F(b_j) \to y].$$

(3) In this case, by Corollary 4.3.1 on page 155, there exists $\bar{c}$ such that $c \overset{\tau}{\Longmapsto} \bar{c} \overset{S(a_i)}{\longmapsto} c'$ and $l(c \overset{\tau}{\Longmapsto} \bar{c}) = l(\hat{c} \overset{\tau}{\Longmapsto} c') > 0$. We may now apply the inductive hypothesis to $\bar{c} \overset{S(a_i)}{\longmapsto} c' \overset{S(b_j)}{\Longmapsto} d$ to obtain that

$$\bar{c}\rho \overset{\tau}{\Longmapsto} \sim d\rho[F(a_i) \to, F(b_j) \to y].$$

By Lemma 4.3.10, $c \overset{\tau}{\Longmapsto} \bar{c}$ implies $c\rho \overset{\tau}{\Longmapsto} \sim \bar{c}\rho$. It is now easy to see that $c\rho \overset{\tau}{\Longmapsto} \sim \bar{c}\rho$ and $\bar{c}\rho \overset{\tau}{\Longmapsto} \sim d\rho[F(a_i) \to, F(b_j) \to y]$ imply $c\rho \overset{\tau}{\Longmapsto} \sim d\rho[F(a_i) \to x, F(b_j) \to y]$. □

The following lemma allows us to derive a move of $c\rho$ from the information provided by clause (4) of Theorem 4.3.5.

Lemma 4.3.13 *Let $c \in \mathbf{L}S^\Gamma$ and ρ be a labelled action refinement compatible with it. Assume that $F(a_i)$ occurs in c and $\rho(F(a_i)) \overset{e}{\Longmapsto} x$. Then $c\rho \overset{e}{\Longmapsto} \sim c\rho[F(a_i) \to x]$.*

Proof: By induction on the structure of c. □

As an immediate corollary of the above lemma we obtain the following useful result.

Corollary 4.3.7 *Let $c \in \mathbf{L}\mathcal{S}^{\Gamma}$ and ρ be a labelled action refinement compatible with it. Assume that $F(a_i)$ occurs in c and $\rho(F(a_i)) \overset{\varepsilon}{\Longmapsto} x$. Then $c\rho \overset{\varepsilon}{\Longmapsto} \sim c\rho[F(a_i) \to x]$.*

The following lemma will find application in the proof of Theorem 4.3.6 on page 182.

Lemma 4.3.14 *Let $c, d \in \mathbf{L}\mathcal{S}^{\Gamma}$ and ρ be a labelled action refinement compatible with c. Then $c \overset{F(a_i)}{\Longmapsto} d$ and $\rho(F(a_i)) \overset{\varepsilon}{\Longmapsto} x\surd$ imply $c\rho \overset{\varepsilon}{\Longmapsto} \sim d\rho$.*

Proof: Assume that $c \overset{F(a_i)}{\Longmapsto} d$ and $\rho(F(a_i)) \overset{\varepsilon}{\Longmapsto} x\surd$. By Corollary 4.3.2 on page 155, we may assume, without loss of generality, that, for some $\bar{d}$, $c \overset{F(a_i)}{\longmapsto} \bar{d} \overset{\varepsilon}{\Longmapsto} d$. It is now sufficient to prove that

$$c \overset{F(a_i)}{\longmapsto} \bar{d} \text{ and } \rho(F(a_i)) \overset{\varepsilon}{\Longmapsto} x\surd \text{ implies } c\rho \overset{\varepsilon}{\Longmapsto} \sim \bar{d}\rho. \tag{4.7}$$

In fact, by using (4.7), we have that $c\rho \overset{\varepsilon}{\Longmapsto} \sim \bar{d}\rho$. By Corollary 4.3.6, $\bar{d} \overset{\varepsilon}{\Longmapsto} d$ implies $\bar{d}\rho \overset{\varepsilon}{\Longmapsto} \sim d\rho$. It is now easy to see that $c\rho \overset{\varepsilon}{\Longmapsto} \sim \bar{d}\rho$ and $\bar{d}\rho \overset{\varepsilon}{\Longmapsto} \sim d\rho$ imply that

$$c\rho \overset{\varepsilon}{\Longmapsto} \sim d\rho.$$

We are thus left to prove (4.7). The proof of this statement is by induction on the length of the proof of the derivation $c \overset{F(a_i)}{\longmapsto} \bar{d}$. We examine only two of the cases which arise in such a proof and leave the remaining ones to the reader.

- $c = F(a_i) \overset{F(a_i)}{\longmapsto} nil = \bar{d}$. Then $c\rho = \rho(F(a_i)) \overset{\varepsilon}{\Longmapsto} x \sim nil$, because $x\surd$ by the assumptions of the statement.

- $c = c_1 \setminus \alpha \overset{F(a_i)}{\longmapsto} d' \setminus \alpha = \bar{d}$ because $c_1 \overset{F(a_i)}{\longmapsto} d'$ and $a \neq \alpha, \bar{\alpha}$. First of all, recall that

$$(c_1 \setminus \alpha)\rho = (c_1\rho[\alpha \mapsto \beta]) \setminus \beta,$$

where $\beta = choice(new\, \alpha\, c_1\, \rho)$. As $a \neq \alpha, \bar{\alpha}$, we have that $\rho[\alpha \mapsto \beta](F(a_i)) = \rho(F(a_i)) \overset{\varepsilon}{\Longmapsto} x\surd$. We may then apply the inductive hypothesis to $c_1 \overset{F(a_i)}{\longmapsto} d'$ and $\rho[\alpha \mapsto \beta](F(a_i)) \overset{\varepsilon}{\Longmapsto} x\surd$ to obtain that, for some y,

$$c_1\rho[\alpha \mapsto \beta] \overset{\varepsilon}{\Longmapsto} y \sim d'\rho[\alpha \mapsto \beta].$$

By the operational semantics and the substitutivity of $\sim$,

$$(c_1 \setminus \alpha)\rho = (c_1\rho[\alpha \mapsto \beta]) \setminus \beta \overset{\varepsilon}{\Longmapsto} y \setminus \beta \sim (d'\rho[\alpha \mapsto \beta]) \setminus \beta.$$

As $\beta \in new\,\alpha\, c_1\, \rho$, it follows that $\beta \in new\,\alpha\, d'\, \rho$. We may now apply the labelled version of Fact 4.2.2 to obtain that $(d' \setminus \alpha)\rho =_\alpha (d'\rho[\alpha \mapsto \beta]) \setminus \beta$. The claim now follows form the fact that $=_\alpha$ is a strong bisimulation and the transitivity of $\sim$. □

Lemma 4.3.15 *Let $c \in \mathbf{L}\mathcal{S}^\Gamma$ and ρ be a labelled action refinement compatible with it. Assume that $c \overset{S(a_i)}{\Longmapsto} c'$, $F(b_j)$ occurs in c, $\rho(a_i) \overset{a'_h}{\Longmapsto} x$ and $\rho(F(b_j)) \overset{\bar{a'}_k}{\Longmapsto} y$. Then $c\rho \overset{\tau}{\Longmapsto} \sim c'\rho[F(a_i) \rightarrow x, F(b_j) \rightarrow y]$.*

Proof: By induction on the length of the derivation $c \overset{S(a_i)}{\Longmapsto} c'$.

- Base case, $c \overset{S(a_i)}{\longmapsto} c'$. The proof proceeds by a sub-induction on the length of the proof of the derivation $c \overset{S(a_i)}{\longmapsto} c'$. We examine only two of the possible cases, leaving the remaining ones to the reader.
  - $c = c_1|c_2 \overset{S(a_i)}{\longmapsto} c_1'|c_2 = c'$ because $c_1 \overset{S(a_i)}{\longmapsto} c_1'$. We distinguish two possibilities depending on whether $F(b_j)$ occurs in c_1 or c_2.

    If $F(b_j)$ occurs in c_1 then we may apply the sub-inductive hypothesis to obtain $c_1\rho \overset{\tau}{\Longmapsto} \sim c_1'\rho[F(a_i) \rightarrow x, F(b_j) \rightarrow y]$. By the operational semantics and the substitutivity of $\sim$,

    $$(c_1|c_2)\rho = c_1\rho|c_2\rho \overset{\tau}{\Longmapsto} \sim (c_1'\rho[F(a_i) \rightarrow x, F(b_j) \rightarrow y])|c_2\rho.$$

    Moreover, as $F(a_i)$ and $F(b_j)$ do not occur in c_2,

    $$(c_1'\rho[F(a_i) \rightarrow x, F(b_j) \rightarrow y])|c_2\rho = (c_1'|c_2)\rho[F(a_i) \rightarrow x, F(b_j) \rightarrow y].$$

    If $F(b_j)$ occurs in c_2 then, by Lemma 4.3.11, $c_1 \overset{S(a_i)}{\longmapsto} c_1'$ and $\rho(a_i) \overset{a'_h}{\Longmapsto} x$ imply $c_1\rho \overset{a'_h}{\Longmapsto} \sim c_1'\rho[F(a_i) \rightarrow x]$. By Lemma 4.3.13, $F(b_j)$ occurs in c_2 and $\rho(F(b_j)) \overset{\bar{a'}_k}{\Longmapsto} y$ imply $c_2 \overset{\bar{a'}_k}{\Longmapsto} c_2\rho[F(b_j) \rightarrow y]$. By the operational semantics and the substitutivity of $\sim$,

    $$(c_1|c_2)\rho = c_1\rho|c_2\rho \overset{\tau}{\Longmapsto} \sim (c_1'\rho[F(a_i) \rightarrow x])|(c_2\rho[F(b_j) \rightarrow y]).$$

    Moreover, as $F(a_i)$ and $F(b_j)$ do not occur in c_2 and c_1', respectively, we have that

    $$(c_1'\rho[F(a_i) \rightarrow x])|(c_2\rho[F(b_j) \rightarrow y]) = (c_1'|c_2)\rho[F(a_i) \rightarrow x, F(b_j) \rightarrow y].$$

  - $c = c_1 \setminus \beta \overset{S(a_i)}{\longmapsto} c_1' \setminus \beta = c'$ because $c_1 \overset{S(a_i)}{\longmapsto} c_1'$ and $a \neq \beta, \bar{\beta}$. First of all, note that

    $$(c_1 \setminus \beta)\rho = (c_1\rho[\beta \mapsto \beta']) \setminus \beta',$$

    where $\beta' = choice(new\,\beta\, c_1\, \rho)$. As $F(b_j)$ occurs in $c_1 \setminus \beta$, by axiom (CR) we have that also $b \neq \beta, \bar{\beta}$. Moreover $F(b_j)$ occurs in c_1. By these observations we have that $\rho[\beta \mapsto \beta'](a_i) = \rho(a_i) \overset{a'_h}{\Longmapsto} x$ and $\rho[\beta \mapsto \beta'](F(b_j)) = \rho(F(b_j)) \overset{\bar{a'}_k}{\Longmapsto} y$. We may then apply the sub-inductive hypothesis to $c_1 \overset{S(a_i)}{\longmapsto} c'_1$ to obtain

$$c_1\rho[\beta \mapsto \beta'] \overset{\tau}{\Longmapsto} \sim c'_1\rho[\beta \mapsto \beta'][F(a_i) \to x, F(b_j) \to y].$$

By the operational semantics and the substitutivity of $\sim$, we then have that

$$(c_1 \setminus \beta)\rho = (c_1\rho[\beta \mapsto \beta']) \setminus \beta' \overset{\tau}{\Longmapsto} \sim (c'_1\rho[\beta \mapsto \beta'][F(a_i) \to x, F(b_j) \to y]) \setminus \beta'.$$

It is now sufficient to prove that

$$(c'_1 \setminus \beta)\rho[F(a_i) \to x, F(b_j) \to y] \;=_\alpha\; (c'_1\hat{\rho}) \setminus \beta', \tag{4.8}$$

where $\hat{\rho} = \rho[\beta \mapsto \beta'][F(a_i) \to x, F(b_j) \to y]$. The claim will, in fact, then follow by (4.8), the fact that $=_\alpha$ is a strong bisimulation and the transitivity of $\sim$. In order to prove statement (4.8), we note first of all that

$$(c'_1 \setminus \beta)\rho[F(a_i) \to x, F(b_j) \to y] = (c'_1\rho[F(a_i) \to x, F(b_j) \to y][\beta \mapsto \alpha]) \setminus \alpha,$$

where $\alpha = \mathit{choice}(\mathit{new}\,\beta\, c'_1\, \rho[F(a_i) \to x, F(b_j) \to y])$. We prove (4.8) by distinguishing two cases, depending on whether $\beta' = \alpha$ or not. If $\alpha = \beta'$ then

$$\begin{aligned}(c'_1 \setminus \beta)\rho[F(a_i) \to x, F(b_j) \to y] &= (c'_1\rho[F(a_i) \to x, F(b_j) \to y][\beta \mapsto \beta']) \setminus \beta' \\ &= (c'_1\rho[\beta \mapsto \beta'][F(a_i) \to x, F(b_j) \to y]) \setminus \beta'.\end{aligned}$$

The last equality depends on the fact that $\beta \neq a, \bar{a}, b, \bar{b}$. The claim now follows by the reflexivity of $=_\alpha$.

If $\alpha \neq \beta'$ then it is easy to see that

$$\beta' \notin FC(c'_1\rho[F(a_i) \to x, F(b_j) \to y][\beta \mapsto \alpha]).$$

Now, by the substitution lemma, we have that

$$\begin{aligned}&(c'_1\rho[F(a_i) \to x, F(b_j) \to y][\beta \mapsto \alpha])\iota[\alpha \mapsto \beta'] &=\\ &c'_1((\iota[\alpha \mapsto \beta']) \circ (\rho[F(a_i) \to x, F(b_j) \to y][\beta \mapsto \alpha])) &=\\ &c'_1\rho[F(a_i) \to x, F(b_j) \to y][\beta \mapsto \beta'] &=\\ &c'_1\rho[\beta \mapsto \beta'][F(a_i) \to x, F(b_j) \to y].\end{aligned}$$

The last equality in the above verification depends on the fact that $\beta \neq a, \bar{a}, b, \bar{b}$. The claim now follows by rule (α) of the definition of $=_\alpha$.

- Inductive step, $c \overset{S(a_i)}{\Longmapsto} c'$ and $l(c \overset{S(a_i)}{\Longmapsto} c') > 1$. By Corollary 4.3.1, we may assume, wlog, that $c \overset{\varepsilon}{\Longmapsto} c'' \overset{S(a_i)}{\longmapsto} c'$ for some c''. By Corollary 4.3.6, $c \overset{\varepsilon}{\Longmapsto} c''$ implies $c\rho \overset{\varepsilon}{\Longmapsto} \sim c''\rho$. By the base case, $c'' \overset{S(a_i)}{\longmapsto} c'$ implies that $c''\rho \overset{\tau}{\Longmapsto} \sim c'\rho[F(a_i) \to x, F(b_j) \to y]$. By combining the two derivations we then have that $c\rho \overset{\tau}{\Longmapsto} \sim c'\rho[F(a_i) \to x, F(b_j) \to y]$. □

Lemma 4.3.16 *Let $c \in \mathbf{L}\mathcal{S}^{\Gamma}$ and ρ be a labelled action refinement compatible with c. Assume that $a_i \neq b_j$, $F(a_i)$ and $F(b_j)$ occur in c, $\rho(a_i) \overset{\alpha_h}{\Longmapsto} x$ and $\rho(F(b_j)) \overset{\bar{\alpha}_k}{\Longmapsto} y$. Then $c\rho \overset{\tau}{\Longmapsto} \sim c\rho[F(a_i) \to x, F(b_j) \to y]$.*

Proof: By structural induction on c. Omitted. □

We now have most of the technical material needed to prove the main theorem of this chapter, namely the refinement theorem stated on page 157. The refinement theorem will be a corollary of a similar statement for labelled states and labelled action refinements. In the following definition we shall make use of the machinery developed in Definition 4.3.7 at page 161.

Definition 4.3.10 **(1)** *For each $\phi \in \mathcal{H}$, let $\mathcal{R}^{sub}_{\phi}$ be defined as follows:*

$$\mathcal{R}^{sub}_{\phi} =_{def} \{\langle c\rho, d\rho'\rangle \mid \begin{array}{ll} \text{(i)} & \textit{there exists } \varphi \in \mathcal{H} \textit{ such that } c \approx_{\varphi} d, \\ \text{(ii)} & c\rho, d\rho' \in \mathbf{L}\mathcal{S}^{\Gamma}, \textit{ and} \\ \text{(iii)} & \textit{there exists a } \varphi\textit{-vector } \sigma \textit{ such that } \rho =_{(\varphi,\sigma)} \rho' \textit{ and} \\ & \phi = \bigcup_{(a,i,j)\in\varphi} \sigma^{(a,i,j)}\}. \end{array}$$

Using $\{\mathcal{R}^{sub}_{\phi} \mid \phi \in \mathcal{H}\}$ we define, for each $\phi \in \mathcal{H}$, the relation

$$\sim \circ \mathcal{R}^{sub}_{\phi} \circ \sim =_{def} \{\langle c, d\rangle \mid c \sim x \mathcal{R}^{sub}_{\phi} y \sim d \textit{ for some } x, y\}.$$

(2) *For each $\phi \in \mathcal{H}$, $\mathcal{S}_{\phi}$ is given by*

$$\mathcal{S}_{\phi} =_{def} \begin{cases} (\sim \circ \mathcal{R}^{sub}_{\phi} \circ \sim) \cup \Pi & \textit{if } \phi = \emptyset \\ \sim \circ \mathcal{R}^{sub}_{\phi} \circ \sim & \textit{otherwise,} \end{cases}$$

where Π is defined as follows:

$$\Pi =_{def} \{\langle \pi_1\rho + a_i, \pi_2\rho' + a_j\rangle \mid \begin{array}{ll} \text{(i)} & \pi_1 =_{\emptyset} \pi_2 \textit{ and, for each } n,\ a_n \textit{ does not occur in} \\ & \pi_1\rho, \pi_2\rho' \\ \text{(ii)} & \pi_1\rho, \pi_2\rho' \in \mathbf{L}\mathcal{S}^{\Gamma}, \textit{ and} \\ \text{(iii)} & \rho =_{(\emptyset,\emptyset)} \rho'\}. \end{array}$$

Our aim is to prove that the $\mathcal{H}$-indexed family of relations $\{\mathcal{S}_{\phi} \mid \phi \in \mathcal{H}\}$ is a weak refine bisimulation. In the proof of this statement we shall need to relate the termination capabilities of configurations $c\rho$ and $d\rho'$ such that $\langle c\rho, d\rho'\rangle \in \mathcal{S}_{\phi}$. This is accomplished in the following theorems.

Definition 4.3.11 *For all $c_i, d_i \in \mathbf{L}\mathcal{S}^{\Gamma}$, $\phi_i \in \mathcal{H}$ and labelled action refinements ρ_i, ρ'_i, $i=1,2$, we write $\langle c_1\rho_1, c_2\rho_2\rangle \prec_r \langle d_1\rho'_1, d_2\rho'_2\rangle$ iff $c_1 \approx_{\phi_1} c_2$, $d_1 \approx_{\phi_2} d_2$ and there exist a ϕ_1-vector σ_1 and a ϕ_2-vector σ_2 such that $\rho_1 =_{(\phi_1,\sigma_1)} \rho_2$ and $\rho'_1 =_{(\phi_2,\sigma_2)} \rho'_2$ and the following conditions hold:*

(1) $|c_1| + |c_2| < |d_1| + |d_2|$ *or*

(2) $|c_1| + |c_2| = |d_1| + |d_2|$ *and* $|c_1\rho_1| + |c_2\rho_2| < |d_1\rho'_1| + |d_2\rho'_2|$,

where, for each $c \in \mathbf{L}\mathcal{S}^\Gamma$, $|c|$ *denotes the depth of its derivation tree.*

The following theorem relates the termination capabilities of $c\rho$ and $d\rho'$, where $c \approx_\phi d$ and $\rho =_{(\phi,\sigma)} \rho'$ for some $\phi \in \mathcal{H}$ and ϕ-vector σ. Unfortunately, the proof of the theorem is rather involved because, in general, there is no simple relationship between the termination potential of a labelled state c and that of $c\rho$. For instance, it is easy to see that *not* $a_i + b_j\surd\!\!\!/$, but $(a_i + b_j)\rho\surd\!\!\!/$, for any ρ such that $\rho(a_i) = \tau$ and $\rho(b_j) = b_j$. On the other hand, $a_i + \tau\surd\!\!\!/$, but *not* $(a_i + \tau)\rho\surd\!\!\!/$ for any labelled action refinement ρ such that $\rho(a_i) = \tau;\delta$.

Theorem 4.3.6 *Let* $c, d \in \mathbf{L}\mathcal{S}^\Gamma$, $\phi \in \mathcal{H}$ *and* ρ, ρ' *be labelled action refinements such that* $\rho =_{(\phi,\sigma)} \rho'$ *for some* ϕ*-vector* σ. *Assume that* $c \approx_\phi d$. *Then* $c\rho\surd\!\!\!/$ *implies* $d\rho'\surd\!\!\!/$.

Proof: By induction on the relation $\prec_r$ defined above. We assume, as inductive hypothesis, that the claim holds for all $\langle c_1\rho_1, c_2\rho_2\rangle \prec_r \langle c\rho, d\rho'\rangle$ and prove that it holds for $\langle c\rho, d\rho'\rangle$, where $c \approx_\phi d$ and $\rho =_{(\phi,\sigma)} \rho'$, for some ϕ-vector σ. Assume that $c\rho\surd\!\!\!/$. We shall prove that $d\rho\surd\!\!\!/$ by distinguishing two possibilities:

(A) $c\rho \not\xrightarrow{\tau}$ and $d\rho' \not\xrightarrow{\tau}$, i.e. both $c\rho$ and $d\rho'$ are *stable*, or

(B) either $c\rho$ or $d\rho'$ is not stable.

We examine the two possibilities in turn.

(A) Assume that $c\rho$ and $d\rho'$ are stable. As $c\rho\surd\!\!\!/$ by the hypothesis of the theorem, we then have that $c\rho\surd$. We shall prove that $d\rho'\surd$. First of all, note that, by Fact 4.3.3, $c\rho\surd$ implies that $\rho(F(a_i))\surd$ for all $F(a_i)$ occurring in c. Moreover, by Lemmas 4.3.10, 4.3.11 and 4.3.13, we have that $d\rho' \not\xrightarrow{\tau}$ implies

(A.1) $d \not\xrightarrow{\tau}$ and

(A.2) $\rho'(F(a_i)) \not\xrightarrow{\tau}$, for all $F(a_i)$ occurring in d'.

By Fact 4.3.3, $c\rho\surd$ implies that, for each $(X, c') \in \eta(c)$, $\rho(X)\surd$ and $c'\rho\surd$. Let $(X, c') \in \eta(c)$. (Note that such a pair always exists because $\eta(c)$ is nonempty) We distinguish two possibilities depending on whether X is empty or not.

If $X = \emptyset$ then, by Lemma 4.3.6.(1), c is a labelled process. By Fact 4.3.7, this implies that $c\surd$. As $c \approx_\phi d$ by the hypothesis of the theorem, we have that $d\surd\!\!\!/$. Hence, by (A.1), $d\surd$ and this implies that $d\rho'\surd$.

Assume now that $X \neq \emptyset$. Then there exist $F(a_i) \in X$. By Lemma 4.3.6.(3), there exists $\bar{c}$ such that $(\{F(a_i)\}, \bar{c}) \in \eta(c)$. By Lemma 4.3.7, $c \overset{F(a_i)}{\longmapsto} \bar{c}$. As $\rho(F(a_i))\surd$, we may apply Lemma 4.3.14 to obtain that

$$c\rho \overset{\varepsilon}{\Longmapsto} x \sim \bar{c}\rho, \text{ for some } x.$$

As $c\rho$ is stable, we have that $c\rho \sim \bar{c}\rho$ and $\bar{c}\rho\surd$. As $c \approx_\phi d$, we have that $c \overset{F(a_i)}{\longmapsto} \bar{c}$ and d stable imply that $d \overset{F(a_j)}{\longmapsto} d'$ for some j and d' such that $\phi_a(i) = j$ and $\bar{c} \approx_{\phi'} d'$, where $\phi' = \phi - \{(a, i, j)\}$. As $\rho =_{(\phi,\sigma)} \rho'$ by the hypotheses of the theorem and $\rho(F(a_i))\surd$, we have that $\rho'(F(a_j))\surd\!\!\!/$. Moreover, by (A.2), it follows that $\rho'(F(a_j))\surd$. We may then apply Lemma 4.3.14 to obtain that, as $d\rho'$ is stable, $d\rho' \sim d'\rho'$. Note now that $\rho =_{(\phi,\sigma)} \rho'$ implies that $\rho =_{(\phi',\sigma')} \rho'$, where σ' is the restriction of σ to a ϕ'-vector. Moreover, as $|\bar{c}| + |d'| < |c| + |d|$, $\langle \bar{c}, d' \rangle \prec_r \langle c, d \rangle$. We may then apply the inductive hypothesis to $\bar{c} \approx_{\phi'} d'$, $\rho =_{(\phi',\sigma')} \rho'$ and $\bar{c}\rho\surd$ to obtain that $d'\rho'\surd\!\!\!/$. As $d\rho' \sim d'\rho'$ and $d\rho$ is stable, we then derive that $d\rho'\surd$. This completes the proof of case (A).

(B) Without loss of generality, we restrict ourselves to examining the case in which $d\rho'$ is not stable. (The case in which $d\rho'$ is stable and $c\rho$ is not can be dealt with in a similar fashion) In order to prove that $d\rho'\surd\!\!\!/$, it is sufficient to show that $d\rho' \overset{\tau}{\longmapsto} \bar{y} \overset{\varepsilon}{\Longmapsto} y$ and y stable imply $y\surd$. By Theorem 4.3.5 and the fact that $=_\alpha$ is a strong bisimulation, $d\rho' \overset{\tau}{\longmapsto} \bar{y}$ implies

(1) there exist $a_i \in LAct(\Lambda)$ and $d', x \in \mathbf{L}\mathcal{S}^\Gamma$ such that $d \overset{S(a_i)}{\longmapsto} d'$, $\rho'(a_i) \overset{\tau}{\longmapsto} x$ and $\bar{y} \sim d'\rho'[F(a_i) \to x]$, or

(2) there exists $(X, d') \in \eta(d)$ such that $X \neq \emptyset$, $\rho'(X)\surd$ and $d'\rho' \overset{\tau}{\longmapsto} \bar{y}$, or

(3) there exist $a_i, b_j \in LAct(\Lambda)$, $\alpha \in \Lambda$ and $\bar{d}, d', x_1, x_2 \in \mathbf{L}\mathcal{S}^\Gamma$ such that $d \overset{S(a_i)}{\longmapsto} \bar{d} \overset{S(b_j)}{\longmapsto} d'$, $\rho'(a_i) \overset{\alpha_h}{\longmapsto} x_1$, $\rho'(b_j) \overset{\bar{\alpha}_k}{\longmapsto} x_2$ and $\bar{y} \sim d'\rho'[F(a_i) \to x_1, F(b_j) \to x_2]$, or

(4) there exist $a_i \in LAct(\Lambda)$ and $x \in \mathbf{L}\mathcal{S}^\Gamma$ such that $F(a_i)$ occurs in d, $\rho'(F(a_i)) \overset{\tau}{\longmapsto} x$ and $\bar{y} \sim d\rho'[F(a_i) \to x]$, or

(5) there exists $d' \in \mathbf{L}\mathcal{S}^\Gamma$ such that $d \overset{\tau}{\longmapsto} d'$ and $\bar{y} \sim d'\rho'$, or

(6) there exist $a_i, b_j \in \mathbf{L}\mathcal{S}^\Gamma$ and $d', x_1, x_2 \in \mathbf{L}\mathcal{S}^\Gamma$ such that $d \overset{S(a_i)}{\longmapsto} d'$, $F(b_j)$ occurs in d, $\rho'(a_i) \overset{a'_h}{\longmapsto} x_1$, $\rho'(F(b_j)) \overset{\bar{a'}_k}{\longmapsto} x_2$ and $\bar{y} \sim d'\rho'[F(a_i) \to x_1, F(b_j) \to x_2]$, or

(7) there exist $a_i, b_j \in LAct(\Lambda)$ and $x_1, x_2 \in \mathbf{L}\mathcal{S}^\Gamma$ such that $F(a_i)$, $F(b_j)$ occur in d, $\rho'(F(a_i)) \overset{\alpha_h}{\longmapsto} x_1$ and $\rho'(F(b_j)) \overset{\bar{\alpha}_k}{\longmapsto} x_2$ and $\bar{y} \sim d\rho'[F(a_i) \to x_1, F(b_j) \to x_2]$.

We proceed by showing that $y\surd$ in each of the following possibilities. We shall only examine three representative cases, leaving the remaining ones to the reader.

(1) Assume that $d\rho' \overset{\tau}{\longmapsto} \bar{y}$ because $d \overset{S(a_i)}{\longmapsto} d'$, $\rho'(a_i) \overset{\tau}{\longmapsto} x$ and $\bar{y} \sim d'\rho'[F(a_i) \to x]$. As $c \approx_\phi d$, we then have that $c \overset{S(a_{i_1})}{\Longmapsto} c'$ for some i_1 and c' such that $c' \approx_{\phi'} d'$, where $\phi' = \phi \cup \{(a, i_1, i)\}$. As $\rho =_{(\phi,\sigma)} \rho'$, we have that $\rho(a_{i_1}) =_\emptyset \rho'(a_i)$. Hence there exists x' such that $\rho(a_{i_1}) \overset{\tau}{\Longmapsto} x'$ and $x' \approx_\emptyset x$. By Lemma 4.3.11, $c \overset{S(a_{i_1})}{\Longmapsto} c'$ and $\rho(a_{i_1}) \overset{\tau}{\Longmapsto} x'$ imply $c\rho \overset{\tau}{\Longmapsto} \bar{x} \sim c'\rho[F(a_{i_1}) \to x']$, for some $\bar{x}$. As $c\rho \not\surd$ we then have that $c'\rho[F(a_{i_1}) \to x'] \not\surd$.

It is easy to see that, as $\rho =_{(\phi,\sigma)} \rho'$, we have that $\rho[F(a_{i_1}) \to x'] =_{(\phi',\sigma')} \rho'[F(a_i) \to x]$, where σ' is the ϕ'-vector given by:

$$\sigma'^{(b,h,k)} =_{def} \begin{cases} \emptyset & \text{if } (b,h,k) = (a, i_1, i) \\ \sigma^{(b,h,k)} & \text{otherwise.} \end{cases}$$

As $|c'| + |d'| < |c| + |d|$, we may now apply the inductive hypothesis to $c' \approx_{\phi'} d'$, $\rho[F(a_{i_1}) \to x'] =_{(\phi',\sigma')} \rho'[F(a_i) \to x]$ and $c'\rho[F(a_{i_1}) \to x'] \not\surd$ to obtain that $d'\rho'[F(a_i) \to x] \not\surd$. As $\bar{y} \sim d'\rho'[F(a_i) \to x]$, we then have that $\bar{y} \not\surd$. This implies that $y\surd$.

(2) Assume that $d\rho' \overset{\tau}{\longmapsto} \bar{y}$ because there exists $(X, d') \in \eta(c)$ such that $X \neq \emptyset$, $\rho'(X)\surd$ and $d'\rho' \overset{\tau}{\longmapsto} \bar{y}$. As $X \neq \emptyset$, there exists $F(a_j) \in X$. By Lemma 4.3.6.(3), $(\{F(a_j)\}, \bar{d}) \in \eta(d)$ for some $\bar{d}$ and, as $\rho'(X)\surd$, $\rho'(F(a_j))\surd$. By Proposition 4.3.9, we have that

$$d\rho' =_\surd d'\rho' =_\surd \bar{d}\rho'.$$

Moreover, by Lemma 4.3.7, $d \overset{F(a_j)}{\longmapsto} \bar{d}$. As $c \approx_\phi d$, there exist i and c' such that $c \overset{F(a_i)}{\Longmapsto} c'$, $\phi_a(i) = j$ and $c' \approx_{\phi'} \bar{d}$, where $\phi' = \phi - \{(a, i, j)\}$. As $\rho =_{(\phi,\sigma)} \rho'$ and $\rho'(F(a_j))\surd$, we have that

- $\rho(F(a_i)) \approx_\emptyset \rho'(F(a_j))$ and $\rho(F(a_j)) \not\surd$, and
- $\rho =_{(\phi',\sigma')} \rho'$, where σ' is the restriction of σ to a ϕ'-vector.

As $\rho(F(a_i)) \not\surd$, there exists x such that $\rho(F(a_i)) \overset{\varepsilon}{\Longmapsto} x\surd$. By Lemma 4.3.14, $c \overset{F(a_i)}{\Longmapsto} c'$ and $\rho(F(a_i)) \overset{\varepsilon}{\Longmapsto} x\surd$ imply $c\rho \overset{\varepsilon}{\Longmapsto} \sim c'\rho$. As $c\rho \not\surd$, we have that $c'\rho \not\surd$. We may now apply the inductive hypothesis to $c' \approx_{\phi'} \bar{d}$, $\rho =_{(\phi',\sigma')} \rho'$ and $c'\rho \not\surd$ to obtain that $\bar{d}\rho' \not\surd$. As $d\rho' =_\surd \bar{d}\rho'$ and $=_\surd$ is a strong bisimulation we then have that $d\rho' \not\surd$. Hence $y\surd$.

(4) Assume that $d\rho' \overset{\tau}{\longmapsto} \bar{y}$ because $F(a_i)$ occurs in d, $\rho'(F(a_i)) \overset{\tau}{\longmapsto} x$ and $\bar{y} \sim d\rho'[F(a_i) \to x]$. Then, as $c \approx_\phi d$, there exists j such that $\phi_a(j) = i$ and $F(a_j)$ occurs in c. Moreover, as $\rho =_{(\phi,\sigma)} \rho'$, we have that $\rho(F(a_j)) \approx_{\sigma(a,j,i)} \rho'(F(a_i))$. Thus there exists x' such that $\rho(F(a_j)) \overset{\varepsilon}{\Longmapsto} x'$ and $x' \approx_{\sigma(a,j,i)} x$. By Corollary 4.3.7, we have that $c\rho \overset{\varepsilon}{\Longmapsto} \sim c\rho[F(a_j) \to x']$. As $c\rho \not\surd$, $c\rho[F(a_j) \to x'] \not\surd$. Moreover, as

$\rho =_{(\phi,\sigma)} \rho'$, we have that $\rho[F(a_j) \to x'] =_{(\phi,\sigma)} \rho'[F(a_i) \to x]$. Note now that, as $|d\rho'[F(a_i) \to x]| < |d\rho'|$,

$$|c\rho[F(a_j) \to x']| + |d\rho'[F(a_i) \to x]| < |c\rho| + |d\rho'|.$$

We may then apply the inductive hypothesis to obtain that $d\rho'[F(a_i) \to x]\surd\!\!\!/$. As $\bar{y} \sim d\rho'[F(a_i) \to x]$, we then have that $\bar{y}\surd\!\!\!/$. This implies that $y\surd$.

The proof of (B) can be completed by checking the remaining cases in a similar way. □

In a similar fashion one can show the following theorem, whose proof is omitted.

Theorem 4.3.7 *For each $\pi_1, \pi_2 \in \mathbf{L}\mathcal{S}^\Gamma$ and labelled action refinements ρ and ρ', if $\langle \pi_1\rho + a_i, \pi_2\rho' + a_j\rangle \in \Pi$ and $\pi_1\rho + a_i\surd\!\!\!/$ then $\pi_2\rho' + a_j\surd\!\!\!/$.*

We have now developed all the technical material needed to prove the following refinement theorem for labelled states and labelled action refinements.

Theorem 4.3.8 *The family of relations $\{\mathcal{S}_\phi \mid \phi \in \mathcal{H}\}$ is a weak refine bisimulation.*

Proof: In order to prove this statement, by the definition of $\{\mathcal{S}_\phi \mid \phi \in \mathcal{H}\}$, it is sufficient to show that $\{\mathcal{S}_\phi \mid \phi \in \mathcal{H}\}$ is symmetric and the defining clauses of $\approx_\mathcal{H}$ are met by:

(1) $\langle \pi_1\rho + a_i, \pi_2\rho' + a_j\rangle \in \Pi$, and

(2) $\langle c, d\rangle \in \sim \circ \mathcal{R}_\phi^{sub} \circ \sim$.

Before giving some of the technical details of the proof, we shall examine, for the sake of clarity, the nature of the arguments used in it. The proofs of statements (1) and (2) make an essential use of the composition/decomposition results proved above. For example, given $\langle c\rho, d\rho'\rangle \in \mathcal{R}_\phi^{sub}$ and a move of $c\rho$, we shall use Theorem 4.3.5 to decompose this move into a move from c and move(s) of the components of ρ. The equivalence of c and d with respect to some history and the pointwise equivalence between ρ and ρ' will then be used to derive "matching moves" from d and the components of ρ'. These moves will then be composed by means of the "whither lemmas" to obtain a move of $d\rho'$, which will be shown to match the original move of $c\rho$ with respect to $\{\mathcal{S}_\phi \mid \phi \in \mathcal{H}\}$. We shall now prove that the defining clauses of $\approx_\mathcal{H}$ are met by (1) and (2) above.

(1) Assume that $\langle \pi_1\rho + a_i, \pi_2\rho' + a_j\rangle \in \Pi \subseteq \mathcal{S}_\emptyset$. By the definition of Π this is because

(A.1) $\pi_1 =_\emptyset \pi_2$ and, for all n, a_n does not occur in $\pi_1\rho$ and $\pi_2\rho'$, and

(B.1) $\rho =_{(\emptyset,\emptyset)} \rho'$.

We shall now prove that the defining clauses of $\approx_{\mathcal{H}}$ are met by $\langle \pi_1\rho + a_i, \pi_2\rho' + a_j\rangle$. First of all, note that, as $\pi_1\rho + a_i$ and $\pi_2\rho' + a_j$ are labelled processes, $\emptyset$ is compatible with $\langle \pi_1\rho + a_i, \pi_2\rho' + a_j\rangle$.

- Assume that $\pi_1\rho + a_i \stackrel{S(b_h)}{\longmapsto} \bar{c}$. By the operational semantics, this is because

  (1.a) either $a_i \stackrel{S(a_i)}{\longmapsto} F(a_i) = \bar{c}$ and $b_h = a_i$,

  (1.b) or $\pi_1\rho \stackrel{S(b_h)}{\longmapsto} \bar{c}$.

  We examine the two possibilities in turn.

  (1.a) In this case, by the operational semantics, $\pi_2\rho' + a_j \stackrel{S(a_j)}{\longmapsto} F(a_j)$. Obviously, $F(a_i) \approx_{\{(a,i,j)\}} F(a_j)$. Moreover, by using the identity substitution ι and the $\{(a,i,j)\}$-vector σ given by $\sigma^{(a,i,j)} = \{(a,i,j)\}$, it is easy to see that $\langle F(a_i), F(a_j)\rangle \in \mathcal{R}^{sub}_{\{(a,i,j)\}}$. This implies that $\langle F(a_i), F(a_j)\rangle \in \mathcal{S}_{\{(a,i,j)\}}$.

  (1.b) In this case, we may apply Theorem 4.3.4 to $\pi_1\rho \stackrel{S(b_h)}{\longmapsto} \bar{c}$ to obtain that there exist a'_n and c, x such that

$$\pi_1 \stackrel{S(a'_n)}{\longmapsto} c,\ \rho(a'_n) \stackrel{S(b_h)}{\longmapsto} x \text{ and } \bar{c} \sim c\rho[F(a'_n) \to x].$$

  Now, as $\pi_1 =_\emptyset \pi_2$ by (A.1), there exist m and d such that $\pi_2 \stackrel{S(a'_m)}{\Longmapsto} d$ and $c \approx_{\{(a',n,m)\}} d$. As $\rho =_{(\emptyset,\emptyset)} \rho'$ by (B.1), we have that $\rho(a'_n) =_\emptyset \rho'(a'_m)$. Thus there exist k and y such that $\rho'(a'_m) \stackrel{S(b_k)}{\Longmapsto} y$ and $x \approx_{\{(b,h,k)\}} y$. By Lemma 4.3.11, $\pi_2 \stackrel{S(a'_m)}{\Longmapsto} d$ and $\rho'(a'_m) \stackrel{S(b_k)}{\Longmapsto} y$ imply

$$\pi_2\rho' \stackrel{S(b_k)}{\Longmapsto} \bar{d} \sim d\rho'[F(a'_m) \to y],$$

  for some $\bar{d}$. By the operational semantics, we then have that $\pi_2\rho' + a_j \stackrel{S(b_k)}{\Longmapsto} \bar{d}$. We shall now check that $\langle \bar{c}, \bar{d}\rangle \in \mathcal{S}_{\{(b,h,k)\}}$ as required. It is sufficient to prove that $\langle c\rho[F(a'_n) \to x], d\rho'[F(a'_m) \to y]\rangle \in \mathcal{R}^{sub}_{\{(b,h,k)\}}$. We have already seen that $c \approx_{\{(a',n,m)\}} d$. The fact that $c\rho[F(a'_n) \to x], d\rho'[F(a'_m) \to y] \in \mathbf{L}\mathcal{S}^\Gamma$ follows by the closure of $\mathbf{L}\mathcal{S}^\Gamma$ under derivations. Define now the $\{(a',n,m)\}$-vector σ by $\sigma^{(a',n,m)} = \{(b,h,k)\}$. As $\rho =_{(\emptyset,\emptyset)} \rho'$ and $x \approx_{\{(b,h,k)\}} y$, we have that

$$\rho[F(a'_n) \to x] =_{(\{(a',n,m)\},\sigma)} \rho'[F(a'_m) \to y].$$

  Moreover, we have that

$$\bigcup_{(b',l,l') \in \{(a',n,m)\}} \sigma^{(b',l,l')} = \sigma^{(a',n,m)} = \{(b,h,k)\}.$$

  Hence, $\langle c\rho[F(a'_n) \to x], d\rho'[F(a'_m) \to y]\rangle \in \mathcal{R}^{sub}_{\{(b,h,k)\}}$.

- Assume that $\pi_1\rho + a_i \overset{\tau}{\longmapsto} \bar{c}$. Then it must be the case that $\pi_1\rho \overset{\tau}{\longmapsto} \bar{c}$. By Theorem 4.3.4, $\pi_1\rho \overset{\tau}{\longmapsto} \bar{c}$ implies:

  (1.a) there exist a'_n and c_1, x such that $\pi_1 \overset{S(a'_n)}{\longmapsto} c_1$, $\rho(a'_n) \overset{\tau}{\longmapsto} x$ and $\bar{c} \sim c_1\rho[F(a'_n) \to x]$, or

  (1.b) there exist a'_n, b_h, α and c_1, c_2, x, y such that $\pi_1 \overset{S(a'_n)}{\longmapsto} c_1 \overset{S(b_h)}{\longmapsto} c_2$, $\rho(a'_n) \overset{\alpha_i}{\longmapsto} x$, $\rho(b_h) \overset{\bar{\alpha}_j}{\longmapsto} y$ and $\bar{c} \sim c_2\rho[F(a'_n) \to x, F(b_h) \to y]$, or

  (1.c) there exists π'_1 such that $\pi_1 \overset{\tau}{\longmapsto} \pi'_1$ and $\bar{c} \sim \pi'_1\rho$.

  We examine the three possibilities separately and show that in each case it is possible to find a matching move for $\pi_1\rho \overset{\tau}{\longmapsto} \bar{c}$.

  (1.a) The details of this case are entirely similar to the ones given above and thus omitted.

  (1.b) In this case, as $\pi_1 =_\emptyset \pi_2$, $\pi_1 \overset{S(a'_n)}{\longmapsto} c_1$ implies that $\pi_2 \overset{S(a'_m)}{\Longmapsto} d_1$, for some m and d_1 such that $c_1 \approx_{\{(a',n,m)\}} d_1$. Similarly, $c_1 \overset{S(b_h)}{\longmapsto} c_2$ implies that $d_1 \overset{S(b_k)}{\Longmapsto} d_2$, for some k and d_2 such that $c_2 \approx_\varphi d_2$, where $\varphi = \{(a', n, m), (b, h, k)\}$. Moreover, as $\rho =_{(\emptyset,\emptyset)} \rho'$, we have that $\rho(a'_n) =_\emptyset \rho'(a'_m)$ and $\rho(b_h) =_\emptyset \rho'(b_k)$. Hence there exist transitions $\rho'(a'_m) \overset{\alpha_{i_1}}{\Longmapsto} x'$ and $\rho'(b_k) \overset{\bar{\alpha}_{j_1}}{\Longmapsto} y'$ such that $x \approx_\emptyset x'$ and $y \approx_\emptyset y'$. By Lemma 4.3.12, $\pi_2 \overset{S(a'_m)}{\Longmapsto} d_1 \overset{S(b_k)}{\Longmapsto} d_2$, $\rho'(a'_m) \overset{\alpha_{i_1}}{\Longmapsto} x'$ and $\rho'(b_k) \overset{\bar{\alpha}_{j_1}}{\Longmapsto} y'$ imply

  $$\pi_2\rho' \overset{\tau}{\Longmapsto} \bar{d} \sim d_2\rho'[F(a'_m) \to x', F(b_k) \to y'],$$

  for some $\bar{d}$. We are now left to show that $\langle \bar{c}, \bar{d} \rangle \in \mathcal{S}_\emptyset$. By the definition of $\mathcal{S}_\emptyset$, it is sufficient to show that

  $$\langle c_2\rho[F(a'_n) \to x, F(b_h) \to y], d_2\rho'[F(a'_m) \to x', F(b_k) \to y'] \rangle \in \mathcal{R}^{sub}_\emptyset.$$

  We already know that $c_2 \approx_\varphi d_2$, where $\varphi = \{(a', n, m), (b, h, k)\}$. The fact that $c_2\rho[F(a'_n) \to x, F(b_h) \to y]$ and $d_2\rho'[F(a'_m) \to x', F(b_k) \to y']$ are in $\mathbf{L}\mathcal{S}^\Gamma$ follows from the closure of $\mathbf{L}\mathcal{S}^\Gamma$ under derivation. Consider now the φ-vector σ given by

  $$\sigma^{(a',n,m)} = \sigma^{(b,h,k)} = \emptyset.$$

  Then, as $\rho =_{(\emptyset,\emptyset)} \rho'$, $x \approx_\emptyset x'$ and $y \approx_\emptyset y'$, it is easy to see that

  $$\rho[F(a'_n) \to x, F(b_h) \to y] =_{(\varphi,\sigma)} \rho'[F(a'_m) \to x', F(b_k) \to y'].$$

  Moreover, we have that

  $$\bigcup_{(b',l,l') \in \varphi} \sigma^{(b',l,l')} = \emptyset.$$

  This completes the proof of case (1.b).

(1.c) In this case, as $\pi_1 =_{\emptyset} \pi_2$, $\pi_1 \stackrel{\tau}{\longmapsto} \pi_1'$ implies $\pi_2 \stackrel{\tau}{\Longmapsto} \pi_2'$, for some π_2' such that $\pi_1' \approx_{\emptyset} \pi_2'$. By Lemma 4.3.10, $\pi_2 \stackrel{\tau}{\Longmapsto} \pi_2'$ implies $\pi_2\rho' \stackrel{\tau}{\Longmapsto} \bar{d} \sim \pi_2'\rho'$, for some $\bar{d}$. By the operational semantics, we then have that $\pi_2\rho' + a_j \stackrel{\tau}{\Longmapsto} \bar{d}$. We are thus left to check that $\langle \bar{c}, \bar{d} \rangle \in \mathcal{S}_{\emptyset}$. Again, it is sufficient to show that $\langle \pi_1'\rho, \pi_2'\rho' \rangle \in \mathcal{R}_{\emptyset}^{sub}$. This is easily shown by following the pattern used in the previous case.

Assume now that $\pi_1\rho + a_i\surd\!\!\!/$. Then, by Theorem 4.3.7, we have that $\pi_2\rho' + a_j\surd\!\!\!/$. This completes the proof of statement (1).

(2) This is the more involved of the two statements. In order to prove that the defining clauses of $\approx_{\mathcal{H}}$ are met by $\langle c, d \rangle \in \sim \circ \mathcal{R}_{\phi}^{sub} \circ \sim$, it is sufficient to show the following claim:

Claim (♠): The defining clauses of $\approx_{\mathcal{H}}$ are met by $\langle c\rho, d\rho' \rangle \in \sim \circ \mathcal{R}_{\phi}^{sub} \circ \sim$.

In fact, let us assume that (♠) holds and that $\langle c, d \rangle \in \sim \circ \mathcal{R}_{\phi}^{sub} \circ \sim$. This is because there exist x, y such that $c \sim x \mathcal{R}_{\phi}^{sub} y \sim d$. All the defining clauses of $\approx_{\mathcal{H}}$ can be easily seen to be met by $\langle c, d \rangle$. It is sufficient to use the definitions of $\sim$ and $\mathcal{R}_{\phi}^{sub}$ and claim (♠). Here we just check that $\mathbf{Places}(a, c) = dom(\phi_a)$ and $\mathbf{Places}(a, d) = range(\phi_a)$.

Let $a \in V(\Lambda)$. Then, as $c \sim x$, we have that $\mathbf{Places}(a, c) = \mathbf{Places}(a, x)$. As $x \mathcal{R}_{\phi}^{sub} y$, we have, by claim (♠), that $\mathbf{Places}(a, x) = dom(\phi_a)$. Note that, by the definition of $\mathcal{R}_{\phi}^{sub}$, x and y must be of the form required in order to apply (♠). Similarly, $y \sim d$ implies that $\mathbf{Places}(a, y) = \mathbf{Places}(a, d)$ and, as $x \mathcal{R}_{\phi}^{sub} y$, claim (♠) implies that $\mathbf{Places}(a, y) = range(\phi_a)$.

We are thus left to prove claim (♠).

Proof of (♠): The proof is by induction on the combined size of c and d, for $\langle c\rho, d\rho' \rangle \in \sim \circ \mathcal{R}_{\phi}^{sub} \circ \sim$. We assume, as inductive hypothesis, that the defining clauses of $\approx_{\mathcal{H}}$ are met by all $\langle c'\rho_1, d'\rho_2 \rangle \in \sim \circ \mathcal{R}_{\phi}^{sub} \circ \sim$ such that $|c| + |d| > |c'| + |d'|$. Reasoning as above, it is easy to see that it is sufficient to prove that (♠) holds for $\langle c\rho, d\rho' \rangle \in \mathcal{R}_{\phi}^{sub}$. Claim (♠) will then follow immediately from the definitions of $\sim$ and $\mathcal{R}_{\phi}^{sub}$. Assume that $\langle c\rho, d\rho' \rangle \in \mathcal{R}_{\phi}^{sub}$. Then, by the definition of $\mathcal{R}_{\phi}^{sub}$, we have that

(A.2) $c \approx_{\varphi} d$, for some φ,

(B.2) $c\rho, d\rho' \in \mathbf{L}\mathcal{S}^{\Gamma}$, and

(C.2) there exists a φ-vector σ such that $\rho =_{(\varphi,\sigma)} \rho'$ and

$$\phi = \bigcup_{(a,i,j) \in \varphi} \sigma^{(a,i,j)}.$$

We shall now show that the defining clauses of $\approx_{\mathcal{H}}$ are met by $\langle c\rho, d\rho'\rangle \in \mathcal{R}^{sub}_{\phi}$. The details are very similar to the ones in the proof of statement (1). Hence, we shall limit ourselves to giving the proof for two cases not covered in (1).

- We shall show that, for each $a \in V(\Lambda)$, $\mathbf{Places}(a, c\rho) = dom(\phi_a)$ and $\mathbf{Places}(a, d\rho') = range(\phi_a)$. We can calculate as follows:

$$\begin{array}{rcll} \mathbf{Places}(a, c\rho) & = & \bigcup_{(b,n,m)\in\varphi} \mathbf{Places}(a, \rho(F(b_n))) & \\ & = & \bigcup_{(b,n,m)\in\varphi} dom(\sigma_a^{(b,n,m)}) & \text{as } \rho(F(b_n)) \approx_{\sigma(b,n,m)} \rho'(F(b_m)) \\ & = & dom(\phi_a). & \end{array}$$

  A similar calculation shows that $range(\phi_a) = \mathbf{Places}(a, d\rho')$.

- Assume that $c\rho \overset{F(a_i)}{\longmapsto} \bar{c}$. We shall prove that $d\rho' \overset{F(a_j)}{\Longmapsto} \bar{d}$, for some $\bar{d}$ and j such that $\phi_a(i) = j$ and $\langle \bar{c}, \bar{d}\rangle \in \mathcal{S}_{\phi-\{(a,i,j)\}}$. By Theorem 4.3.5, $c\rho \overset{F(a_i)}{\longmapsto} \bar{c}$ implies

  (2.a) there exists $(X, c') \in \eta(c)$ such that $X \neq \emptyset$, $\rho(X)\surd$ and $c'\rho \overset{F(a_i)}{\longmapsto} \bar{c}$, or

  (2.b) there exist a'_n and x such that $F(a'_n)$ occurs in c, $\rho(F(a'_n)) \overset{F(a_i)}{\longmapsto} x$ and $\bar{c} \sim c\rho[F(a'_n) \to x]$. (None of the other possibilities in Theorem 4.3.5 applies to end-moves)

  We proceed by examining the two possibilities separately.

  (2.a) Assume that there exists $(X, c') \in \eta(c)$ such that $X \neq \emptyset$, $\rho(X)\surd$ and $c'\rho \overset{F(a_i)}{\longmapsto} \bar{c}$. Then, by Proposition 4.3.9, $c\rho =_{\surd} c'\rho$. As $=_{\surd}$ is a strong bisimulation, we have that $c\rho \sim c'\rho$. By the definition of $\sim \circ \mathcal{R}^{sub}_{\phi} \circ \sim$, we have that $\langle c'\rho, d\rho'\rangle \in \sim \circ \mathcal{R}^{sub}_{\phi} \circ \sim$. As the combined size of c' and d is less than that of c and d, we may apply the inductive hypothesis to obtain that $d\rho' \overset{F(a_j)}{\Longmapsto} \bar{d}$ for some $\bar{d}$ such that $\phi_a(i) = j$ and $\langle \bar{c}, \bar{d}\rangle \in \mathcal{S}_{\phi-\{(a,i,j)\}}$.

  (2.b) Assume that there exist a'_n and x such that $F(a'_n)$ occurs in c, $\rho(F(a'_n)) \overset{F(a_i)}{\longmapsto} x$ and $\bar{c} \sim c\rho[F(a'_n) \to x]$. As $c \approx_{\varphi} d$ by (A.2) and $F(a'_n)$ occurs in c, we have that $F(a'_m)$ occurs in d, with $m = \phi_a(n)$. As $\rho =_{(\varphi,\sigma)} \rho'$ by (C.2), we have that $\rho(F(a'_n)) \approx_{\sigma(a',n,m)} \rho'(F(a'_m))$. Thus $\rho(F(a'_n)) \overset{F(a_i)}{\longmapsto} x$ implies $\rho'(F(a'_m)) \overset{F(a_j)}{\Longmapsto} y$, for some y such that $\sigma_a^{(a',n,m)}(i) = j$ and

$$x \approx_{\sigma(a',n,m)-\{(a,i,j)\}} y.$$

  By Lemma 4.3.13, $F(a'_m)$ occurs in d and $\rho'(F(a'_m)) \overset{F(a_j)}{\Longmapsto} y$ imply that $d\rho' \overset{F(a_j)}{\Longmapsto} \bar{d} \sim d\rho'[F(a'_m) \to y]$, for some $\bar{d}$. We are left to prove that $\langle \bar{c}, \bar{d}\rangle \in \mathcal{S}_{\phi-\{(a,i,j)\}}$. By the definition of $\mathcal{S}_{\phi-\{(a,i,j)\}}$, it is sufficient to show that

$$\langle c\rho[F(a'_n) \to x], d\rho'[F(a'_m) \to y]\rangle \in \mathcal{R}^{sub}_{\phi-\{(a,i,j)\}}.$$

We know that $c \approx_{\varphi} d$. Again, $c\rho[F(a'_n) \to x], d\rho'[F(a'_m) \to y] \in \mathbf{L}\mathcal{S}^{\Gamma}$ because $\mathbf{L}\mathcal{S}^{\Gamma}$ is closed under derivation. Consider now the φ-vector σ_1 given by:

$$\sigma_1^{(b',l,l')} = \begin{cases} \sigma^{(a',n,m)} - \{(a,i,j)\} & \text{if } (b',l,l') = (a',n,m) \\ \sigma^{(b',l,l')} & \text{otherwise.} \end{cases}$$

Then, as $\rho =_{(\varphi,\sigma)} \rho'$ and $x \approx_{\sigma^{(a',n,m)}-\{(a,i,j)\}} y$, it is easy to see that

$$\rho[F(a'_n) \to x] =_{(\varphi,\sigma_1)} \rho'[F(a'_m) \to y].$$

Moreover, we have that

$$\begin{aligned} \bigcup_{(b',l,l')\in\varphi} \sigma_1^{(b',l,l')} &= (\bigcup_{(b',l,l')\in\varphi-\{(a',n,m)\}} \sigma^{(b',l,l')}) \cup \sigma_1^{(a',n,m)} \\ &= (\bigcup_{(b',l,l')\in\varphi-\{(a',n,m)\}} \sigma^{(b',l,l')}) \cup (\sigma^{(a',n,m)} - \{(a,i,j)\}) \\ &\stackrel{\star}{=} \bigcup_{(b',l,l')\in\varphi} \sigma^{(b',l,l')} - \{(a,i,j)\} \\ &= \phi - \{(a,i,j)\} \end{aligned}$$

Step $\stackrel{\star}{=}$ in the above verification is justified by the fact that $(a,i,j) \notin \sigma^{(b',l,l')}$ for $(b',l,l') \neq (a',n,m)$. This completes the verification of (2.b).

- Assume that $c\rho\surd\!\!\!/$ and $\langle c\rho, d\rho'\rangle \in \mathcal{R}_{\phi}^{sub}$. Then, by Theorem 4.3.6, we have that $d\rho'\surd\!\!\!/$. The verification of the remaining clauses of the definition of $\approx_{\mathcal{H}}$ for statement (2) follows the lines of the similar cases for statement (1) and are thus left to the reader.

This completes the proof of claim (♠) and thus of (2).

In order to prove that $\{\mathcal{S}_{\phi} \mid \phi \in \mathcal{H}\}$ is a weak refine bisimulation, we are thus left to show that it is symmetric. The straightforward, but rather tedious, calculations are omitted. □

The refinement theorem for $\approx_r$ and $\approx_r^+$ is now an obvious consequence of the previous result. The following lemma will be used in the proof of the refinement theorem.

Lemma 4.3.17 *Let π be a labelled process and ρ be a labelled action refinement compatible with π. Assume that, for each $a \in V(\Lambda)$, $\mathbf{un}(\rho(a_i)) = \mathbf{un}(\rho(a_j))$, for all i, j. Let $\rho_u : V(\Lambda) \to \mathbf{P}^{\Gamma}$ be given by $\rho_u(a) = \mathbf{un}(\rho(a_i))$, which is well-defined by the above assumption. Then $\mathbf{un}(\pi\rho) = \mathbf{un}(\pi)\rho_u$.*

Proof: A straightforward induction on the structure of π. □

Notation 4.3.3 *Let ρ and ρ' be action refinements. Then $\rho \approx_r^+ \rho'$ iff for all $a \in V(\Lambda)$, $\rho(a) \approx_r^+ \rho'(a)$.*

Theorem 4.3.9 (The Refinement Theorem) *Let $p, q \in \mathbf{P}^\Gamma$ and ρ, ρ' be action refinements. Then:*

(1) $p \approx_r q$ *implies* $p\rho \approx_r q\rho$;

(2) $p \approx_r^+ q$ *implies* $p\rho \approx_r^+ q\rho$;

(3) $p \approx_r^+ q$ *and* $\rho \approx_r^+ \rho'$ *imply* $p\rho \approx_r^+ q\rho'$.

Proof: We shall only give the proof of statement (2) and, for the sake of simplicity, we shall restrict ourselves to giving the details of the proof for an action refinement ρ which acts like the identity on all actions but $a \in V(\Lambda)$. Let ρ be such that

$$\begin{aligned} \rho(a) &= r \\ \rho(\bar{a}) &= r' \\ \rho(b) &= b \quad \text{for all } b \neq a, \bar{a}. \end{aligned}$$

The general case is only notationally more cumbersome. Assume that $p, q \in \mathbf{P}^\Gamma$ and $p \approx_r^+ q$. Then, by Proposition 4.3.6, $p =_r q$. Then there exist labelled processes π_1, π_2 such that $\mathbf{un}(\pi_1) = p$, $\mathbf{un}(\pi_2) = q$ and $\pi_1 =_\emptyset \pi_2$. We shall now construct a labelled action refinement ρ' compatible with both π_1 and π_2 (see page 161) such that, for each i, $\mathbf{un}(\rho'(a_i)) = \rho(a)$ and $\mathbf{un}(\rho'(\bar{a}_i)) = \rho(\bar{a})$.

Let $\{i_1, \ldots, i_k\}$ be the set of indices of occurrences of a in π_1 and π_2. Similarly, let $\{j_1, \ldots, j_h\}$ be the set of indices of occurrences of $\bar{a}$ in π_1 and π_2. Correspondingly, let $\{\pi_{i_1}, \ldots, \pi_{i_k}, \pi\}$ and $\{\bar{\pi}_{j_1}, \ldots, \bar{\pi}_{j_h}, \bar{\pi}\}$ be sets of processes with pairwise disjoint labellings such that

(a) for all $1 \leq n \leq k$, $\mathbf{un}(\pi_{i_n}) = \mathbf{un}(\pi) = r$, and

(b) for all $1 \leq n \leq h$, $\mathbf{un}(\bar{\pi}_{j_n}) = \mathbf{un}(\bar{\pi}) = \bar{r}$.

The labelled action refinement $\rho' : LAct(\Lambda) \cup \{F(a_i) \mid a_i \in LAct(\Lambda)\} \to \mathbf{L}\mathcal{S}^\Gamma$ is now given by:

- $\rho'(a_n) = \pi_{im}$ if $n = i_m$, π otherwise;
- $\rho'(\bar{a_n}) = \bar{\pi}_{jm}$ if $n = j_m$, $\bar{\pi}$ otherwise;
- ρ' is the identity everywhere else.

By the construction of ρ' and the fact that ρ is an action refinement, it is easy to see that ρ' is a labelled action refinement and that $\rho' =_{(\emptyset,\emptyset)} \rho'$. By construction, ρ' is compatible with both π_1 and π_2. By the definition of $\{\mathcal{S}_\phi \mid \phi \in \mathcal{H}\}$, we have that $\langle \pi_1\rho' + b_i, \pi_2\rho' + b_j \rangle \in \mathcal{S}_\emptyset$, for some b not occurring in $\pi_1\rho'$ and $\pi_2\rho'$. By Theorem 4.3.8, $\pi_1\rho' + b_i \approx_\emptyset \pi_2\rho' + b_j$ and, by Corollary 4.3.6, $\pi_1\rho' =_\emptyset \pi_2\rho'$. By the definition of $=_r$, we then have that

$$\mathbf{un}(\pi_1\rho') =_r \mathbf{un}(\pi_2\rho').$$

By the previous lemma and Corollary 4.3.6, we have that $p\rho \approx_r^+ q\rho$. □

4.4 An Example Distinguishing $\approx_t$ from $\approx_r$

In this section, we shall present an example showing that $\approx_t$ and $\approx_r$ are different equivalences over the language $\mathbf{P}^\Gamma$. The example is based on a similar one devised by R. Van Glabbeek to show that, in general, splitting an action into three gives rise to a different equivalence from $\approx_t$. What van Glabbeek's example, called the *owl-example* by its author, essentially shows is that $\sim_t$ and $\approx_t$ are *not* preserved by action refinements over a language which is rich enough to describe the processes used in it. It turns out that a slightly modified version of the processes used in the owl-example can indeed be described in the language $\mathbf{P}^\Gamma$ considered in this chapter. As a corollary of van Glabbeek's result we then have that:

Fact 4.4.1 *$\approx_t$ is not preserved by action refinement over $\mathbf{P}^\Gamma$.*

In the remainder of this section we shall use van Glabbeek's owl-example to prove that $\approx_t$ and $\approx_r$ are different equivalences over $\mathbf{P}^\Gamma$. The version of the owl-example which is presented below is due to Frits Vaandrager, [Va90], who translated the original example into $\mathbf{CCS}^2$. Let P be the process given by

$$P = (a.(\bar{\beta}|(\bar{\alpha'} + c.(\alpha.e + \gamma.P_1)))|b.(\bar{\alpha}|(\bar{\gamma} + c.(\beta.d + \alpha'.P_2)))) \setminus \alpha \setminus \alpha' \setminus \beta \setminus \gamma,$$

where $P_1 = d|c.\tau.e$ and $P_2 = e|c.\tau.d$. Let Q be identical to P, but with the rôles of d and e interchanged. We shall show that $P \approx_t Q$, but $P \not\approx_r Q$. The arguments given below will hopefully be made more comprehensible by working with a semantic representation of the processes P and Q. The labelled transition systems denoted by the processes P and Q are given in Figure 4.3. Intuitively, in this representation of processes, confluence in the process graph represents concurrency between actions, whilst absence of confluence stands for conflict. Note that only the labels of the transitions along the edges of the graph are explicitly given; the labels of the inner transitions can be obtained by assigning the same label to "parallel edges" in the graph.

We shall now prove that $\mathcal{G}(P) \approx_t \mathcal{G}(Q)$, where $\mathcal{G}(P)$ and $\mathcal{G}(Q)$ denote the process graphs for P and Q, respectively. In order to make our argument clearer, we refer the reader to Figure 4.4,

[2] I thank Frits Vaandrager for making this example available to me.

where the interesting parts of the process graphs associated with P and Q with respect to the timed operational semantics are given. There the dotted arrows stand for beginings and endings of the actions labelling the transitions which will be used in the arguments to follow. Before arguing informally that $\mathcal{G}(P) \approx_t \mathcal{G}(Q)$, it will be useful to give a bit of notation.

Notation 4.4.1 *Given a graph $\mathcal{G}$ and a node n of $\mathcal{G}$, $\mathcal{G}_n$ will denote the sub-graph of $\mathcal{G}$ whose root is n. Isomorphism between graphs will be denoted by $\cong$.*

We shall now informally sketch the construction of a timed bisimulation between $\mathcal{G}(P)$ and $\mathcal{G}(Q)$. The core of such a bisimulation is given by the relation

$$\mathcal{R} = \{\langle n, m\rangle \mid n \text{ is a node of } \mathcal{G}(P),\ m \text{ is a node of } \mathcal{G}(Q) \text{ and } \mathcal{G}(P)_n \cong \mathcal{G}(Q)_m\}.$$

Note that, for instance, $\mathcal{R}$ relates the node labelled x in $\mathcal{G}(P)$ to the one labelled z in $\mathcal{G}(Q)$ and, conversely, the node labelled z in $\mathcal{G}(P)$ to the labelled x in $\mathcal{G}(Q)$. Moreover, $\langle y, y\rangle$ and $\langle w, w\rangle$ are in $\mathcal{R}$. It is easy to see that $\mathcal{R}$ is a timed bisimulation between $\mathcal{G}(P)_1$ and $\mathcal{G}(Q)_1$. We shall now extend $\mathcal{R}$ to a timed bisimulation between $\mathcal{G}(P)$ and $\mathcal{G}(Q)$. Note that *any* timed bisimulation between $\mathcal{G}(P)$ and $\mathcal{G}(Q)$ *must* relate the nodes labelled x and those labelled z in the two graphs, i.e. must contain the pairs $\langle x, x\rangle$ and $\langle z, z\rangle$. This is because:

- the nodes labelled x in $\mathcal{G}(P)$ and $\mathcal{G}(Q)$ are the unique ones reachable from the roots of these graphs via the sequence of sub-actions $S(b)F(b)S(c)F(c)S(a)F(a)$ in which the start of an e-action and of a d-action are both possible, and
- the nodes labelled z in $\mathcal{G}(P)$ and $\mathcal{G}(Q)$ are the unique ones reachable from the roots of these graphs via the sequence of sub-actions $S(a)F(a)S(c)F(c)S(b)F(b)$ in which the start of an e-action and of a d-action are both possible.

An exhaustive analysis of all the possible transitions originating from the nodes labelled x and z in the two graphs shows that $\mathcal{G}(P)_x \approx_t \mathcal{G}(Q)_x$ and $\mathcal{G}(P)_z \approx_t \mathcal{G}(Q)_z$. We are thus left to extend the bisimulation relation to the nodes in $\mathcal{G}(P)$ and $\mathcal{G}(Q)$ lying above those labelled x, y and z. Again, an exhaustive argument shows that "nodes in the same position" in the two graphs are indeed timed bisimilar. We have thus informally hinted at how to construct a timed bisimulation between $\mathcal{G}(P)$ and $\mathcal{G}(Q)$. Hence $\mathcal{G}(P) \approx_t \mathcal{G}(Q)$, as claimed.

We now turn our attention to proving that $\mathcal{G}(P) \not\approx_r \mathcal{G}(Q)$. In order to clarify our argument, we refer the reader to Figure 4.5, where the occurrences of action c relevant to proving the claim are labelled. Note, first of all, that *any* weak refine bisimulation between $\mathcal{G}(P)$ and $\mathcal{G}(Q)$ must relate the nodes labelled n and t in Figure 4.5 with respect to the history $\{(c, i, j)\}$. This is because n and t are the unique nodes in $\mathcal{G}(P)$ and $\mathcal{G}(Q)$, respectively, reachable from

Process Graph for P

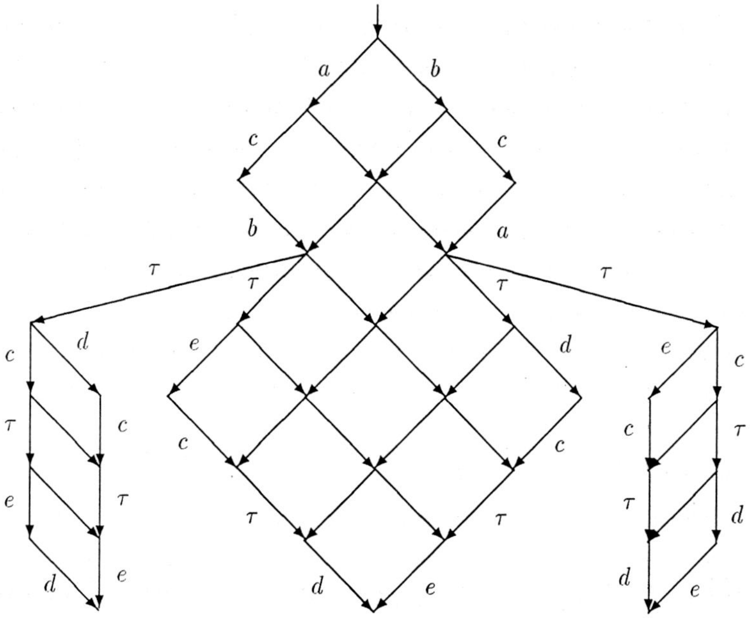

Process Graph for Q

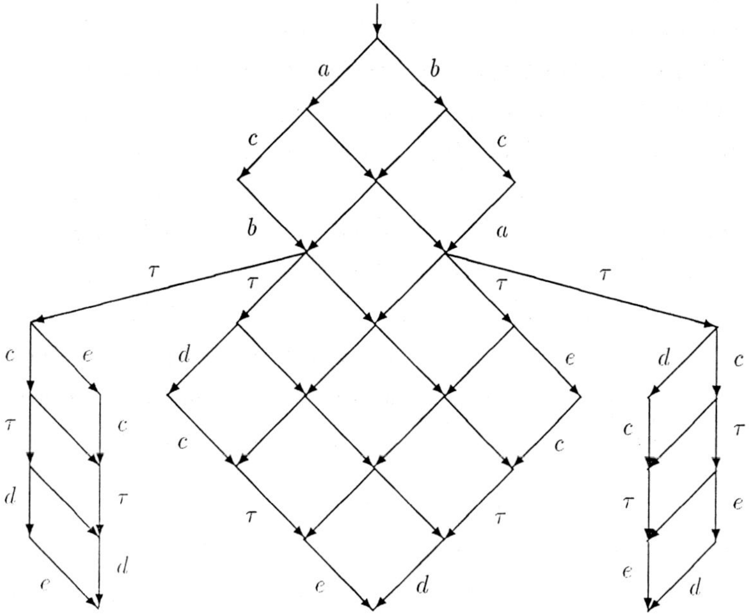

Figure 4.3. The Process Graphs for P and Q

the roots by performing the sequences $S(b)F(b)S(c_i)S(a)F(a)$ and $S(b)F(b)S(c_j)S(a)F(a)$, respectively. When in the state corresponding to node n, the process denoted by $\mathcal{G}(P)$ can perform $S(c_h)$ and enter state x. Similarly, t can perform $S(c_k)$ and enter state y. However, x and y are *not* equivalent with respect to the history $\phi = \{(c,i,j),(c,h,k)\}$. In fact, we can get from state x to state w in $\mathcal{G}(P)$ by peforming the end of action c_i, $F(c_i)$, and in state w it is possible to start a weak d-move. However, none of the states reachable from y in $\mathcal{G}(Q)$ by performing the end of action c_j, the one associated with c_i by the history ϕ, can start a d-move. Hence there is no refine bisimulation between $\mathcal{G}(P)$ and $\mathcal{G}(Q)$, i.e. $\mathcal{G}(P) \not\approx_r \mathcal{G}(Q)$. It is easy to see that $\approx_r$ is contained in $\approx_t$. We then have that:

Fact 4.4.2 $\approx_r$ *is strictly contained in* $\approx_t$ *over* $\mathbf{P}^\Gamma$.

The same example can be used to prove that the definition of timed equivalence is not robust, in the sense that slight modifications in its definition give rise to different notions of equivalence. In §4.3, Proposition 4.3.4, we proved that adding a clause requiring the matching of complete actions to the definition of $\approx_r$ does not change the resulting notion of equivalence. A natural question to ask is whether the same is true of $\approx_t$. We shall now argue that the addition of a clause requiring the matching of complete actions in the definition of $\approx_t$ gives rise to a different notion of equivalence by showing that the resulting equivalence distinguishes the processes P and Q above. Again, it will be convenient to use the process graph representation of P and Q. Consider the state w' in Figure 4.4 reachable from the root in $\mathcal{G}(P)$ via the the sequence $\sigma = bS(c)ac$. Any timed bisimulation requiring in addition the matching of complete actions should relate w' to some state with equivalent potential reachable from the root of $\mathcal{G}(Q)$ via σ. There are only two states reachable from the root of $\mathcal{G}(Q)$ via a weak σ-transition. These states are labelled w and v in Figure 4.4. We claim that w' can be equivalent to neither w nor v. In fact, w' can perform a weak e-move whilst w and v cannot do so. This shows that there can be *no* timed bisimulation between $\mathcal{G}(P)$ and $\mathcal{G}(Q)$ which requires the matching of complete actions.

The situation described above does not present itself with respect to $\approx_t$. In fact, in the definition of $\approx_t$, complete actions are not allowed to occur atomically and the transition σ from the root of $\mathcal{G}(P)$ to the state w' considered above corresponds to performing the sequence $\sigma' F(c)$, where $\sigma' = S(b)F(b)S(c)S(a)F(a)S(c)$. The only state reachable from the root of $\mathcal{G}(P)$ by performing σ' is state y in $\mathcal{G}(P)$. Similarly, the only state reachable from the root of $\mathcal{G}(Q)$ by performing σ' is state y in $\mathcal{G}(Q)$. We have already seen that $\mathcal{G}(P)_y \cong \mathcal{G}(Q)_y$. Hence the transition from y to w' in $\mathcal{G}(P)$ can indeed be matched, up to $\approx_t$, by a transition in $\mathcal{G}(Q)$.

Process Graph for P

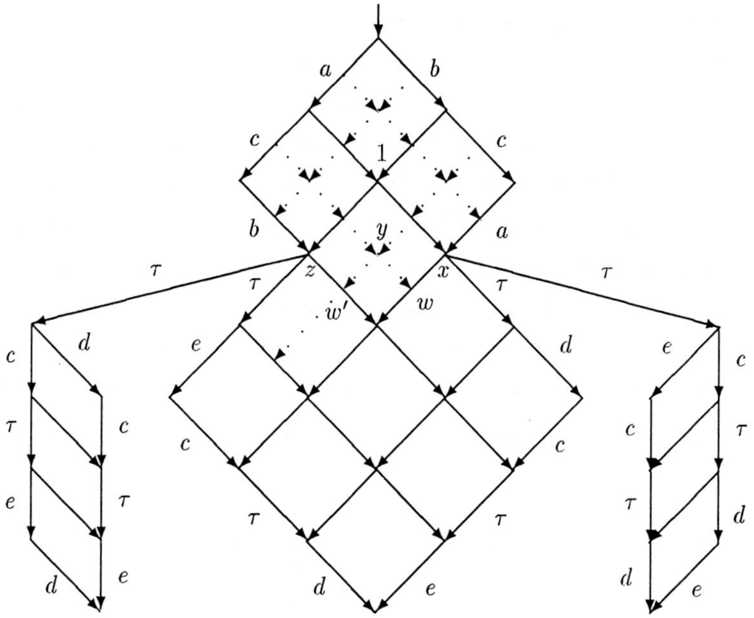

Process Graph for Q

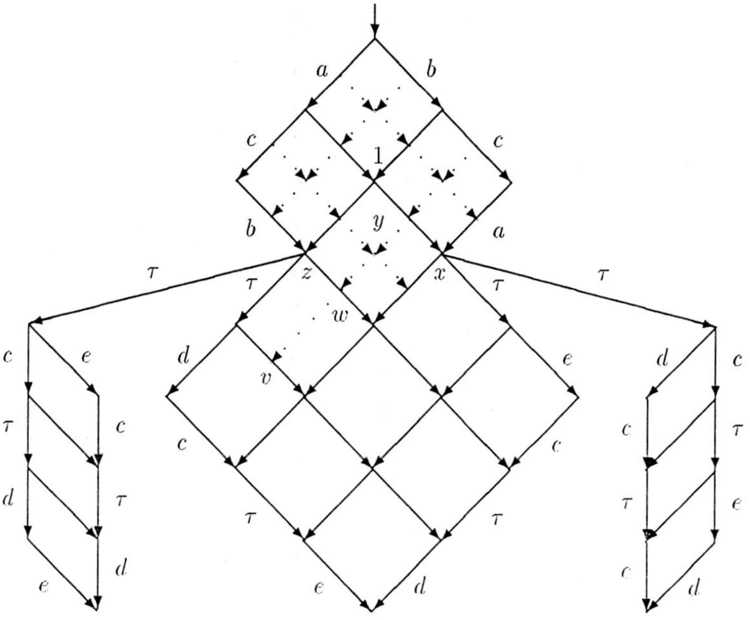

Figure 4.4. The Timed process Graphs for P and Q

4.5 Concluding Remarks

This chapter has been devoted to the development of a process algebra for the specification of concurrent, communicating systems, which incorporates an operator for the refinement of actions by processes, and to the study of a suitable semantic equivalence between specifications written in this language. As we have seen, the syntactic and semantic treatment of action refinement for such a language is much more delicate than it was the case for the simple language considered in Chapter 3. In particular, the synchronization structure of processes, the internal nature of the τ action and the "scoping" of channel names induced by the restriction operator have to be taken into account both in the definition of a suitable notion of action refinement for the richer language considered in this chapter and in the application of action refinements to processes.

At the semantic level, a suitable notion of equivalence for the language we have considered can still be obtained by assuming that the actions processes perform during their evolution are not atomic. However, due to the expressive power of the language under consideration, the formalization of a suitable notion of bisimulation equivalence for it based on this intuition has turned out to be more involved than for the language considered in Chapter 3. In particular, the natural weak version of the refine equivalence introduced in Chapter 3 turns out to play a more fundamental rôle than weak timed equivalence.

The treatment of action refinement for a language with communication, an internal action and restriction presented in this chapter is, at least to the author's knowledge, new. Most of the studies of action refinement in the literature deal with this operator at the *semantic* level and do not address *syntactic* issues like the treatment of binders. Moreover, synchronization between concurrent activities does not require a special treatment when dealt with at the semantic level only. As an example, consider the processes $p = a|\bar{a}$ and $q = p + \tau$. As remarked in this chapter, suitable notions of syntactic refinement for the language P^{Γ} must take into account the extra structure on the set of actions induced by the synchronization algebra underlying the parallel composition operator. In particular, action refinements should not interfere with the synchronization structure of processes. This requirement has led us to restrict ourselves to considering action refinements which satisfy axiom (ComPres) on page 141. However, semantically, process p above has a "τ-*summand*" and most semantic representations for p and q would be *identical*. In our treatment of action refinement, the action refinement operators are applied to syntactic descriptions of processes and thus have had to be restricted to those which, in some formal sense, preserve the intended semantics of processes.

The refinement theorem presented in this chapter is related to the one for ST-bisimulation for Prime Event Structures presented in [Gl90,90a]. In fact, ST-bisimulation and refine equivalence, although defined in slightly different ways and on different domains, are both attempts to formalize the idea of a bisimulation-like relation between processes based on the matching of non-atomic actions. In the above quoted references, the author only considers Prime Event Structures without internal τ-actions. Moreover, because of the chosen system model, he restricts the class of refinements to the finite, conflict-free ones, i.e. for each action a, $\rho(a)$, the process used to refine a, is a finite, conflict-free Event Structure. The ideas underlying our refinement theorem and the one presented in [Gl90] are closely related to the ones upon which the *failures semantics based on interval semiwords* proposed in [Vo90] is based. There the author proposes a notion of failures semantics, [BHR84], for safe Petri Nets, [Rei85], which is the largest congruence with respect to action refinement contained in the standard, interleaving failures equivalence.

The semantic equivalences considered in the last two chapters, which are suitable for process algebras incorporating an action refinement combinator, are rather discriminating. In particular, parallelism is a primitive notion in their semantic theories. However, both the timed equivalence considered in Chapter 3 and weak refine equivalence are not based on explicit observations of the causal structure of systems. It is interesting to ask whether any natural causality based models are fully abstract with respect to either of these equivalences. The following chapter will be entirely devoted to giving a positive answer to the above question for a simple language and a basic model of concurrency based on partial orders: the series-parallel pomsets of Pratt and Gischer, [Gi84].

Process Graph for P

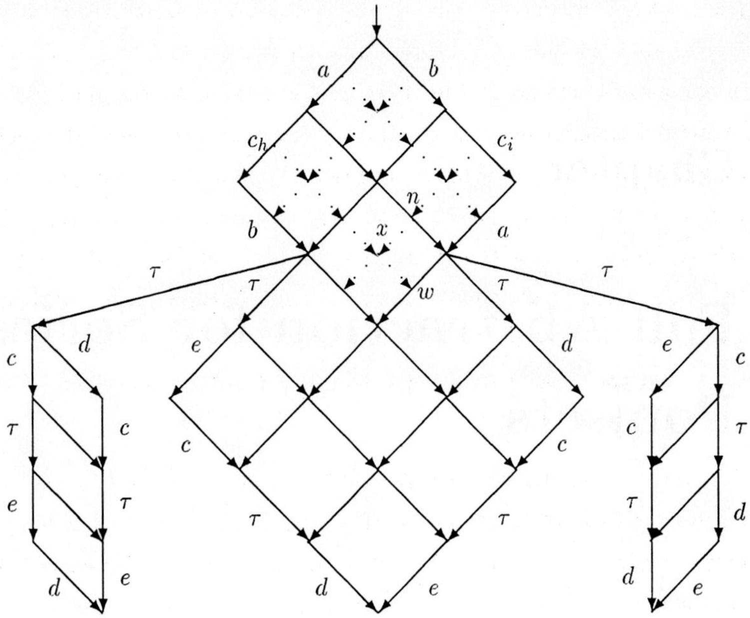

Process Graph for Q

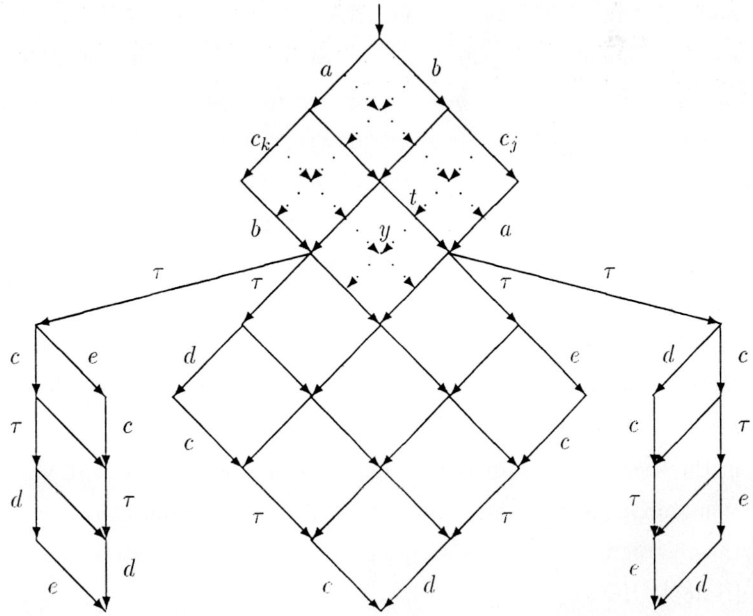

Figure 4.5. (Parts of) The Labelled Process Graphs for P and Q

Chapter 5

Full Abstraction for Series-Parallel Pomsets

5.1 Introductory Remarks

In recent years, many models of concurrent computation based upon partial orders have been proposed in the literature, e.g. *Petri Nets* [Rei85], *Event Structures* [Win80,87], *Pomsets* [Pr86] and, more recently, *Causal Trees* [DD89a]. These models are based upon the idea that concurrent, communicating systems are characterized by their causal structure, i.e. by the computational events a system performs during its evolution together with the causal dependencies amongst them, and that its proper description is necessary in accounting for the nonsequential behaviour of distributed systems. The mathematical tractability of causality-based models has been investigated in the literature by providing operational and denotational semantics for process algebras, such as **CCS** [Mil80,89], **CSP** [Hoare85] and **ACP** [BK85], in terms of the above mentioned models. Partial order operational semantics for standard process algebras have been presented in e.g. [BC88], [DDM88], [DD89a], and denotational semantics are given in e.g. [Win82], [Go88], [Tau89]. Several notions of equivalence over the above mentioned models, which allow to abstract from the way processes evolve, have recently been proposed in the literature (the interested reader is invited to consult [GV87] and [Gl90] for a comparison among some of the proposals), thus importing into the partial ordering setting some of the abstraction techniques supported by the standard interleaving equivalences and preorders [HM85], [H88a], [BHR84].

However, not much work has been carried out in studying reasonable testing scenarios, [DH84], which justify the use of these models in giving semantics to concurrent programming languages. Notions of observability play a fundamental rôle in the study of suitable semantics for programming languages. Following Milner and Plotkin's paradigm, mathematical models for programming languages should be justified by comparing them with some natural notion of behaviourally defined equivalence between processes. Models that are in complete agreement with the coarsest equivalence over processes induced by the chosen notion of observability are called *fully abstract* in [Mil77], [Pl77], [HP79], [Sto88]. As fully abstract semantic models are the most abstract ones which are consistent with the chosen notion of observability, it is natural to try to justify the choice of a model for a language by showing that it induces exactly all the distinctions that can be made by means of some natural notion of observation.

The main aim of this chapter is to provide such a behavioural justification for a simple model based on partial orders, namely the class of *series-parallel* or *N-free pomsets* [Gi84], [Pr86]. Series-parallel pomsets have been extensively studied in the literature, see e.g. [Gi84], [BC88], [Ts88], and have a pleasing algebraic and order-theoretic structure that will be exploited in the proofs of the main results of this chapter. Following Gischer, the algebraic structure of the class of series-parallel pomsets will allow us to relate it to a simple process algebra, whose terms are built from a set of generators by means of the operators of sequential and parallel composition. This gives us a syntax for denoting such partially ordered structures and will allow us to give a standard LTS semantics for the resulting language. We shall define a standard notion of observational equivalence over processes by means of the bisimulation technique [Pa81], [Mil83]. The notion of observability underlying such a notion of equivalence has been thoroughly investigated in [Ab87a] and is called "tightly-controlled testing" in [H88b]. Series-parallel pomsets do not give rise to a fully abstract model with respect to standard bisimulation equivalence; however, by enriching the language with a *refinement operator* like the one used in Chapters 3 and 4 and closing bisimulation equivalence with respect to all the contexts built using this new language construct we shall be able to make series-parallel pomsets fully observable. In other words, series-parallel pomsets are fully abstract with respect to the coarsest congruence obtained by applying Abramsky's testing scenario in all refinement contexts. By relying on results from Chapter 3, we shall be able to provide a natural behavioural characterization of series-parallel pomsets in terms of Hennessy's *timed equivalence* [H88c]. These results will cast the observability of series-parallel pomsets in a well-known interleaving setting.

A natural notion of basic observation which is widely used in the interleaving models for concurrency is that of *trace*, [Hoare85]. Indeed, some natural models for concurrency, e.g. Hennessy's *Acceptance Trees*, [H85], and the *Failures model*, [BHR84], have been shown to be fully abstract with respect to behavioural equivalences which are intrinsically based on such a notion of observability, [Main88], [H88a]. A natural question to ask is whether series-parallel pomsets can be made fully observable by assuming a trace-based notion of observation. In this chapter, we shall provide a partial answer to this question by showing that the order structure of a series-parallel pomset is totally revealed by its set of *ST-traces*, [Gl90]. ST-semantics has been recently proposed in [GV87], [Gl90] as a refinement of *split-semantics*, [H88c], in which an explicit link is required between the beginning and the end of any event. It will be shown that series-parallel pomsets give rise to a fully abstract model with respect to ST-trace equivalence over the simple process algebra considered in this chapter. As a corollary of this result, we shall be able to give a complete axiomatization of ST-trace equivalence over the class of series-parallel pomsets.

We now give a brief outline of the remainder of the chapter. Section 5.2 is devoted to a review of mostly standard material in the theory of pomsets. Two behavioural semantics for series-parallel pomsets, based on the notion of bisimulation equivalence, are presented in §5.3. We shall show that series-parallel pomsets are fully abstract with respect to the finer of the two behavioural semantics, which may be seen as arising by applying Abramsky's testing scenario for bisimulation equivalence in all refinement contexts. The proof of this result is algebraic in nature and relies on Gischer's axiomatization of the theory of series-parallel pomsets, [Gi84], and on work presented in Chapter 3. Section 5.4 is entirely devoted to providing another behavioural characterization for series-parallel pomsets. We shall prove that ST-trace equivalence coincides with equality over the class of SP pomsets, thus giving a trace-theoretic understanding of this simple model based on partial orders. We end with a section of concluding remarks.

5.2 Series-Parallel Pomsets

This section will be devoted to a brief review of some basic notions of the theory of *partially ordered multisets* (or *pomsets* in Pratt and Gischer's terminology) which will find application in the remainder of the chapter. The interested reader is referred to [Gr81], [Gi84], [Pr86], for more information on pomsets and further references. The following definition introduces the main objects of study of the chapter.

Definition 5.2.1

(1) *A labelled poset $\mathbb{P}$ over a label set L is a triple $\mathbb{P} = (P, <, l)$, where*

- *P is a finite set of events,*
- *$<$ is a binary, transitive and acyclic relation over P, and*
- *$l : P \to L$ is a labelling function.*

Two L-labelled posets $\mathbb{P}_i = (P_i, <_i, l_i)$, $i = 1, 2$, are isomorphic, written $\mathbb{P}_1 \cong \mathbb{P}_2$, iff there exists a bijective function $h : P_1 \to P_2$ such that, for all $u, v \in P_1$, $u <_1 v$ iff $h(u) <_2 h(v)$ and $l_1(u) = l_2(h(u))$.

(2) *A pomset over* L, $\alpha = [P, <, l]$, *is an isomorphism class of* L*-labelled posets. For a label set* L, **Pom[L]** *will denote the set of pomsets over* L *and will be ranged over by* $\alpha, \beta \ldots$.

Several operations on pomsets have been defined in the above given references. Since pomsets are isomorphism classes of labelled posets, it will be convenient to define operations on them by using arbitrary representatives of the isomorphism class. For each operation it will be straightforward to establish that the result of the operation is independent of the chosen representative and such verifications will be omitted. Let **A** be a set of *observable actions* ranged over by $a, b, a' \ldots$. In the remainder of the chapter we shall only need the following operations over **Pom[A]**.

- *Empty pomset.* **I** will denote the isomorphism class of the **A**-labelled poset $(\emptyset, \emptyset, \emptyset)$.

- *Atomic actions.* For each $a \in \mathbf{A}$, a will denote, with abuse of notation, the isomorphism class of the one element poset labelled with a. In what follows, **A** will be used to denote, with abuse of notation, the set of all such atomic pomsets.

- *Sequential and parallel composition.* Let $\alpha = [P_1, <_1, l_1]$ and $\beta = [P_2, <_2, l_2]$ be pomsets on **A** and assume, wlog, that $P_1 \cap P_2 = \emptyset$. Then $\alpha; \beta$, the *sequential composition* of α and β is given by

$$\alpha; \beta = [P_1 \cup P_2, <_1 \cup <_2 \cup (P_1 \times P_2), l_1 \cup l_2]$$

  and $\alpha | \beta$, the *parallel composition* of α and β, is given by

$$\alpha | \beta = [P_1 \cup P_2, <_1 \cup <_2, l_1 \cup l_2].$$

Following Gischer [Gi84], the class SP of *series-parallel pomsets* over **A** may now be defined to be the closure of **A** and **I** with respect to the operations of sequential and parallel compositon. The definition of the class of pomsets SP has a pleasing algebraic flavour; indeed, the class of pomsets SP is in close correspondence with the set of terms **SP** built from the set of observable actions **A** by means of the operators of sequential and parallel composition. More formally, let **SP** be the set of terms generated by the syntax

$$p ::= nil \mid a \mid p;p \mid p|p,$$

where $a \in \mathbf{A}$. **SP** will be ranged over by $p, q, p' \ldots$. Following Gischer, the set of terms **SP** may be interpreted as series-parallel pomsets by defining the semantic map $[\![\cdot]\!] : \mathbf{SP} \to SP$ as follows:

- $[\![nil]\!] = \mathbf{I}$,

$$\begin{array}{llcl}
\textbf{(PAR1)} & x|nil & = & x \\
\textbf{(PAR2)} & x|y & = & y|x \\
\textbf{(PAR3)} & (x|y)|z & = & x|(y|z)
\end{array}$$

$$\begin{array}{llclcl}
\textbf{(SEQ1)} & x;nil & = & x & = & nil;x \\
\textbf{(SEQ2)} & (x;y);z & = & x;(y;z) & &
\end{array}$$

Figure 5.1. The set of axioms E

- $[\![a]\!] = a$,
- $[\![p;q]\!] = [\![p]\!];[\![q]\!]$ and
- $[\![p|q]\!] = [\![p]\!]|[\![q]\!]$.

The following theorem, which formalizes the close connection between **SP** and *SP* and gives a complete axiomatization of the congruence on **SP** induced by the above given denotational semantics, has been proven in [Gi84] (Theorem 5.2, page 23). Let $=_E$ denote the least **SP**-congruence which satisfies the set of axioms E in Figure 5.1.

Theorem 5.2.1 (Gischer) *For each* $p, q \in \mathbf{SP}$, $[\![p]\!] = [\![q]\!]$ *iff* $p =_E q$.

The algebraic characterization of the theory of *SP* pomsets given by the above theorem will provide the key to their behavioural characterization, which will be presented in the following section; namely, we shall give a behavioural view of the processes in **SP**, based on a well-understood testing scenario familiar from the theory of bisimulation semantics, and prove that the denotational semantics for **SP** given by the map $[\![\cdot]\!]$ is fully abstract with respect to it. Following Milner and Plotkin's paradigm, this will justify the choice of *SP* as a denotational model for **SP** by showing that *SP* is the most abstract model for **SP** which is consistent with the chosen testing scenario.

We end this review of standard material on series-parallel pomsets with an order-theoretic characterization of the class of pomsets *SP*. It is well-known that the class of pomsets *SP* coincides with that of the so-called *N-free pomsets*, see e.g. [Gi84] (Theorem 3.2. pp. 14-15) and [BC88], where a more general result is proven for Event Structures. Here we only present a result from [Ts88] giving a characterization of *SP* pomsets in terms of their order structure which will find application in §5.4.

Proposition 5.2.1 (Tschantz) *A pomset* $[P, <, l]$ *is series-parallel iff the following property holds:*

$$(N)\ \forall w, x, y, z \in P \quad w < y, w < z \text{ and } x < z \text{ imply } y < z \text{ or } w < x \text{ or } x < y.$$

In what follows, a labelled poset $\mathbb{P}$ will be said to be series-parallel (SP) iff it satisfies the above-given property (N). **Pos[A]** will denote the class of SP posets labelled on **A**.

5.3 Full-Abstraction for Series-Parallel Pomsets

This section will be entirely devoted to a discussion of a behavioural semantics for the simple language **SP** and to a proof of full abstraction of the denotational semantics given by the map $[\![\cdot]\!]$ with respect to it. Following Milner and Plotkin's approach, the behavioural view of processes we shall present, which is based on the notion of testing characterizing standard bisimulation semantics studied in [Ab87a], will justify the denotational semantics in terms of series-parallel pomsets. In what follows, we shall introduce two operational semantics for the set of processes **SP** and two notions of observational equivalence for it. Relying on results from Chapter 3, we shall study the relationships between the two behavioural theories of processes and prove that the denotational semantics is fully abstract with respect to the finer one, the timed equivalence proposed in [H88c] and studied in Chapter 3 for a super-language of **SP**.

Operationally, following the work presented in Chapter 3, the constructs in the language for processes **SP** will be interpreted in a fairly standard way; following Milner [Mil80,89], *nil* will be interpreted as the process that cannot perform any move. A generator $a \in \mathbf{A}$ will be interpreted as a process which is capable of performing the task represented by a and terminate in doing so. The combinators ; and | will stand for sequential composition and parallel composition (without communication), respectively. Both the operational semantics for the language **SP** consist of two ingredients:

(1) a termination predicate $\surd$, used in giving an operational account of the sequential composition operator, and

(2) a standard LTS semantics for **SP** given using Plotkin's SOS method, [Pl81].

The termination predicate $\surd$ for **SP** is the restriction to this language of the one defined for the language **P** in Chapter 3. (See Definition 3.2.4) For the sake of clarity, we recall that $\surd$ is the least subset of **SP** which satisfies the following axiom and rule:

- $nil \in \surd$,

$$\begin{array}{llll}
(1) & a \xrightarrow{a} nil & & \\
(2) & p \xrightarrow{a} p' & \text{implies} & p;q \xrightarrow{a} p';q \\
(3) & p\surd, q \xrightarrow{a} q' & \text{imply} & p;q \xrightarrow{a} q' \\
(4) & p \xrightarrow{a} p' & \text{implies} & p|q \xrightarrow{a} p'|q \\
 & & & q|p \xrightarrow{a} q|p'
\end{array}$$

Figure 5.2. Axiom and rules for $\xrightarrow{a}$

- $p \in \surd$ and $q \in \surd$ imply $p; q \in \surd$ and $p|q \in \surd$.

As usual, $p \in \surd$ will be often written as $p\surd$. Using this termination predicate we may now give the first Labelled Transition System semantics for **SP**; this semantics will be based on the assumption that processes evolve by performing actions which are atomic. For each $a \in \mathbf{A}$, $\xrightarrow{a}$ will denote the restriction to **SP** of the transition relations given for the super-language **P** in Definition 3.2.5. The interested reader may find the defining axiom and rules for $\xrightarrow{a}$ collected in Figure 5.2. As usual, we shall use $\sim$ to denote bisimulation equivalence over the LTS $\langle \mathbf{SP}, \mathbf{A}, \longrightarrow \rangle$. (See Definition 3.2.2) The following proposition is then an immediate consequence of Proposition 3.2.2.

Proposition 5.3.1 *$\sim$ is an* **SP***-congruence.*

The testing scenario which is needed to characterize $\sim$ as a testing equivalence has been spelt out by S. Abramsky in [Ab87a]; a tutorial exposition of Abramsky's testing characterization of the equivalence $\sim$ may be found in [H88b]. The main import of Abramsky's results is that, by using $\sim$ as our basic notion of equivalence, we automatically have a testing scenario justifying it; in the remainder of this section we shall behaviourally characterize the class of *SP* pomsets by means of the testing scenario presented in [Ab87a]. However, as it is stated in the following proposition, there is still a mismatch between the denotational semantics for **SP** given by $[\![\cdot]\!]$ and the behavioural semantics given in terms of $\sim$. In fact, the denotational semantics is sound, but *not* complete, with respect to the behavioural one.

Proposition 5.3.2 (i) *For all $p, q \in$* **SP**, *$[\![p]\!] = [\![q]\!]$ implies $p \sim q$.*

(ii) *$a; a \sim a|a$, but $[\![a; a]\!] \neq [\![a|a]\!]$.*

Proof:

(i) By Gischer's theorem and the above proposition, it is sufficient to prove that all the equations in E are sound with respect to $\sim$. The straightforward verifications are omitted.

(ii) Trivial. □

The import of the above proposition is that, not surprisingly, series-parallel pomsets do not give rise to a fully-abstract model with respect to standard bisimulation equivalence. The remainder of this section is devoted to showing how to define a behavioural semantics for the language **SP** with respect to which the denotational model *SP* is fully abstract. Following the system-testing approach discussed in [H88b], the discriminating power of the testing scenario which induces the equivalence $\sim$ over **SP** may be increased by enriching the language with some computationally meaningful constructs and by applying the basic tests presented in [Ab87a] to processes in all language contexts built using the new combinators. In what follows, we shall apply this philosophy by enriching the language **SP** with a *refinement operator* ρ like the ones considered in e.g. [NEL88], [GG88] and Chapters 3 and 4 of this thesis.

Definition 5.3.1

(i) *A* refinement map *is a function* $\rho : \mathbf{A} \to \mathbf{SP}$.

(ii) *The closure of* $\sim$ *with respect to all refinement contexts,* $\sim^\rho$*, is given by*

$$p \sim^\rho q \textit{ iff, for all refinement maps } \rho,\ p\rho \sim q\rho,$$

where $p\rho$ *and* $q\rho$ *are the terms (in* **SP***) obtained by syntactically replacing* $\rho(a)$ *for each occurrence of* a *in* p *and* q*, respectively.*

By construction, $\sim^\rho$ is the largest **SP**-congruence contained in $\sim$ which is preserved by all refinements of actions by processes. As pointed out before, this notion of equivalence may be seen as arising by applying Abramsky's testing scenario in all refinement contexts. We shall now show that $\sim^\rho$ is indeed the behavioural counterpart of the denotational model *SP*, i.e. that series-parallel pomsets are fully abstract with respect to the behavioural semantics induced by $\sim^\rho$. The proof of this claim proceeds in two steps. First of all, relying on work presented in Chapter 3, we shall give a behavioural characterization of $\sim^\rho$ in terms of Hennessy's timed equivalence [H88c], $\sim_t$. Secondly, we shall prove that the set of equations E in Figure 5.1 completely axiomatize $\sim_t$ over **SP**. The result will then follow as $\sim_t$ and the congruence induced over **SP** by the denotational semantics have a common axiomatization.

It is easy to see that $\sim$ is strictly weaker than $\sim^\rho$. For instance, as previously remarked, $a;a \sim a|a$; however, $(a;a)\rho \not\sim (a|a)\rho$, where ρ is any refinement map such that $\rho(a) = b;c$. In fact, $(a|a)\rho = (b;c)|(b;c)$ can perform two b-moves in a row, whilst $(a;a)\rho = (b;c);(b;c)$ can not. Hence $a;a \not\sim^\rho a|a$ and this implies that $\sim$ itself is not preserved by the refinement combinator

(1) $a \xrightarrow{S(a)}_t F(a)$

(2) $F(a) \xrightarrow{F(a)}_t nil$

(3) $s \xrightarrow{e}_t s'$ implies $s;p \xrightarrow{e}_t s';p$

(4) $s\surd_{\mathcal{S}}, p \xrightarrow{e}_t s'$ imply $s;p \xrightarrow{e}_t s'$

(5) $s_1 \xrightarrow{e}_t s_1'$ implies $s_1|s_2 \xrightarrow{e}_t s_1'|s_2$, $s_2|s_1 \xrightarrow{e}_t s_2|s_1'$

Figure 5.3. Axioms and rules for $\xrightarrow{e}_t$

over **SP**. As pointed out in Chapter 3, this is not at all surprising as the definition of $\sim$ is based on the assumption that processes evolve from state to another by performing actions which are atomic. This behavioural view of processes becomes inadequate in the presence of a refinement operator like ρ and a more refined behavioural description of the processes in **SP** is needed. In Chapter 3, we presented a behavioural view of processes based on the assumption that beginnings and endings of actions are distinct events and that they may be observed. Let us recall that, for each $a \in \mathbf{A}$, $S(a)$ and $F(a)$ are used to denote the beginning and the termination of action a, respectively, and that $\mathbf{A_S} =_{def} \{S(a), F(a) \mid a \in \mathbf{A}\}$ will be the new set of observable events and will be ranged over by e.

As pointed out in [H88c] and Chapter 3, the language for processes is not sufficiently expressive to describe a possible state a process may reach by executing the beginning of an action. To overcome this problem, a new symbol $F(a)$ for each $a \in \mathbf{A}$ is introduced into the language. $F(a)$ will denote the state in which action a is being executed but is not terminated yet. The set of *process states* $\mathcal{S}$ over **SP** is then defined by considering **SP** in place of **P** in Definition 3.3.1. The operational semantics for process states may be defined along the lines followed in Chapter 3. For each $e \in \mathbf{A_S}$, the transition relation $\xrightarrow{e}_t$ over $\mathcal{S}$ is the restriction to **SP** of the one given in Definition 3.3.3 for the super-language **P**. The interested reader may find the defining axioms and rules for $\xrightarrow{e}_t$ in Figure 5.3. The defining rules of $\xrightarrow{e}_t$ use a termination predicate on process states, $\surd_{\mathcal{S}}$, which is induced on $\mathcal{S}$ by the one previously defined on $\mathcal{SP}$; namely, $s\surd_{\mathcal{S}}$ iff $s \in \mathcal{SP}$ and $s\surd$. A standard behavioural equivalence over process states may now be defined using the notion of timed bisimulation introduced in the previous two chapters. We recall that a relation $\mathcal{R} \subseteq \mathcal{S}^2$ is a *timed bisimulation* iff it is symmetric and, for each $(s_1, s_2) \in \mathcal{R}$, $e \in \mathbf{A_S}$,

$$s_1 \xrightarrow{e}_t s_1' \text{ implies, for some } s_2',\ s_2 \xrightarrow{e}_t s_2' \text{ and } (s_1', s_2') \in \mathcal{R}.$$

As in Chapter 3, we let $\sim_t$ denote the maximum timed bisimulation. The following theorem from Chapter 3 states that $\sim_t$ gives a behavioural characterization of the relation $\sim^\rho$ defined previously by purely algebraic means. (See Theorem 3.3.4 on page 86)

Theorem 5.3.1 *For all $p, q \in \mathbf{SP}$, $p \sim_t q$ iff $p \sim^\rho q$.*

The behavioural characterization of $\sim^\rho$ given by the above-stated theorem will be the touchstone for relating $\sim^\rho$ to the denotational semantics for **SP** in terms of series-parallel pomsets. The proof of full-abstractness of the denotational semantics with respect to $\sim^\rho$ relies on Gischer's axiomatization of the congruence induced by $[\![\cdot]\!]$ over **SP** stated in Theorem 5.2.1. Let us recall, for the sake of clarity, that $=_E$ denotes the least congruence over **SP** that satisfies the set of equations E in Figure 5.1. The key to the full-abstraction result is then provided by the following theorem.

Theorem 5.3.2 (Equational characterization of $\sim_t$) *For all $p, q \in \mathbf{SP}$, $p \sim_t q$ iff $p =_E q$.*

The following section will be entirely devoted to a proof of Theorem 5.3.2. The full-abstractness of series-parallel pomsets with respect to $\sim^\rho$ now follows fairly straightforwardly from the results stated above.

Theorem 5.3.3 (Full-abstraction for series-parallel pomsets) *For all $p, q \in \mathbf{SP}$, $[\![p]\!] = [\![q]\!]$ iff $p \sim^\rho q$.*

Proof: Assume $p, q \in \mathbf{SP}$. Then:

$$
\begin{aligned}
[\![p]\!] = [\![q]\!] &\iff p =_E q \quad \text{by Theorem 5.2.1} \\
&\iff p \sim_t q \quad \text{by Theorem 5.3.2} \\
&\iff p \sim^\rho q \quad \text{by Theorem 5.3.1.} \ \Box
\end{aligned}
$$

In order to establish the above theorem, it is then sufficient to prove Theorem 5.3.2 and the following section will be entirely devoted to a detailed proof of this result. We end this section with a few comments on the equational characterization of $\sim_t$ provided by Theorem 5.3.2. It is interesting to remark that the equational characterization of $\sim_t$, and consequently of $\sim^\rho$, is finite and does not make use of any auxiliary operator. This is not in contrast with F. Moller's results on the non-finite axiomatizability of "strong bisimulation"-like equivalences because $\sim_t$, when considered over SP, does not satisfy his "reasonableness criterion". See [Mol89] for more details. Moreover, we can prove a stronger version of Theorem 5.3.2 stating that the above-given equational characterization of $\sim_t$ is also ω-complete [Mol89], i.e. complete for the open term theory.

We shall now present a proof of the ω-completeness of the set of equations E with respect to $\sim_t$. Let Var be a countable set of *variables* ranged over by x, y, z. **SP**(Var) will denote the set of expressions built by adding the clause

- $x \in \texttt{Var}$ implies $x \in \mathbf{SP}(\texttt{Var})$

to the formation rules for **SP**. **SP**(Var) will be ranged over by $t, t', t_1, \ldots$. The equivalence $\sim_t$ can now be extended to **SP**(Var) in the standard way as follows:

Definition 5.3.2 *Let $t, t' \in \mathbf{SP}(\texttt{Var})$. Then $t \sim_t t'$ iff for all closed substitutions $\sigma : \texttt{Var} \to \mathbf{SP}$, $t\sigma \sim_t t'\sigma$. An equational theory EQ over the signature of* **SP** *is then called* ω-complete *with respect to $\sim_t$ iff for all open terms $t, t' \in \mathbf{SP}(\texttt{Var})$, $t \sim_t t'$ iff $EQ \vdash t = t'$. $EQ \vdash t = t'$ will also be written as $t =_{EQ} t'$.*

We shall now prove that the set of axioms E presented in Figure 1 is indeed ω-complete with respect to $\sim_t$ over **SP**(Var). In the proof we shall make use of a novel technique for proving the ω-completeness of a set of equations developed by J.F. Groote in [Gro90]. For the sake of clarity, we shall now briefly outline Groote's proof-technique for showing the ω-completeness of a set of equations. Assume that t and t' are open terms in **SP**(Var) and $t \sim_t t'$, i.e., by Theorem 5.3.2, $t\sigma =_E t'\sigma$ for all closed substitutions σ. The application of Groote's technique requires the isolation of a closed substitution $\rho : \texttt{Var} \to \mathbf{SP}$, mapping each variable occurring in t and t' to a distinguished closed term representing this variable, and of a translation map $R : \mathbf{SP} \to \mathbf{SP}(\texttt{Var})$, which replaces each subterm representing a variable by the variable itself. This pair of functions is required to satisfy the following conditions:

(1) $t =_E R(\rho(t))$ and $t' =_E R(\rho(t'))$,

(2) for each $\odot \in \{;, |\}$ and $p_1, p_2, q_1, q_2 \in \mathbf{SP}$, $R(p_1 \odot p_2) =_{E'} R(q_1 \odot q_2)$, where $E' = E \cup \{R(p_i) = R(q_i) \mid i = 1, 2\}$, and

(3) for each axiom $t_1 = t_2$ in E and closed substitution σ, $R(\sigma(t_1)) =_E R(\sigma(t_2))$.

Having found such a pair of maps ρ and R satisfying conditions (1)-(3) above, we could then obtain the ω-completeness of E with respect to $\sim_t$ by applying the following instance of Theorem 3.1 from [Gro90], page 317.

Theorem 5.3.4 *If for each $t, t' \in \mathbf{SP}(\texttt{Var})$ such that $t\sigma =_E t'\sigma$, for all closed substitutions σ, there exist a closed substitution $\rho : \texttt{Var} \to \mathbf{SP}$ and a map $R : \mathbf{SP} \to \mathbf{SP}(\texttt{Var})$ satisfying (1)-(3) above then E is ω-complete.*

We shall now apply the technique described above to prove that E is indeed ω-complete with respect to $\sim_t$ over **SP**.

Theorem 5.3.5 (ω-Completeness) *For each $t, t' \in \mathbf{SP}(\mathtt{Var})$, $t \sim_t t'$ iff $t =_E t'$.*

Proof: Let $t, t' \in \mathbf{SP}(\mathtt{Var})$ be such that $t \sim_t t'$. By Theorem 5.3.4, in order to prove that E is ω-complete, it is sufficient to find $\rho : \mathtt{Var} \to \mathbf{SP}$ and $R : \mathbf{SP} \to \mathbf{SP}(\mathtt{Var})$ satisfying conditions (1)-(3) above. Define $\rho : \mathtt{Var} \to \mathbf{SP}$ by $\rho(x) = a_x \in \mathbf{A}$, where, for each $x, y \in \mathtt{Var}$,

- a_x does not occur in t and t', and
- $a_x = a_y$ implies $x = y$. (Note that such a map can be found because $\mathbf{A}$ is infinite)

The translation map $R : \mathbf{SP} \to \mathbf{SP}(\mathtt{Var})$ is defined by induction on the structure of $p \in \mathbf{SP}$ as follows:

- $R(nil) = nil$,
- $R(a) = \begin{cases} x & \text{if } a = a_x \\ a & \text{otherwise,} \end{cases}$
- $R(p \odot q) = R(p) \odot R(q)$, for $\odot \in \{;, |\}$.

We are now left to prove that ρ and R satisfy conditions (1)-(3). We examine each of the conditions in turn.

(1) We prove that, for all $\bar{t} \in \mathbf{SP}(\mathtt{Var})$ not containing actions of the form a_x, $\bar{t} =_E R(\rho(\bar{t}))$. The proof is by structural induction on $\bar{t}$. We only examine two of the cases.

- $\bar{t} = a$. Then $R(\rho(a)) = R(a) = a$ because $a \neq a_x$, for all x. The claim now follows by the reflexivity of $=_E$.
- $\bar{t} = t_1; t_2$. Then we have that

$$\begin{aligned} R(\rho(t_1; t_2)) &= R(\rho(t_1); \rho(t_2)) \\ &= R(\rho(t_1)); R(\rho(t_2)) \\ &=_E t_1; t_2 && \text{by induction.} \end{aligned}$$

(2) Let $\odot \in \{;, |\}$ and $p_1, p_2, q_1, q_2 \in \mathbf{SP}$. Then, letting $E' = E \cup \{R(p_i) = R(q_i) \mid i = 1, 2\}$, we have that

$$\begin{aligned} R(p_1 \odot p_2) &= R(p_1) \odot R(p_2) \\ &=_{E'} R(q_1) \odot R(q_2) \\ &= R(q_1 \odot q_2). \end{aligned}$$

(3) Let $t_1 = t_2$ be an equation in E and σ be a closed substitution. Then it is easy to see that $R(\sigma(t_1)) =_E R(\sigma(t_2))$. For instance,

$$\begin{array}{rcll} R(\sigma((x|y)|z)) & = & R((\sigma(x)|\sigma(y))|\sigma(z)) & \\ & = & (R(\sigma(x))|R(\sigma(y)))|R(\sigma(z)) & \\ & =_E & R(\sigma(x))|(R(\sigma(x))|R(\sigma(y))) & \text{by (PAR1)} \\ & = & R(\sigma(x|(y|z))). & \end{array}$$

As ρ and R satisfy conditions (1)-(3), by Theorem 5.3.4 we have that E is indeed ω-complete. □

5.3.1 Proof of the Equational Characterization

This section is entirely devoted to a proof of the equational characterization of the relation $\sim_t$ over **SP** (Theorem 5.3.2). The proof is based upon standard techniques used in the literature to axiomatize non-interleaving equivalences, see, e.g., [CH89] and [H88c], and follows the one of the analogous result presented in Chapter 3. Again, let us recall that $=_E$ denotes the least congruence over **SP** which satisfies the set of axioms E in Figure 5.1. The following proposition, whose proof is standard and thus omitted, states that the axioms are indeed sound with respect to $\sim_t$.

Proposition 5.3.3 (Soundness) *For each $p, q \in$ **SP**, $p =_E q$ implies $p \sim_t q$.*

In what follows we will concentrate on the more challenging proof of completeness of the set of axioms E with respect to $\sim_t$. In order to prove the completeness theorem, it will be convenient to reduce the processes in **SP** to what will be called *reduced forms* or *rf's*. Before giving the definition of this class of terms, let us introduce some useful notation.

Notation 5.3.1 *In what follows, for each process $p \in$ **SP**,*

$$p \longrightarrow_t \text{ iff there exist } a \in \mathbf{A}, s \in \mathcal{S} : p \xrightarrow{S(a)}_t s.$$

Note that $p \not\longrightarrow_t$ iff $p \sim_t nil$.

Definition 5.3.3 (Reduced Forms) *The sets* RF *(*reduced forms*),* RF_{Seq} *(*sequential reduced forms*) and* RF_{Par} *(*parallel reduced forms*) are defined simultaneously as the least sets satisfying:*

- *$nil \in$ RF;*

- $a; p \in \mathtt{RF}_{Seq}$ *if* $p \in \mathtt{RF}$;
- $p; q \in \mathtt{RF}_{Seq}$ *if* $p \in \mathtt{RF}_{Par}$, $q \in \mathtt{RF}$ *and* $q \longrightarrow_t$;
- $\prod_{i \in I} p_i \in \mathtt{RF}_{Par}$ *if* I *is a finite index set such that* $|I| > 1$ *and* $p_i \in \mathtt{RF}_{Seq}$, *for all* $i \in I$;
- $p \in \mathtt{RF}$ *if* $p \in \mathtt{RF}_{Seq}$ *or* $p \in \mathtt{RF}_{Par}$.

The notation $\prod_{i \in I} p_i$ *stands for* $p_{i_1} | \cdots | p_{i_k}$ $(I = \{i_1, \ldots, i_k\}$, $k \geq 0)$ *and is justified by axioms (PAR2)-(PAR3). By convention, if* $I = \emptyset$ *then* $\prod_{i \in \emptyset} p_i \equiv nil$.

In what follows, the *size* of a process p, $|p|$, will be taken to be the length of the longest sequence of sub-actions it can perform. (See page 111) This notion of size is generalized in the obvious way to $s \in \mathcal{S}$. The next lemma states that it is possible to reduce each process in **SP** to a reduced form using the axioms in E.

Lemma 5.3.1 (Reduction Lemma) *For each* $p \in$ **SP**, *there exists a reduced form* $h(p)$ *such that* $p =_E h(p)$ *and* $|p| = |h(p)|$.

Proof: By induction on the size of p. The proof proceeds by a further induction on the structure of p and will use all of the axioms in E.

- $p \equiv nil$. Then p is already a *rf*.
- $p \equiv a$. Then $a =_E a; nil$, which is a sequential *rf*, by axiom (SEQ1).
- $p \equiv q; r$. By the inner inductive hypothesis, $q =_E h(q)$ for some *rf* $h(q)$ such that $|q| = |h(q)|$. The proof proceeds by a case analysis on the possible form of $h(q)$.
  - If $h(q) \equiv nil$ then $p \equiv q; r =_E nil; h(r)$, by the inner inductive hypothesis. By (SEQ1), we then have that $p =_E h(r)$.
  - If $h(q)$ is of the form $a; q'$, with $q' \in \mathtt{RF}$, then

$$\begin{aligned} p \equiv q; r &=_E (a; q'); r && \text{by substitutivity} \\ &=_E a; (q'; r) && \text{by (SEQ2)} \\ &=_E a; h(q'; r) && \text{by the outer inductive hypothesis.} \end{aligned}$$

  - If $h(q)$ is of the form $q_1; q_2$, with $q_1 \in \mathtt{HRF}_{Par}$, $q_2 \in \mathtt{RF}$ and $q_2 \longrightarrow_t$, then

$$\begin{aligned} p \equiv q; r &=_E (q_1; q_2); r && \text{by substitutivity} \\ &=_E q_1; (q_2; r) && \text{by (SEQ2)} \\ &=_E q_1; h(q_2; r) && \text{by the outer inductive hypothesis.} \end{aligned}$$

  It is now easy to see that $q_1; h(q_2; r) \in \mathtt{HRF}_{Seq}$ because $|h(q_2; r)| = |q_2; r| \geq |q_2| > 0$.

- If $h(q)$ is of the form $\prod_{i \in I} q_i$, $|I| > 1$ and $q_i \in \mathtt{HRF}_{Seq}$, for all $i \in I$, then we distinguish two possibilities. If $h(r) \equiv nil$ then $p \equiv q; r =_E h(q); nil =_E h(q)$, by (SEQ1). If $h(r) \longrightarrow_t$ then, by substitutivity, $p \equiv q; r =_E (\prod_{i \in I} q_i); h(r)$, which is a sequential *rf*.

- $p \equiv q|r$. By the inner inductive hypothesis, there exist *rf*'s $h(q)$ and $h(r)$ such that $q =_E h(q)$ and $r =_E h(r)$, respectively. If either of $h(q)$ or $h(r)$ is nil then a *rf* may be obtained by applying axioms (PAR1)-(PAR2). Otherwise both $h(q)$ and $h(r)$ are different from nil. In this case it is easy to see that $p \equiv q|r =_E h(q)|h(r)$, which is a *rf*.

This completes the proof of the reduction lemma. □

The proof of the completeness theorem follows the one of the analogous result given in §3.5 of this thesis and is based upon several results stating important decomposition properties. In what follows, for the sake of clarity, we shall state the main results used in the proof of the completeness theorem and refer the reader to the corresponding ones in §3.5 for their proofs.

Proposition 5.3.4 *For each* $s_1, s_2 \in \mathcal{S}$, $s_1 \sim_t s_2$ *implies* $|s_1| = |s_2|$.

The following lemma is the counterpart of Lemma 3.5.2 on page 111.

Lemma 5.3.2 *Let* $s_1, s_2 \in \mathcal{S}$. *Then* $s_1; p \sim_t s_2; p$ *implies* $s_1 \sim_t s_2$.

As in the proof of Theorem 3.5.6, two special subclasses of the set of states $\mathcal{S}$ will play an important rôle in the proof of the completeness theorem. These are the sequential configurations and parallel configurations introduced in Definition 3.5.2. For the sake of clarity, we recall their definition for the language SP.

Definition 5.3.4 (Parallel and Sequential Configurations) *The sets* C_{Seq}, *the set of* sequential configurations, *and* C_{Par}, *the set of* parallel configurations, *are simultaneously defined as the least subsets of* $\mathcal{S}$ *which satisfy the following clauses:*

- $F(a); p \in C_{Seq}$ *if* $p \in$ SP;
- $c \in C_{Par}$, $p \longrightarrow_t$ *imply* $c; p \in C_{Seq}$;
- $\{c_i \mid i \in I\} \subseteq C_{Seq}$, $|I| > 0$ *and* $p \longrightarrow_t$ *imply* $\prod_{i \in I} c_i | p \in C_{Par}$;
- $\{c_i \mid i \in I\} \subseteq C_{Seq}$, $|I| > 1$ *imply* $\prod_{i \in I} c_i \in C_{Par}$.

In the above clauses I is always assumed to stand for a finite *index set.*

The proof of Theorem 3.5.6 has highlighted the importance of "decomposition results" for $\sim_t$ with respect to the operations of sequential and parallel composition. As it might be expected, the proof of the completeness theorem which will be presented in the remainder of this section will make an essential use of similar decomposition results. The sequential and the parallel decomposition theorems for the language **SP** and the set of states $\mathcal{S}$ are just instances of the corresponding results in §3.5.2. Therefore we shall limit ourselves to giving the results without proof. The interested reader is invited to consult §3.5.2 for the technical details. First of all, we give the Sequential Decomposition Theorem for the language considered in this chapter.

Theorem 5.3.6 (Sequential Decomposition Theorem) *Let $c, d \in C_{Par}$. Then $c;p \sim_t d;q$ implies $c \sim_t d$ and $p \sim_t q$.*

Proof: The theorem is just an instance of Theorem 3.5.2 on page 120. □

Our next aim is to show that, for $c, d \in C_{Seq}$, $c|p \sim_t d|q$ implies $c \sim_t d$ and $p \sim_t q$. The following Simplification Lemma, which is the counterpart of Lemma 3.5.8, is used in proving the above mentioned result and will find application in the proof of the completeness theorem.

Lemma 5.3.3 *Let $c, d, f \in \mathcal{S}$. Then:*

(1) *For each $e \in \mathbf{A_S}$, $f' \in \mathcal{S}$, $f \xrightarrow{e}_t f'$ and $c|f \sim_t d|f'$ imply $d \xrightarrow{e}_t d'$, for some d' such that $c \sim_t d'$.*

(2) *$c|f \sim_t d|f$ imply $c \sim_t d$.*

Proof: See Lemma 3.5.8 on page 120. □

In the parallel decomposition theorem we shall use some notation which was introduced in Definition 3.5.5 on page 124.

Theorem 5.3.7 (Parallel Decomposition Theorem) *Let $\{c_i \mid i \in I\}, \{d_j \mid j \in J\} \subseteq C_{Seq}$, $p, q \in \mathbf{SP}$. Then $c \equiv \prod_{i \in I} c_i|p \sim_t \prod_{j \in J} d_j|q \equiv d$ implies $\{c_i \mid i \in I\} \sim_t \{d_j \mid j \in J\}$ and $p \sim_t q$.*

Proof: This theorem is just an instance of Theorem 3.5.4 on page 124. □

In order to prove the promised completeness result, we shall need some further lemmas. Again, we shall omit their proofs and refer the reader to the corresponding results in §3.5.2 for more information.

Lemma 5.3.4 *Let* $c, d \in C_{Seq} \cup C_{Par}$. *Then:*

(i) $c \in C_{Seq}$ *and* $c \sim_t d$ *imply* $d \in C_{Seq}$.

(ii) $c \in C_{Par}$ *and* $c \sim_t d$ *imply* $d \in C_{Par}$.

Proof: See Lemma 3.5.9 on page 126. □

Lemma 5.3.5 *Let* $c, d \in C_{Seq} \cup C_{Par}$. *Then:*

1) $c \equiv S(a); p \sim_t d$ *implies* $d \equiv S(a); q$ *and* $p \sim_t q$.

2) $c \equiv c_1; p \sim_t d$, *with* $c_1 \in C_{Par}$ *and* $|p| > 0$, *implies* $d \equiv d_1; q$ *with* $d_1 \in C_{Par}$, $|q| > 0$, $c_1 \sim_t d_1$ *and* $p \sim_t q$.

3) $c \equiv c_1 | p \sim_t d$, *with* $c_1 \in C_{Seq}$ *and* $|p| > 0$, *implies* $d \equiv d_1 | q$, *with* $d_1 \in C_{Seq}$, $|q| > 0$, $c_1 \sim_t d_1$ *and* $p \sim_t q$.

Proof: See Lemma 3.5.10 on page 127. □

The following lemma studies the effect of start-moves on sequential and parallel reduced forms.

Lemma 5.3.6 *Let* $p \in \mathtt{HRF}_{Seq} \cup \mathtt{HRF}_{Par}$. *Then:*

(1) $p \in \mathtt{HRF}_{Seq}$ *and* $p \xrightarrow{S(a)}_t c$ *imply* $c \in C_{Seq}$.

(2) $p \in \mathtt{HRF}_{Par}$ *and* $p \xrightarrow{S(a)}_t c$ *imply* $c \in C_{Par}$ *and* c *is of the form* $d|q$, *with* $d \in C_{Seq}$ *and* $|q| > 0$.

Proof: See Lemma 3.5.3 on page 113. □

We have now developed all the technical machinery needed in the proof of the completeness theorem.

Theorem 5.3.8 (Completeness) *Let* $p, q \in \mathrm{SP}$. *Then* $p \sim_t q$ *implies* $p =_E q$.

Proof: The proof is by induction on the combined size of the terms. By the reduction lemma and the soundness of $=_E$ we may assume, wlog, that p and q are reduced forms. The proof proceeds by an analysis of the possible structure of p. The claim is trivial if $p \equiv nil$.

- $p \equiv a;r$, $r \in \mathtt{RF}$. Then $p \xrightarrow{S(a)}_t S(a);r$. As $p \sim_t q$, there exists c such that $q \xrightarrow{S(a)}_t c \sim_t S(a);r$. By lemma 5.3.5, $c \sim_t S(a);r$ implies $c \equiv S(a);t$ with $r \sim_t t$. By the inductive hypothesis, $r =_E t$. Moreover, $q \xrightarrow{S(a)}_t S(a);t$ and $q \in \mathtt{HRF}_{Seq} \cup \mathtt{HRF}_{Par}$ imply $q \equiv a;t$. Hence, by substitutivity, $p \equiv a;r =_E a;t \equiv q$.

- $p \equiv r_1;r_2$, with $r_1 \in \mathtt{HRF}_{Par}$, $r_2 \in \mathtt{RF}$ and $|r_2| > 0$. Assume $p \xrightarrow{S(a)}_t c$. Then c must be of the form $c';r_2$ with $r_1 \xrightarrow{S(a)}_t c'$. By lemma 5.3.6, $c' \in C_{Par}$ and thus $c \in C_{Seq}$. As $p \sim_t q$, there exists d such that $q \xrightarrow{S(a)}_t d \sim_t c';r_2$. By lemma 5.3.5, $d \sim_t c';r_2$ implies d has the form $d';t_2$, with $d' \in C_{Par}$, $|t_2| > 0$, $c' \sim_t d'$ and $r_2 \sim_t t_2$. By the inductive hypothesis, $r_2 =_E t_2$.

  It is now easy to see that $q \xrightarrow{S(a)}_t d';t_2$ and $q \in \mathtt{HRF}_{Seq} \cup \mathtt{HRF}_{Par}$ imply $q \equiv t_1;t_2$, with $t_1 \in \mathtt{HRF}_{Par}$ and $t_1 \xrightarrow{S(a)}_t d'$. By lemma 5.3.2 and substitutivity, $p \equiv r_1;r_2 \sim_t t_1;t_2 \equiv q$ and $r_2 \sim_t t_2$ imply $r_1 \sim_t t_1$. Hence, by the inductive hypothesis, $r_1 =_E t_1$ and, by substitutivity, $p \equiv r_1;r_2 =_E t_1;t_2 \equiv q$.

- $p \equiv \prod_{i \in I} p_i$, with $|I| > 1$ and $p_i \in \mathtt{HRF}_{Seq}$, for all $i \in I$. Assume $p \xrightarrow{S(a)}_t c$. Then, wlog, we may assume that c is of the form $c_1|r_2$, with $p_1 \xrightarrow{S(a)}_t c_1$ and $r_2 \equiv \prod_{i \in I - \{1\}} p_i$. By lemma 5.3.6, $c_1 \in C_{Seq}$. As $p \sim_t q$, there exists d such that $q \xrightarrow{S(a)}_t d \sim_t c_1|r_2$. By lemma 5.3.5, it must be the case that $d \equiv d_1|t_2$, with $d_1 \in C_{Seq}$, $|t_2| > 0$, $c_1 \sim_t d_1$ and $r_2 \sim_t t_2$. By the inductive hypothesis, $r_2 =_E t_2$.

  As $q \xrightarrow{S(a)}_t d_1|t_2$ and $q \in \mathtt{HRF}_{Seq} \cup \mathtt{HRF}_{Par}$, q must be of the form $t_1|t_2$, with $t_1 \in \mathtt{HRF}_{Seq}$ and $t_1 \xrightarrow{S(a)}_t d_1$. As $p \sim_t q$ and $r_2 \sim_t t_2$, we then have, by lemma 5.3.3 and substitutivity, that $p_1 \sim_t t_1$. By the inductive hypothesis, $p_1 =_E t_1$. By substitutivity, $p \equiv p_1|r_2 =_E t_1|t_2 \equiv q$.

This completes the proof of the theorem. □

5.4 Series-Parallel Pomsets and ST-Traces

In the previous section we showed that series-parallel pomsets are fully-abstract with respect to the equivalence obtained by applying Abramsky's testing scenario for bisimulation in all refinement contexts. The observability of SP pomsets was then cast in a well-known interleaving setting. The aim of this section is to investigate to what extent the model SP can be made fully observable by assuming a *trace-based* basic notion of observation. It will be shown that the causal structure of an N-free pomset is totally revealed by its set of *ST-traces* [Gl90], i.e. that SP pomsets are fully abstract with respect to ST-trace equivalence over the set of processes **SP**. ST-semantics has recently been proposed in [GV87], [Gl90] as a refinement of the timed behavioural view of processes outlined in §5.3. This more refined view of processes is obtained by requiring a link between the beginning and the end of any event; this allows one to express that a start-action $S(a)$ and an end-action $F(a)$ represent the beginning and the end of the same occurrence of action a. Notions of ST-bisimulation and ST-trace equivalence have been proposed and studied in [GV87], [Gl90] for Petri Nets and Event Structures, respectively, and the interested reader is invited to consult these references for more details on ST-semantics.

In what follows we shall mainly work with labelled posets rather than pomsets; this will make the technical development slightly less cumbersome. All the results will be lifted to pomsets and the process language **SP** in a straightforward way. Our first aim is to give the class of labelled SP posets the structure of a labelled transition system following the intuitions underlying the timed view of processes described in [H88c] and §5.3. In order to provide an LTS semantics for the class of **A**-labelled posets, we shall have to extend the class of labelled posets in order to express those intermediate stages in the evolution of a process in which some actions have started but have not yet terminated.

Definition 5.4.1 *Let* $\mathbf{A_S} = \mathbf{A} \cup \{S(a) \mid a \in \mathbf{A}\}$. *An* $\mathbf{A_S}$*-labelled poset* $I\!P = (P, <, l)$ *is* sensible *iff, for all* $u \in P$, $l(u) = F(a)$, *for some* $a \in \mathbf{A}$, *implies* u *is minimal in* $I\!P$.

$\mathbf{Pos}[\mathbf{A_S}]$ *will denote the class of sensible, series-parallel* $\mathbf{A_S}$*-labelled posets.*

Note that each $I\!P \in \mathbf{Pos}[\mathbf{A}]$ is a sensible, series-parallel $\mathbf{A_S}$-labelled poset. Intuitively, $\mathbf{A_S}$-labelled posets $(P, <, l)$ in which $l(u) \in \mathbf{A}$, for all $u \in P$, are the model-theoretic counterpart of the processes in **SP** and those with at least a minimal element labelled $F(a)$, for some $a \in \mathbf{A}$, correspond to *proper states* in $\mathcal{S}$, i.e. states in which some actions will have started, but have not yet terminated. The following definition introduces the transition relations over $\mathbf{Pos}[\mathbf{A_S}]$.

Definition 5.4.2 (Transition relations for posets) *Let* $I\!P = (P, <, l) \in \mathbf{Pos}[\mathbf{A_S}]$. *Then:*

(i) $I\!P \xrightarrow{(S(a),u)} I\!P'$ *iff*

(a) *u is minimal in* $I\!P$,

(b) $l(u) = a$, *and*

(c) $I\!P' = (P, <, l')$, *where, for each* $v \in P$,

$$l'(v) = \begin{cases} F(a) & \text{if } u = v \\ l(u) & \text{otherwise.} \end{cases}$$

(ii) $I\!P \stackrel{\langle F(a),u\rangle}{\longrightarrow} I\!P_1$ *iff*

(a) u *is minimal in* $I\!P$,

(b) $l(u) = F(a)$, *and*

(c) $I\!P_1 = (P_1, <_1, l_1)$, *where* $P_1 = P - \{u\}$ *and* $<_1$, l_1 *are the restrictions of* $<$ *and* l *to* P_1, *respectively.*

The following fact, whose proof follows easily from the definition of the transition relations $\stackrel{\langle e,u\rangle}{\longrightarrow}$, $e \in Ev$, and is thus omitted, states that $\mathbf{Pos}[\mathbf{A_S}]$ is indeed closed under derivation.

Fact 5.4.1 (Closure under derivation) *Let* $I\!P \in \mathbf{Pos}[\mathbf{A_S}]$ *and* $e \in Ev$. *Then* $I\!P \stackrel{\langle e,u\rangle}{\longrightarrow} I\!P'$ *implies* $I\!P' \in \mathbf{Pos}[\mathbf{A_S}]$.

Using the above-given operational semantics for $\mathbf{Pos}[\mathbf{A_S}]$, it is now possible to define a natural notion of complete trace of a poset $I\!P \in \mathbf{Pos}[\mathbf{A_S}]$. Intuitively, a complete trace γ of a poset $I\!P \in \mathbf{Pos}[\mathbf{A_S}]$ records a possible linear history of the evolution of the process denoted by $I\!P$, i.e. the set of events the process involves in together with their relative order of execution. In what follows we shall only be interested in this notion and the ones derived from it for SP posets $I\!P \in \mathbf{Pos}[\mathbf{A}]$.

Definition 5.4.3 (Complete traces) *Let* $I\!P = (P, <, l) \in \mathbf{Pos}[\mathbf{A}]$. *A sequence* $\gamma = \langle e_1, u_1\rangle \cdots \langle e_k, u_k\rangle \in (Ev \times P)^*$, $k \geq 0$, *is a* complete trace *of* $I\!P$ *iff there exist* $I\!P_0, \ldots, I\!P_k$ *in* $\mathbf{Pos}[\mathbf{A_S}]$ *such that*

(i) $I\!P_0 = I\!P$, $I\!P_k = (\emptyset, \emptyset, \emptyset)$ *and*

(ii) $I\!P_i \stackrel{\langle e_{i+1}, u_{i+1}\rangle}{\longrightarrow} I\!P_{i+1}$, *for all* $i < k$.

$CT(I\!P)$ will denote the set of complete traces of $I\!P$. The projection maps will be homomorphically extended to strings over $(Ev \times P)^*$, i.e. for $\gamma = \langle e_1, u_1\rangle \cdots \langle e_k, u_k\rangle$, $\pi_1(\gamma) = e_1 \cdots e_k$ and $\pi_2(\gamma) = u_1 \cdots u_k$.

It is easy to see that if $\gamma = \langle e_1, u_1\rangle \cdots \langle e_k, u_k\rangle$ is a complete trace of $I\!P = (P, <, l) \in \mathbf{Pos}[\mathbf{A}]$ then $P = \{u_1, \ldots, u_k\}$, $k = 2m$ where $|P| = m$ and, for all $u \in P$, there exist unique i, j

such that $1 \le i < j \le k$, $\langle e_i, u_i \rangle = \langle S(a), u \rangle$ and $\langle e_j, u_j \rangle = \langle F(a), u \rangle$, with $a = l(u)$. By using the above notion of complete trace it is now possible to define two key notions of trace equivalence over **Pos[A]**. The first one, which is based on the operational intuition underlying the timed operational semantics defined in §5.3, is (complete) *split trace* equivalence [Va88], [Gl90]. Split trace equivalence is just standard interleaving trace equivalence, [Hoare85], but based on interleavings of beginnings and endings. *ST-trace* equivalence, [Gl90], will then be defined as a refinement of split trace equivalence by requiring that beginnings and endings of the same occurrence of an action $a \in \mathbf{A}$ are explicitly connected in a complete trace.

Definition 5.4.4 (Split and ST-trace equivalence) *Let* $\mathbb{P}, \mathbb{P}_1, \mathbb{P}_2 \in \mathbf{Pos[A]}$.

(i) *A sequence* $\sigma \in Ev^\star$ *is a* split trace *of* $\mathbb{P}$ *iff there exists* $\gamma \in CT(\mathbb{P})$ *such that* $\pi_1(\gamma) = \sigma$. $S(\mathbb{P})$ *will denote the set of split traces of* $\mathbb{P}$. *Then* $\mathbb{P}_1$ *and* $\mathbb{P}_2$ *are split trace equivalent,* $\mathbb{P}_1 \sim_{2t} \mathbb{P}_2$, *iff* $S(\mathbb{P}_1) = S(\mathbb{P}_2)$.

(ii) $\mathbb{P}_1$ *and* $\mathbb{P}_2$ *are* ST-trace *equivalent,* $\mathbb{P}_1 \sim_{ST} \mathbb{P}_2$, *iff the following conditions hold:*

(a) *for each* $\gamma_1 = \langle e_1, u_1 \rangle \cdots \langle e_k, u_k \rangle \in CT(\mathbb{P}_1)$ *there exists* $\gamma_2 = \langle f_1, v_1 \rangle \cdots \langle f_h, v_h \rangle \in CT(\mathbb{P}_2)$ *such that*

- $h = k$, $\pi_1(\gamma_1) = \pi_1(\gamma_2)$, *and*
- *for all* $1 \le i < j \le k$, $u_i = u_j$ *iff* $v_i = v_j$ *(ST-condition);*

(b) *viceversa, with the rôles of* $\mathbb{P}_1$ *and* $\mathbb{P}_2$ *interchanged.*

The following fact states two basic properties of the above-given notions of equivalence over **Pos[A]**. The first justifies our choice of working with labelled posets rather than pomsets by showing that the notions of equivalence given above may be consistently lifted to pomsets. The second states that $\sim_{ST}$ is at least as strong as split trace equivalence. For the sake of clarity, we recall that isomorphism between labelled posets is denoted by $\cong$ (see Definition 5.2.1).

Fact 5.4.2 *Let* $\mathbb{P}_1$, $\mathbb{P}_2 \in \mathbf{Pos[A]}$. *Then:*

(i) $\mathbb{P}_1 \cong \mathbb{P}_2$ *implies* $\mathbb{P}_1 \sim_{2t} \mathbb{P}_2$ *and* $\mathbb{P}_1 \sim_{ST} \mathbb{P}_2$.

(ii) $\mathbb{P}_1 \sim_{ST} \mathbb{P}_2$ *implies* $\mathbb{P}_1 \sim_{2t} \mathbb{P}_2$.

In the light of the statement above, $\sim_{2t}$ and $\sim_{ST}$ may be now extended to SP in a rather straightforward way.

Definition 5.4.5 *Let* $\alpha = [P_1, <_1, l_1], \beta = [P_2, <_2, l_2] \in SP$. *Then* $\alpha \sim_{2t} \beta$ ($\alpha \sim_{ST} \beta$) *iff* $(P_1, <_1, l_1) \sim_{2t} (P_2, <_2, l_2)$ ($(P_1, <_1, l_1) \sim_{ST} (P_2, <_2, l_2)$).

The remainder of this section will be entirely devoted to showing that ST-trace equivalence coincides with isomorphism over **Pos[A]** (and thus with equality over SP). This implies that SP pomsets can be made fully observable by assuming a trace-like notion of observation, albeit one in which beginnings and endings of the same occurrence of an action are explicitly linked. The following standard example shows that $\sim_{ST}$ does not coincide with isomorphism over general labelled posets and pomsets.

Example 5.4.1 *Let* α *and* β *denote the following pomsets:*

- $\alpha = [\![(a; b)|(a; b)]\!]$ *and*
- $\beta = [P, <, l]$, *where* $P = \{1, 2, 3, 4\}$, $1 < 2$, $3 < 4$ *and* $1 < 4$, $l(1) = l(3) = a$ *and* $l(2) = l(4) = b$. *This pomset is just Gischer's* $N(a, b, a, b)$, *[Gi84].*

Then $\alpha \sim_{ST} \beta$, *but obviously* $\alpha \neq \beta$. *Note that* β *is not a series-parallel pomset.*

The following lemmas, which analyze basic properties of the transition relations $\xrightarrow{\langle e,u\rangle}$, $e \in Ev$, will be useful in the proof of the main result of this section. The following lemma states that sequences of start-moves are made up of independent transitions; such transitions may then be performed in any order without influencing the resulting target state. A similar property holds for end-moves.

Lemma 5.4.1 (Commuting start and end moves) *Let* $\mathbb{P} = (P, <, l) \in \mathbf{Pos}[\mathbf{A_S}]$. *Then the following properties hold.*

(i) $\mathbb{P} \xrightarrow{\langle S(a),u\rangle} \mathbb{P}' \xrightarrow{\langle S(b),v\rangle} \mathbb{P}''$ *implies* $\mathbb{P} \xrightarrow{\langle S(b),v\rangle} \mathbb{P}_1 \xrightarrow{\langle S(a),u\rangle} \mathbb{P}''$, *for some* $\mathbb{P}_1 \in \mathbf{Pos}[\mathbf{A_S}]$.

(ii) $\mathbb{P} \xrightarrow{\langle F(a),u\rangle} \mathbb{P}' \xrightarrow{\langle F(b),v\rangle} \mathbb{P}''$ *implies* $\mathbb{P} \xrightarrow{\langle F(b),v\rangle} \mathbb{P}_1 \xrightarrow{\langle F(a),u\rangle} \mathbb{P}''$, *for some* $\mathbb{P}_1 \in \mathbf{Pos}[\mathbf{A_S}]$.

Proof: We shall only prove (ii) as the proof of (i) is similar. Assume that $\mathbb{P} \xrightarrow{\langle F(a),u\rangle} \mathbb{P}' \xrightarrow{\langle F(b),v\rangle} \mathbb{P}''$, with $\mathbb{P} = (P, <, l) \in \mathbf{Pos}[\mathbf{A_S}]$. Then, by the definition of the transition relations over $\mathbf{Pos}[\mathbf{A_S}]$ and the sensibility requirement for $\mathbb{P}$, it is easy to see that u and v are both minimal in $\mathbb{P}$ and that $\mathbb{P}'' = (P'', < \lceil P''^2, l\lceil P'')$, where $P'' = P - \{u, v\}$. Consider now $\mathbb{P}_1 = (P_1, < \lceil P_1^2, l\lceil P_1)$, where $P_1 = P - \{v\}$. Then it is easy to see that $\mathbb{P}_1 \in \mathbf{Pos}[\mathbf{A_S}]$ and that it is the required mediating state, i.e.

$$\mathbb{P} \xrightarrow{\langle F(b),v\rangle} \mathbb{P}_1 \xrightarrow{\langle F(a),u\rangle} \mathbb{P}''. \quad \square$$

The following lemma states that end-moves and start-moves corresponding to events which are not causally related may be performed in any order without influencing the resulting target state.

Lemma 5.4.2 *Let* $I\!P = (P, <, l) \in \mathbf{Pos}[\mathbf{A_S}]$. *Then* $I\!P \stackrel{\langle F(a),u\rangle}{\longrightarrow} I\!P' \stackrel{\langle S(b),v\rangle}{\longrightarrow} I\!P''$ *and* $u \not< v$ *imply* $I\!P \stackrel{\langle S(b),v\rangle}{\longrightarrow} I\!P_1 \stackrel{\langle F(a),u\rangle}{\longrightarrow} I\!P''$, *for some* $I\!P_1 \in \mathbf{Pos}[\mathbf{A_S}]$.

Proof: First of all, note that $I\!P \stackrel{\langle F(a),u\rangle}{\longrightarrow} I\!P' \stackrel{\langle S(b),v\rangle}{\longrightarrow} I\!P''$ if, and only if,

(1) u is minimal in $I\!P$, $l(u) = F(a)$, $I\!P' = (P', < \lceil P'^2, l\lceil P'^2)$ with $P' = P - \{u\}$, and

(2) v is minimal in $I\!P'$, $l(v) = b$ and $I\!P'' = (P', < \lceil P'^2, l')$, where

$$l'(x) = \begin{cases} F(b) & \text{if } x = v \\ l(x) & \text{otherwise.} \end{cases}$$

As $u \not< v$ and v is minimal in $I\!P'$, we have that v is also minimal in $I\!P$. Consider now $I\!P_1 = (P, <, \bar{l})$, where

$$\bar{l}(x) = \begin{cases} F(b) & \text{if } x = v \\ l(x) & \text{otherwise.} \end{cases}$$

Then it is easy to see that $I\!P_1 \in \mathbf{Pos}[\mathbf{A_S}]$ and that it is the required mediating state, i.e.

$$I\!P \stackrel{\langle S(b),v\rangle}{\longrightarrow} I\!P_1 \stackrel{\langle F(a),u\rangle}{\longrightarrow} I\!P''. \ \square$$

The following result presents a basic consistency requirement on the complete traces of a poset $I\!P = (P, <, l) \in \mathbf{Pos}[\mathbf{A}]$; namely that, for each $v \in P$, the end of each event $u < v$ must precede the start of v in every linear history of $I\!P$.

Lemma 5.4.3 *Let* $I\!P = (P, <, l) \in \mathbf{Pos}[\mathbf{A}]$. *Assume that* $u, v \in P$ *and* $u < v$. *Then* $\gamma = \langle e_1, u_1\rangle \cdots \langle e_k, u_k\rangle \in CT(I\!P)$, $\langle F(a), u\rangle = \langle e_i, u_i\rangle$ *and* $\langle S(b), v\rangle = \langle e_j, u_j\rangle$, *with* $l(u) = a$ *and* $l(v) = b$, *imply* $i < j$.

Proof: Assume $\gamma = \langle e_1, u_1\rangle \cdots \langle e_k, u_k\rangle \in CT(I\!P)$. Then, for some $I\!P_1, \ldots I\!P_{k-1}$,

$$I\!P \stackrel{\langle e_1,u_1\rangle}{\longrightarrow} I\!P_1 \stackrel{\langle e_2,u_2\rangle}{\longrightarrow} \cdots I\!P_{k-1} \stackrel{\langle e_k,u_k\rangle}{\longrightarrow} (\emptyset, \emptyset, \emptyset).$$

Assume, towards a contradiction, that $I\!P_i \stackrel{\langle S(b),v\rangle}{\longrightarrow} I\!P_{i+1}$ and that, for no $j < i$, $I\!P_j \stackrel{\langle F(a),u\rangle}{\longrightarrow} I\!P_{j+1}$. Then, by the definition of the transition relations, it is easy to see that v is minimal in $I\!P_i = (P_i, <_i, l_i)$ and that

$$P_i = P - \{x \in P \mid x = u_j, \text{ for some } j < i \text{ such that } e_j = F(y), \ y \in \mathbf{A}\}.$$

Moreover, $<_i = < \lceil P_i^2$. By the assumption that $\langle e_j, u_j \rangle = \langle F(a), u \rangle$ for no $j < i$, we have that $u \in P_i$. Hence, by the above characterization of $<_i$ and the minimality of v in $I\!P_i$, we have that $u \not< v$; this contradicts the assumption that $u < v$. □

We now have all the technical material which is needed to prove the main theorem of this section, namely that $\sim_{ST}$ coincides with isomorphism over **Pos**[**A**].

Theorem 5.4.1 (ST-trace equivalence = isomorphism over Pos[A])

Let $I\!P_i = (P_i, <_i, l_i) \in \mathbf{Pos}[\mathbf{A}]$, $i = 1, 2$. *Then* $I\!P_1 \cong I\!P_2$ *iff* $I\!P_1 \sim_{ST} I\!P_2$.

Proof: The "only if" implication follows by Fact 5.4.2. We shall now concentrate on the proof of the "if" implication. Assume that $I\!P_1, I\!P_2 \in \mathbf{Pos}[\mathbf{A}]$ and that $I\!P_1 \sim_{ST} I\!P_2$; we will show that $I\!P_1 \cong I\!P_2$. The proof proceeds in two steps:

(1) first of all, we shall show that $I\!P_i$, $i = 1, 2$, may be recovered from a particular $\gamma_i \in CT(I\!P_i)$;

(2) secondly, we shall construct an isomorphism between $I\!P_1$ and $I\!P_2$ by making use of the information on the order structure of the posets obtained in the previous step and the fact that $I\!P_1 \sim_{ST} I\!P_2$.

Following [Ts88], let $\ll$ be the ordering relation over $CT(I\!P_1)$ obtained by lexicographically extending the one over $Ev \times P_1$ given by

$$\langle S(a), u \rangle \ll \langle F(b), v \rangle, \text{ for all } a, b \in \mathbf{A} \text{ and } u, v \in P_1.$$

Let $\gamma_1 = \langle e_1, u_1 \rangle \cdots \langle e_k, u_k \rangle$, $k \geq 0$, be minimal in $CT(I\!P_1)$ with respect to $\ll$. Then there exist $I\!P_1, \ldots, I\!P_{k-1}$ such that

$$\Gamma = I\!P \stackrel{\langle e_1, u_1 \rangle}{\longrightarrow} I\!P_1 \stackrel{\langle e_2, u_2 \rangle}{\longrightarrow} \cdots I\!P_{k-1} \stackrel{\langle e_k, u_k \rangle}{\longrightarrow} (\emptyset, \emptyset, \emptyset).$$

As previously remarked, $P_1 = \{u_1, \ldots, u_k\}$; we shall now show that the ordering relation $<_1$ and the labelling function l_1 of P_1 may be recovered from γ_1. By the definition of the transition relations over $\mathbf{Pos}[\mathbf{A_S}]$, it is easy to see that, for each $u \in P_1$, $l_1(u) = a$ iff $\langle S(a), u \rangle = \langle e_i, u_i \rangle$ for some $1 \leq i \leq k$. We shall now concentrate on showing how $<_1$ may be recovered from γ_1. As we are dealing with finite partial orders, $<_1$ is completely determined by the *covering relation* over $I\!P_1$; for all $u, v \in P_1$, u is covered by v iff $u <_1 x <_1 v$, for no $x \in P_1$. Intuitively, γ_1 begins with a block of start-moves followed alternately by blocks of end-moves and blocks of start-moves and then ends with a final block of end-moves. Then the events in P_1 appearing in the first block of start-moves will correspond to the minimal elements in $I\!P_1$, those appearing in the last block to the maximal elements of $\mathbb{P}_1$ and, for each intervening block of end-moves followed by a block of start-moves, the events appearing in the block of end-moves will correspond to events in $\mathbb{P}_1$ covered by those appearing in the block of start-moves. We shall thus be able to recover from γ_1 the covering relation in $\mathbb{P}_1$; this is sufficient to recover $<_1$.

Let $\prec_1$ be the relation over P_1 such that, for all $u, v \in P_1$, $u \prec_1 v$ iff

(FS) there exists a subword $\langle e_h, u_h \rangle \cdots \langle e_{h+r}, u_{h+r} \rangle \langle e_{h+r+1}, u_{h+r+1} \rangle \cdots \langle e_l, u_l \rangle$ of γ_1, with $h < l$ and $r \geq 0$, such that

(i) $u_h = u$ and $u_l = v$,

(ii) for all $i \leq r$, $e_{h+i} = F(a_i)$ for some $a_i \in \mathbf{A}$, and

(iii) for all $h + r + 1 \leq j \leq l$, $e_j = S(b_j)$ for some $b_j \in \mathbf{A}$.

Let $\prec_1^+$ denote the transitive closure of $\prec_1$. We claim that

$$u \prec_1^+ v \iff u <_1 v. \tag{5.1}$$

- We prove, first of all, that $u \prec_1^+ v$ implies $u <_1 v$, i.e. that $\prec_1^+$ is *sound* with respect to $<_1$. Assume that $u, v \in P_1$ and $u \prec_1 v$. Then there exists a subsequence of Γ

$$\mathbb{P}_h \xrightarrow{\langle F(a_{h+1}), u_{h+1} \rangle} \mathbb{P}_{h+1} \xrightarrow{\langle F(a_{h+2}), u_{h+2} \rangle} \cdots \mathbb{P}_{h+r+1} \xrightarrow{\langle S(b_{h+r+2}), u_{h+r+2} \rangle} \cdots \xrightarrow{\langle S(b_{l+1}), u_{l+1} \rangle} \mathbb{P}_{l+1},$$

with $h < l$, $r \geq 0$, such that $u_{h+1} = u$, $u_{l+1} = v$ and

$$\langle F(a_{h+1}), u_{h+1} \rangle \cdots \langle F(a_{h+r+1}), u_{h+r+1} \rangle \langle S(b_{h+r+2}), u_{h+r+2} \rangle \cdots \langle S(b_{l+1}), u_{l+1} \rangle$$

satisfies the property (FS). Assume, towards a contradiction, that $u \not<_1 v$. Then, by repeatedly applying Lemma 5.4.1, we have that, for some $\mathbb{P}'$ and $\mathbb{P}''$,

$$\mathbb{P}_h \xrightarrow{\omega_1} \mathbb{P}' \xrightarrow{\langle F(a_{h+1}), u \rangle} \mathbb{P}_{h+r+1} \xrightarrow{\langle S(b_{l+1}), v \rangle} \mathbb{P}'' \xrightarrow{\omega_2} \mathbb{P}_{l+1},$$

with $\omega_1 = \langle F(a_{h+2}), u_{h+2} \rangle \cdots \langle F(a_{h+r+1}), u_{h+r+1} \rangle$ and $\omega_2 = \langle S(b_{h+r+2}), u_{h+r+2} \rangle \cdots \langle S(b_l), u$ As $u \not<_1 v$, by Lemma 5.4.2 there exists $\bar{\mathbb{P}}$ such that

$$\mathbb{P}' \xrightarrow{\langle S(b_{l+1}), v \rangle} \bar{\mathbb{P}} \xrightarrow{\langle F(a_{h+1}), u \rangle} \mathbb{P}''.$$

Thus $\gamma = \langle e_1, u_1 \rangle \cdots \langle e_h, u_h \rangle \omega_1 \langle S(b_{l+1}), v \rangle \langle F(a_{h+1}), u \rangle \omega_2 \langle e_{l+2}, u_{l+2} \rangle \cdots \langle e_k, u_k \rangle \in CT(\mathbb{P}_1)$ and $\gamma \ll \gamma_1$. However, this contradicts the minimality of γ_1 in $CT(\mathbb{P}_1)$ with respect to $\ll$. Hence $u \prec_1 v$ implies $u <_1 v$; by transitivity, $u \prec_1^+ v$ implies $u <_1 v$.

- We now prove that $u <_1 v$ implies $u \prec_1^+ v$, i.e. that $\prec_1^+$ is *complete* with respect to $<_1$. The proof of this fact will depend on the assumption that $\mathbb{P}_1$ is a series-parallel poset. Assume that u is covered by v in $\mathbb{P}_1$. Then, by Lemma 5.4.3, $\langle F(a), u\rangle = \langle e_h, u_h\rangle$ and $\langle S(b), v\rangle = \langle e_l, u_l\rangle$, for some h, l such that $h < l$. If the subword of γ

$$\langle e_h, u_h\rangle \cdots \langle e_l, u_l\rangle$$

has the (FS) property then we have that $u \prec_1 v$. Otherwise, there exist h_1 and h_2, with $h < h_1 < h_2 < l$, such that $e_{h_1} = S(a_{h_1})$, $e_{h_2} = F(a_{h_2})$, for some $a_{h_1}, a_{h_2} \in \mathbf{A}$, and $\langle e_h, u_h\rangle \cdots \langle e_{h_1}, u_{h_1}\rangle$ and $\langle e_{h_2}, u_{h_2}\rangle \cdots \langle e_l, u_l\rangle$ have the (FS) property. By the definition of $\prec_1$, we have that $u = u_h \prec_1 u_{h_1}$ and $u_{h_2} \prec_1 u_l = v$. By the soundness of $\prec_1$ with respect to $<_1$, we have that $u <_1 u_{h_1}$ and $u_{h_2} <_1 v$. Hence we have that

$$u <_1 v,\ u <_1 u_{h_1} \text{ and } u_{h_2} <_1 v.$$

As $\mathbb{P}_1$ is a series-parallel poset, by applying the (N) property in Proposition 5.2.1 we derive that

(a) $u_{h_1} <_1 v$ or

(b) $u <_1 u_{h_2}$ or

(c) $u_{h_2} <_1 u_{h_1}$.

We examine each possibility in turn, showing that each of them leads to a contradiction. If (a) holds then we have that $u <_1 u_{h_1} <_1 v$, contradicting the hypothesis that u is covered by v in $\mathbb{P}_1$. Similarly, if (b) holds. If (c) holds then $u_{h_2} <_1 u_{h_1}$, but this contradicts the assumption that $h_1 < h_2$, i.e. that the start of event u_{h_1} occurs before the end of event u_{h_2} in γ_1. Hence we have shown that if u is covered by v in $\mathbb{P}_1$ then $u \prec_1 v$; by transitivity, $u <_1 v$ implies $u \prec_1^+ v$.

Thus we have shown that $u <_1 v$ iff $u \prec_1^+ v$, for all $u, v \in P_1$. As $\mathbb{P}_1 \sim_{ST} \mathbb{P}_2$, there exists $\gamma_2 = \langle f_1, v_1\rangle \cdots \langle f_n, v_n\rangle \in CT(\mathbb{P}_2)$ such that:

(i) $k = n$, $\pi_1(\gamma_1) = \pi_1(\gamma_2)$, and

(ii) for all $1 \leq i < j \leq k$, $u_i = u_j$ iff $v_i = v_j$.

Again, we have that $P_2 = \{v_1, \ldots, v_n\}$. We may now define $\prec_2$ from γ_2 as we did for $\prec_1$ from γ_1 and, as γ_2 is also minimal in $CT(\mathbb{P}_2)$, by symmetry and (5.1) we obtain that $x <_2 y$ iff $x \prec_2^+ y$, for all $x, y \in P_2$. Let us now define $\phi : P_1 \to P_2$ by $\phi(u) = x$ iff there exists $1 \leq i \leq k$ such that $u_i = u$ and $v_i = x$. Then, ϕ is a well-defined function by clause (ii) above and it is label-preserving by clause (i) and the definition of the transition relations. It is easy to see that ϕ is also bijective by clause (ii). Moreover, by construction, ϕ is such that $u \prec_1 v$ iff $\phi(u) \prec_2 \phi(v)$, for all $u, v \in P_1$. Hence, by claim (5.1) and transitivity, we have that $u <_1 v$ iff $\phi_1 <_2 \phi(v)$, for all $u, v \in P_1$. Thus $I\!P_1 \cong I\!P_2$. □

The following result is an immediate corollary of the above theorem.

Corollary 5.4.1 *Let $\alpha, \beta \in SP$. Then $\alpha = \beta$ iff $\alpha \sim_{ST} \beta$.*

ST-trace equivalence can be inherited by **SP** via the semantic map $[\![\cdot]\!]$ in a straightforward way; for each $p, q \in \mathbf{SP}$, we write $p \sim_{ST} q$ iff $[\![p]\!] \sim_{ST} [\![q]\!]$. By using the results presented in §5.3 and the above theorem and corollary, it is now possible to provide a complete axiomatization of ST-trace equivalence over **SP**. Moreover, as stated by the following theorem, $\sim_{ST}$ gives yet another characterization of the largest congruence over **SP** which is preserved by refinement and is contained in $\sim$.

Theorem 5.4.2 *For all $p, q \in \mathbf{SP}$, $p \sim_{ST} q$ iff $p \sim^{r} q$ iff $p =_E q$.*

Proof: The claim follows by the above corollary, Theorem 5.2.1 and theorem 5.3.3. □

ST-trace equivalence, $\sim_{ST}$, could be defined directly on the language **SP** without much difficulty; however, the proof of the main result of this section has been greatly simplified by working with labelled posets rather than with terms in **SP**.

5.5 Concluding Remarks

In this chapter, we have presented a behavioural characterization of the class of series-parallel pomsets, [Gi84], based on a natural interleaving testing scenario. This has been obtained by showing that the model of series-parallel pomsets is fully-abstract with respect to the behavioural equivalence obtained by applying Abramsky's testing scenario for bisimulation equivalence, [Ab87a], in all refinement contexts. Following Milner and Plotkin's paradigm, this result justifies the use of this simple mathematical model based on partial orders in giving semantics to the basic process algebra studied in this chapter. Moreover, we have shown that identity over the class of SP pomsets coincides with ST-trace equivalence, [Gl90]. Thus SP pomsets can be made fully abstract by assuming a trace-based notion of observation, albeit one in which beginnings and ends of the same occurrence of an action are explicitly linked. This retrievability result has allowed us to give a complete axiomatic characterization of ST-trace equivalence over the class of SP pomsets. A natural question to ask is whether SP pomsets are completely characterized by their set of *split traces*, see [Va88], [Gl90] and §5.4. The following conjecture naturally suggests itself.

Conjecture: For all $\alpha, \beta \in SP$, $\alpha = \beta$ iff $\alpha \sim_{2t} \beta$.

All the author's attempts to prove or disprove the above conjecture have so far failed. It is interesting to note that the validity of the above conjecture would have some striking consequences. First of all, it would imply that, for all $p, q \in \mathbf{SP}$, $p \sim_t q$ iff $p \sim_{2t} q$, i.e. that timed-bisimulation and split trace equivalence coincide over **SP**. As it is well-known, this result is *not* true of standard strong bisimulation and trace equivalence because the processes in **SP** are not *deterministic*, [Mil89], [Va88]. Moreover, by following the proof of the results presented in Chapter 3, it would be possible to show that equality between SP pomsets is the largest congruence contained in standard interleaving trace equivalence which is preserved by refinement.

The work presented in this chapter may be seen as an embryonic attempt at defining a natural testing scenario which justifies the use of partial order semantics without assuming any notion of "causal observation". We have shown that such a testing scenario does exist for the simple model considered in this chapter; however, as work by R. van Glabbeek on ST-bisimulation semantics shows, [Gl90], a notion of system testing based on the refinement operator does not suffice to reveal the full-distinguishing power of partial order semantics. The search for a testing scenario which justifies models like *Event Structures* and *Causal Trees* seems to be a very interesting topic for future research.

We end this section with a brief discussion of related work. Precursors of the work presented in §5.3 are [Gi84], [Ts88], where language equivalence for pomsets and series-parallel pomsets are studied in detail, and recent papers in the literature studying notions of equivalence for concurrent systems which are perserved by refinement of actions, [GG88], [NEL88], [Gl90]. In all these references, the authors present semantic theories for processes which support refinement of actions. The reference [GG88] gives a good survey of the work in this area; [NEL88] gives a natural fully abstract model for a language incorporating a refinement operator. In [Gl90], the author studies notions of ST-bisimulation and ST-trace equivalence over prime Event Structures [Win87] and proves that they are both preserved by refinement.

Retrievability results like the one presented in §5.4 for SP pomsets have been shown in, e.g., [Va88]. There the author shows that deterministic Event Structures are characterized, up to isomorphism, by their set of step-sequences. A similar result is shown for split-traces; this implies that the causal structure of a deterministic concurrent system can be reconstructed by observers which are capable of observing the beginning and the end of events.

Chapter 6

On Relating Concurrency and Nondeterminism

6.1 Introductory Remarks

As pointed out in the previous chapters, notions of equivalence between process descriptions are an important component of *Process Algebras*, such as **CCS** [Mil80,89], **CSP** [Hoare85] and **ACP** [BK85]. The interest in equivalences and preorders, which relate descriptions of concurrent systems in terms of these languages, stems from the fact that process algebras are used not only for describing actual systems, but also their specifications. Notions of equivalence between descriptions are thus an important component of these languages as they allow to formally state when (the description of) a system is a correct implementation of a given specification. Roughly, the proposals presented in the literature may be divided into two broad categories:

(1) the equivalences and preorders which semantically reduce parallelism to sequential nondeterminism, and

(2) those whose semantic theories treat parallelism as a primitive notion.

Let us recall that the equivalences which semantically reduce parallelism to sequential nondeterminism are usually called *interleaving equivalences* as in their associated theories concurrency between events is interpreted as their arbitrary interleaving (i.e. their possibility to occur in any temporal order). Several different notions of equivalence, based upon the interleaving approach, have been proposed in the literature and a list of some of the most widely used amongst them may be found in Chapter 1. Although their theories are very different from each other, all of the behavioural equivalences based on the interleaving approach have much in common; the only form of abstraction they support is related to *nondeterminism*. Processes can only be equivalent if their behaviour is the same *modulo nondeterminism*.

As pointed out in Chapter 1, several researchers have recently argued that, although they allow a faithful description of the functional behaviour of processes, the assumptions underlying the interleaving semantic models are inadequate in accounting for the nonsequential behaviour of distributed systems and that semantic theories which consider parallelism as a primitive notion are best suited for this purpose. Consequently, several equivalences for process algebras which distinguish concurrency from sequential nondeterminism have recently been proposed in the literature, e.g. *distributed bisimulation* [CH89], *timed equivalence* [H88c], *pomset bisimulation* [BC87,88], *NMS-bisimulation* [DDM86]. A variety of equivalences has been discussed in [GV87] in the context of Petri Nets [Rei85]. All of these equivalences are based on adaptations of the standard notion of bisimulation equivalence and draw a sharp line between concurrent execution of actions and their interleavings, between causal and temporal dependencies among computational events.

Although the above-mentioned equivalences differ in the degree in which they model the interplay between causality and the branching structure of processes, they identify descriptions of processes only if they can exhibit the same degree of parallelism. However, it may be argued that, whereas the interleaving equivalences are too coarse in that they forget too much of the structure of processes, the above-mentioned equivalences are perhaps too discriminating in that they do not allow us to relate "concurrent implementations" and "nondeterministic specifications" at all. As it is frequently more natural to specify the behaviour of a system in terms of a sequential nondeterministic process and more efficient to implement it in a parallel fashion, it would be helpful to have a semantic theory of processes which allows us to relate these two notions without semantically reducing parallelism to sequential nondeterminism. One possible use of this feature of the theory is in requiring that all the parallelism which is present in a specification be maintained in the implementation.

The main aim of this chapter is to provide such a semantic theory for a simple CCS-like language and to show how standard tools used in defining interleaving and non-interleaving equivalences for this language may be adapted in order to reconcile both philosophies to the semantics of concurrency.

Intuitively, our proposal is based upon a preorder $\sqsubseteq_C$ over processes such that, for processes p and q, $p \sqsubseteq_C q$ if, and only if, p and q have the same "functional behaviour", but q is "at least as parallel as" p. In order to formalize this idea, one needs to make precise the notion of "functional behaviour" over processes and to give a formal way of measuring the degree of parallelism a process may exhibit during its evolution. In this chapter, following the work presented in the previous chapters of this thesis, we take the view that a reasonable notion of equivalence between the functional behaviour of two concurrent, nondeterministic processes is captured by *bisimulation equivalence*, [Pa81], [Mil83]. Of course there is some arbitrariness in the choice of such an interleaving equivalence as the touchstone of our approach. However, bisimulation equivalence has a simple and elegant mathematical theory, [Mil88], and its properties have been investigated in a number of papers in the literature, [HM85], [Ab87a,b]. Moreover, the simplicity of its definition will allow us to concentrate on the issues which are more relevant to the aim of this chapter.

The preorder $\sqsubseteq_C$ will be defined by means of a bisimulation-like relation which will be dependent on the second parameter mentioned above: the measure of the degree of parallelism processes may exhibit during their evolution. In order to formalize this notion, following [BC87,88], we shall drop the requirement of atomicity over the actions performed by a concurrent process and shall axiomatize a preorder over the set of computations, $\leq_C$. Intuitively, for computations u and v, $u \leq_C v$ is intended to capture the fact that u and v correspond to the performance of the same atomic events, but these events are performed in a possibly more parallel fashion in v. The reasonableness of the preorder $\leq_C$ will be justified by showing that it coincides with a simple model-theoretic relation over an interpretation of computations as finite labelled posets. It will be shown how, by using the preorder $\leq_C$ over computations in a bisimulation-like relation, it is possible to obtain, in a rather simple way, a preorder over processes $\sqsubseteq_C$ which gives rise to a semantic theory in which concurrency is related to nondeterminism, but is not reduced to it.

We now give a brief outline of the remainder of the chapter. In §6.2 we introduce the language studied in the chapter, essentially **CCS** without synchronization, relabelling and restriction, and give both a standard labelled transition system semantics, [Kel76], and a pomset transition system semantics, [BC87,88], for it. This section also introduces the notion of computation studied in detail in §6.3. The syntactic and model-theoretic properties of a preorder on the set of computations are investigated in §6.3. We axiomatize a preorder $\leq_C$ over the set of computations, with the intent of capturing their relative degree of parallelism, and show that it coincides with an intuitive preorder over an interpretation of computations as finite labelled posets. The preorder $\leq_C$ on computations is used in §6.4 to induce one over processes, $\sqsubseteq_C$, by means of a bisimulation-like protocol. We show that $\sqsubseteq_C$ is a precongruence with respect to all the operators of the calculus and study its relationships with bisimulation equivalence and the pomset bisimulation equivalence of [BC87,88]. An equational characterization of $\sqsubseteq_C$ over the set of recursion-free processes is presented in §6.5. Finally, §6.6 presents a simple application of the semantic theory developed in the previous sections to the specification of concurrent systems. We end with a section of concluding remarks.

6.2 The Language and its Operational Semantics

This section is devoted to the presentation of a simple process algebra, essentially a subset of Milner's **CCS** [Mil80,89], which will be used to introduce the formalism and the intuitions underlying the semantic theory for processes presented in this chapter. The process algebra, which is parameterized over a set of atomic actions A, consists of a facility for recursive definitions and the following operators:

- nil, a constant used to denote *inaction*, the process that cannot perform any action, [Mil80,89];
- $a.\_$, a unary operator, one for each $a \in A$, used to prefix an action to a process. Intuitively, $a.p$ will denote a process capable of performing action a and behaving like p thereafter;
- $+$, for *nondeterministic choice*;
- $|$, for *parallel composition* (without communication).

Formally:

Definition 6.2.1 *Let A be a countable set of* atomic actions, *ranged over by $a, b \ldots$, and X be a countable set of* variables, *ranged over by $x, y \ldots$. The set of terms over A and X is generated by the following syntax:*

$$t ::= nil \mid a.t \mid t + t \mid t|t \mid x \mid rec\, x.\ t,$$

where $a \in A$ and $x \in X$. We assume the usual notions of free and bound variables in terms, with $rec\, x.\ \_$ as the binding operator. The set of closed terms (or processes*) will be denoted by $\mathcal{P}$ $(p, q \ldots \in \mathcal{P})$ and that of* finite processes, *those not containing occurrences of $rec\, x.\ \_$, will be denoted by $\mathcal{P}_{Fin}$.*

Following Milner [Mil80,89], a standard operational semantics may be given to $\mathcal{P}$ by means of Plotkin's SOS, [Pl81]. This can be done by defining a binary transition relation $\xrightarrow{a}$ for each $a \in A$. A standard way of defining $\xrightarrow{a}$, $a \in A$, is to stipulate that $\xrightarrow{a}$ is the least binary relation on $\mathcal{P}$ which satisfies the axiom and rules in Figure 6.1. Several interpretations of the relations $\xrightarrow{a}$ are possible; a standard one, [Mil80], is the following:

$$p \xrightarrow{a} p' \text{ if } p \text{ may perform action } a \text{ and thereby become } p'.$$

This interpretation of the transition relation $\xrightarrow{a}$ closely corresponds to Milner's experimental approach to the semantics of concurrent processes, [Mil80]. Roughly, experimenting on a

$$
\begin{array}{llll}
(1) & a.p \xrightarrow{a} p & & \\
(2) & p \xrightarrow{a} p' & \text{implies} & p+q \xrightarrow{a} p' \\
 & & & q+p \xrightarrow{a} p' \\
(3) & p \xrightarrow{a} p' & \text{implies} & p|q \xrightarrow{a} p'|q \\
 & & & q|p \xrightarrow{a} q|p' \\
(4) & t[rec\,x.\,t/x] \xrightarrow{a} p & \text{implies} & rec\,x.\,t \xrightarrow{a} p
\end{array}
$$

Figure 6.1. Axiom and rules for $\xrightarrow{a}$

process is taken to mean communicating with it and the binary relations $\xrightarrow{a}$, $a \in A$, are used to formalize the changes of state caused by successful a-experiments on processes by their environment. Informally, two processes that exhibit the same observable behaviour are deemed to be equivalent.

Throughout this thesis, bisimulation equivalence has been used to formalize the above notion of equivalence between processes. Again, we shall use $\sim$ to denote bisimulation equivalence over the labelled transition system $\langle \mathcal{P}, A, \longrightarrow \rangle$. (See Definition 3.2.2) For the sake of completeness, we recall the following result from [Mil80,89]:

Proposition 6.2.1 $\sim$ *is a congruence over* $\mathcal{P}$.

The aim of this chapter is to investigate whether it is possible to carry out the programme sketched in the introduction employing the tools of Milner's approach to the semantics of concurrency: structural operational semantics and bisimulation.

In order to relate concurrency and nondeterminism in a way that does not semantically reduce parallelism to sequential nondeterminism, we shall take a more liberal view of the computational steps of a concurrent system than the one underlying the operational semantics given in terms of the labelled transition system $\langle \mathcal{P}, A, \longrightarrow \rangle$. Underlying an interleaving description of the semantics of concurrent processes is the idea that processes evolve from one (global) state to another by performing atomic actions. Following [BC87,88], we shall assume that concurrent processes evolve by performing actions which are no longer atomic in space and time. [CaS88]. The computational steps considered in what follows will be transitions of the form $p \xrightarrow{u}_p q$, where u is a syntactically deterministic process (a syntactic notation for a *partially ordered multiset*, or *pomset* in Pratt and Gischer's terminology [Pr86], [Gi84]). The operational semantics for the set of processes $\mathcal{P}$ will thus be defined by means of what G. Boudol and I. Castellani call a *pomset transition system*.

Definition 6.2.2 *The set of* computations, *Comp, is generated by the following syntax:*

$$u ::= nil \mid a.u \mid u|u,$$

where $a \in A$. Comp will be ranged over by $u, v, w. \ldots$

As already mentioned above, a computation is thus a syntactically deterministic process. This idea of computational step is already implicit in Winskel's notion of *configuration* for Event Structures, [Win80,87], and has been studied at length in [BC87,88]. We can now define a pomset transition system semantics for $\mathcal{P}$ following [BC87,88].

Definition 6.2.3 *For each $u \in Comp$, $\xrightarrow{u}_p$ is the least binary relation over $\mathcal{P}$ which satisfies the following axiom and rules:*

$$\begin{array}{llll}
(1) & a.p \xrightarrow{a.nil}_p p & & \\
(2) & p \xrightarrow{u}_p p' & \textit{implies} & a.p \xrightarrow{a.u}_p p' \\
(3) & p \xrightarrow{u}_p p' & \textit{implies} & p + q \xrightarrow{u}_p p' \\
 & & & q + p \xrightarrow{u}_p p' \\
(4) & p \xrightarrow{u}_p p' & \textit{implies} & p|q \xrightarrow{u}_p p'|q \\
 & & & q|p \xrightarrow{u}_p q|p' \\
(5) & p \xrightarrow{u}_p p', q \xrightarrow{v}_p q' & \textit{imply} & p|q \xrightarrow{u|v}_p p'|q' \\
(6) & t[rec\, x.\, t/x] \xrightarrow{u}_p p & \textit{implies} & rec\, x.\, t \xrightarrow{u}_p p.
\end{array}$$

Comments about these rules may be found in the above-quoted references. Notationally, occurrences of *nil* will be often omitted in the context *a.nil*, both in processes and computations.

The pomset operational semantics defined above allows us to clearly distinguish causal dependencies from temporal ones, as the following examples show.

Example 6.2.1

$$\begin{array}{lllll}
(1) & a.b + b.a & \xrightarrow{a}_p b & \xrightarrow{b}_p nil \\
 & & \xrightarrow{b}_p a & \xrightarrow{a}_p nil \\
 & & \xrightarrow{a.b}_p nil & \\
 & & \xrightarrow{b.a}_p nil. & \\
(2) & a|b & \xrightarrow{a}_p nil|b & \xrightarrow{b}_p nil|nil \\
 & & \xrightarrow{b}_p a|nil & \xrightarrow{a}_p nil|nil \\
 & & \xrightarrow{a|b}_p nil|nil. &
\end{array}$$

Note how the process $a|b$ is capable of performing actions a and b in any temporal order (represented in this setting by the composition of the relations $\xrightarrow{a}_p$ and $\xrightarrow{b}_p$), but it is not capable of performing them in a single computational step labelled $a.b$ or $b.a$ (which corresponds to a causal dependence between the two actions). Thus there is a whole set of transitions which allow one to differentiate the processes $a|b$ and $a.b + b.a$, which are typically equated by interleaving semantic theories, [Mil80], [DH84].

The strength of a pomset transition system view of processes has been exploited by G. Boudol and I. Castellani, who define a standard bisimulation equivalence on $\langle \mathcal{P}, Comp, \longrightarrow_p \rangle$ up to a simple equational theory of computations. The resulting equivalence $\approx_{Pom}$, which they call *pomset bisimulation*, clearly distinguishes concurrency from sequential nondeterminism and its properties are investigated in [BC87,88]. A similar notion of equivalence has been presented in [GV87] in the setting of Petri Nets. However, as pointed out in the introduction, it might be argued that these equivalences are too discriminating. In fact, they do not allow us to relate sequential nondeterministic processes with concurrent ones at all, which is a drawback as it is frequently more natural to specify a system in a nondeterministic fashion, but it might be more efficient to implement it in a parallel way. The aim of this work is to show how, using the notions of pomset transition system and bisimulation, it is possible to define a preorder on $\mathcal{P}$ which is compatible with both an interleaving view of processes and the idea that concurrency should not be reduced to sequential nondeterminism.

6.3 A Preorder on Computations

This section is devoted to the introduction of the basic tool used in this chapter for relating nondeterminism and concurrency and to a study of its elementary properties. The tool for relating nondeterminism and concurrency in the behaviour of the processes in $\mathcal{P}$ will be a preorder over computations $\leq_C$. Intuitively, for computations u and v, $u \leq_C v$ is intended to capture the fact that u and v correspond to the performance of the same atomic actions, but these actions are performed in a more parallel fashion in the computation v.

Formally, let $\leq_C$ denote the least precongruence over *Comp* which satisfies the following set of axioms C:

$$
\begin{array}{lrcl}
\text{(PAR1)} & x|nil & = & x \\
\text{(PAR2)} & x|y & = & y|x \\
\text{(PAR3)} & (x|y)|z & = & x|(y|z) \\
\text{(SEQ)} & a.(x|y) & \leq & a.x|y.
\end{array}
$$

Of the above axioms, axiom (SEQ) is the interesting one; it expresses the intuitive fact that a computation in which action a causes both x and y is "more sequential" than a computation in which action a only causes x. Its relevance is probably best illustrated by means of an example.

Example 6.3.1 *Using the axioms in C it is possible to give a formal proof of the intuitive statement "a.b is more sequential than a|b" as follows:*

$$\begin{array}{llll} a.b.nil & =_C & a.b.(nil|nil) & \textit{by (PAR1) and substitutivity} \\ & \leq_C & a.(nil|b) & \textit{by (SEQ) and (PAR2)} \\ & \leq_C & a|b & \textit{by (SEQ).} \end{array}$$

The reasonableness of the preorder $\leq_C$ over computations will be demonstrated by showing that the set of axioms C completely axiomatizes a natural preorder on an interpretation of computations as finite posets labelled on A.

It is clear from the definition of the pomset transition system semantics for $\mathcal{P}$ that there are computations $u \in Comp$ such that $\xrightarrow{u}_p = \emptyset$. For instance, $p \xrightarrow{a|nil}_p$ or $p \xrightarrow{nil}_p$ for no p and a. The computations $u \in Comp$ such that $p \xrightarrow{u}_p$ for some $p \in \mathcal{P}$ will be referred to as *relevant*.

Definition 6.3.1 *Let RComp be the subset of Comp generated by the following syntax:*

$$u ::= a.nil \mid a.u \mid u|u,$$

where $a \in A$.

The following lemma states that every relevant computation u is a member of $RComp$.

Lemma 6.3.1 *For each* $p \in \mathcal{P}$, $u \in Comp$, $p \xrightarrow{u}_p$ *implies* $u \in RComp$.

It is now easy to see that $RComp$ is indeed the set of relevant computations mentioned above. This follows from the above lemma and the observation that, for each $u \in RComp$, $u \xrightarrow{u}_p$. The next definition and lemma express some properties of $\leq_C$ over some important subclasses of computations.

Definition 6.3.2 (Sequential and Maximally Parallel Computations) *A computation* $u \in Comp$ *is said to be:*

(1) sequential *if* $u =_C a_0.\ldots.a_n$ *for some* $n \geq 0$ *and* $\{a_0, \ldots, a_n\} \subseteq A$;

(2) maximally parallel *if* $u =_C \prod_{i \in I} a_i$, *where I is a finite nonempty index set and* $\{a_i \mid i \in I\} \subseteq A$. *The notation* $\prod_{i \in I} u_i$ *stands for* $u_{i_1}|\ldots|u_{i_k}$ $(I = \{i_1, \ldots, i_k\}, k \geq 0)$ *and is justified by axioms (PAR2)-(PAR3). By convention* $\prod_{i \in \emptyset} u_i \equiv nil$.

Lemma 6.3.2 *Let* $u, v \in Comp$ *be such that* $u \leq_C v$. *Then:*

(i) *if v is sequential then $u =_C v$;*

(ii) *if $v \equiv a_0.\dots.a_n$ and $u \in RComp$ then $u \equiv v$;*

(iii) *if u is maximally parallel then $u =_C v$;*

(iv) *if $u \equiv a.nil$ and $v \in RComp$ then $u \equiv v$.*

The properties of (relevant) computations stated in the above lemma will be very useful in relating the notion of preorder developed in §6.4 with Milner's strong bisimulation equivalence $\sim$. Before giving the definition of our preorder which relates concurrency to nondeterminism, we shall show that the preorder $\leq_C$ on computations, defined in this section by purely syntactical means, has indeed an intuitive model-theoretic characterization. The remaining part of this section is devoted to the study of such a model-theoretic characterization of $\leq_C$ over *Comp*. Computations will be interpreted as finite posets labelled on A. A preorder, denoted by $\lhd$, which is intended to capture the intuition conveyed by the defining axioms of $\leq_C$, is defined over the interpretation of *Comp* and it is shown to coincide with $\leq_C$.

6.3.1 Labelled Posets

The following definition introduces the main object of study of this section. (See also Definition 5.2.1)

Definition 6.3.3 (A-posets) *A finite* A-labelled partially ordered set (A-poset) *is a triple $\mathcal{E} = (E, \leq, l)$ where:*

- *E is a finite set of* events *(the vertices of the poset),*
- *$\leq$ is a partial order on E (the* causality relation*), and*
- *$l : E \longrightarrow A$ is a labelling function.*

In what follows, A-posets will be always considered up to isomorphism. Isomorphism classes of A-posets (for suitable label sets A) are often met under different names in the literature: for example, Pratt calls them *pomsets* (an acronym for partially ordered multisets), [Gi84], [Pr86], and Grabowski calls them *partial words*, [Gr81].

The following notions about posets will be useful in what follows.

Definition 6.3.4 *Let $\mathcal{E} = (E, \leq, l)$ be an A-poset. Then:*

- $\smile_{\mathcal{E}} =_{def} E^2 - (\leq \cup \leq^{-1})$ *is the* concurrency relation *over* $\mathcal{E}$. *The subscript* $\mathcal{E}$ *will be often dropped whenever the A-poset we are referring to is clear from the context;*
- $Min(\mathcal{E}) =_{def} \{e \in E \mid \forall e' \in E\ e' \leq e \text{ implies } e' = e\}$ *will denote the set of* minimal events *in* $\mathcal{E}$;
- *for each* $e \in E$, $\uparrow e$ *will denote, with abuse of notation, both*

  (a) $\uparrow e =_{def} \{e' \in E \mid e \leq e'\}$, *and*

  (b) $\uparrow e =_{def} (\uparrow e, \leq \lceil (\uparrow e)^2, l \lceil (\uparrow e))$, *the substructure of* $\mathcal{E}$ *corresponding to* $\uparrow e$.

Due to the restricted form of sequential composition, action-prefixing, allowed in the signature for computations, the A-posets which are interpretations of computations will have a particularly simple order structure.

Definition 6.3.5 (Separable A-Posets) *An A-poset* $\mathcal{E} = (E, \leq, l)$ *is* separable *iff, for each* $e, e' \in E$, $e \smile e'$ *implies* $e \leq e''$ *and* $e' \leq e''$ *for no* $e'' \in E$.

We can now define a preorder over A-posets which will be shown to give a model-theoretic characterization of the relation $\leq_C$ defined syntactically over computations. The definition of the preorder is based upon the notion of *projection* presented in the following definition.

Definition 6.3.6 (The Model-Theoretic Preorder)

(1) *Let* $\mathcal{E}_i = (E_i, \leq_i, l_i)$, $i = 1, 2$, *be two A-posets. A* projection *of* $\mathcal{E}_2$ *onto* $\mathcal{E}_1$ *is a function* $h : E_2 \longrightarrow E_1$ *such that:*

   (i) *h is bijective,*

   (ii) $\forall e, e' \in E_2\ e \leq_2 e'$ *implies* $h(e) \leq_1 h(e')$, *i.e. h is monotonic, and*

   (iii) $\forall e \in E_2\ l_2(e) = l_1(h(e))$, *i.e. h is label-preserving.*

(2) *The relation* $\lhd$ *over the set of A-posets is given by:*

$$\mathcal{E}_1 \lhd \mathcal{E}_2 \textit{ iff there exists a projection } h \textit{ of } \mathcal{E}_2 \textit{ onto } \mathcal{E}_1.$$

Lemma 6.3.3 $\lhd$ *is a preorder over the set of A-posets.*

Intuitively, $\mathcal{E}_1 \lhd \mathcal{E}_2$ means that all the causal dependencies among the events in the A-poset $\mathcal{E}_2$ can be bijectively embedded into causal dependencies in $\mathcal{E}_1$ (i.e. $\mathcal{E}_1$ has "at least as many causal dependencies" as $\mathcal{E}_2$). This is intended to capture the intuition that $\mathcal{E}_1 \lhd \mathcal{E}_2$ iff "$\mathcal{E}_2$ is at least as parallel as $\mathcal{E}_1$".

Example 6.3.2 *A-posets will be drawn, up to isomorphism, as Hasse diagrams growing rightwards. As we are concerned with labelled structures, the events will be left anonymous. The following A-posets are related by* $\lhd$*:*

$$a \rightarrow b \rightarrow c \quad \lhd \quad \begin{array}{c} a \rightarrow b \\ c \end{array} \quad \lhd \quad \begin{array}{c} a \\ b \\ c \end{array}$$

On the other hand, $\begin{array}{c} a \\ b \end{array} \;\not\!\lhd\; a \rightarrow b.$

The following proposition establishes some important properties of projection functions.

Proposition 6.3.1 *Let* $\mathcal{E}_i = (E_i, \leq_i, l_i)$, $i = 1, 2$, *be two A-posets. Assume that* $h : E_2 \longrightarrow E_1$ *is a projection function. Then:*

(1) *for each* $e \in E_2$, $h(e) \in Min(\mathcal{E}_1)$ *implies* $e \in Min(\mathcal{E}_2)$*;*

(2) *assume* $\mathcal{E}_1$ *is separable. Then:*

  (i) $\{\mathcal{E}_2(e) \mid e \in Min(\mathcal{E}_1)\}$, *where* $\mathcal{E}_2(e) =_{def} \{\bar{e} \in E_2 \mid e \leq_1 h(\bar{e})\}$, *is a partition of* E_2*. Moreover, for each* $e \in Min(\mathcal{E}_1)$, $h\lceil\mathcal{E}_2(e)$ *is a projection of* $\mathcal{E}_2(e)$ *onto* $\uparrow e$*.*

  (ii) *For each* $e_1, e_2 \in E_2$, $e_1 \leq_2 e_2 \in \mathcal{E}_2(e)$ *and* $e \in Min(\mathcal{E}_1)$ *imply* $e_1 \in \mathcal{E}_2(e)$*.*

Proof: Assume that $\mathcal{E}_1$ and $\mathcal{E}_2$ are A-posets and that $h : E_2 \longrightarrow E_1$ is a projection.

(1) Assume $e \notin Min(\mathcal{E}_2)$. Then, by the finiteness of E_2, there exists $\bar{e} \in Min(\mathcal{E}_2)$ such that $\bar{e} <_2 e$. By the monotonicity and the injectivity of h, $h(\bar{e}) <_1 h(e)$. Thus $h(e) \notin Min(\mathcal{E}_1)$.

(2) Assume that $\mathcal{E}_1$ is separable.

  (i) We prove, first of all, that $\{\mathcal{E}_2(e) \mid e \in Min(\mathcal{E}_1)\}$ is a partition of E_2.

  - We show that $E_2 = \bigcup\{\mathcal{E}_2(e) \mid e \in Min(\mathcal{E}_1)\}$. It is sufficient to prove that, for each $e' \in E_2$, there exists $e \in Min(\mathcal{E}_1)$ such that $e' \in \mathcal{E}_2(e)$. Assume $e' \in E_2$. By the finiteness of E_1, $Min(\mathcal{E}_1) \neq \emptyset$ and, for each $\bar{e} \in E_1$, there exists $e \in Min(\mathcal{E}_1)$ such that $e \leq_1 \bar{e}$. Thus $e \leq_1 h(e')$, for some $e \in Min(\mathcal{E}_1)$. This implies $e' \in \mathcal{E}_2(e)$.
  - We show that, for $e, e' \in Min(\mathcal{E}_1)$, $e \neq e'$ implies $\mathcal{E}_2(e) \cap \mathcal{E}_2(e') = \emptyset$. Assume, towards a contradiction, that $e, e' \in Min(\mathcal{E}_1)$, $e \neq e'$ and $\bar{e} \in \mathcal{E}_2(e) \cap \mathcal{E}_2(e')$. By the definition of $\mathcal{E}_2(\cdot)$, we have that $e \leq_1 h(\bar{e})$ and $e' \leq_1 h(\bar{e})$. As $e, e' \in Min(\mathcal{E}_1)$ and $e \neq e'$ imply $e \smile_{\mathcal{E}_1} e'$, the above assumption violates the hypothesis that $\mathcal{E}_1$ is separable. Hence, for each $e, e' \in Min(\mathcal{E}_1)$, $e \neq e'$ implies $\mathcal{E}_2(e) \cap \mathcal{E}_2(e') = \emptyset$.

We have thus shown that $\{\mathcal{E}_2(e) \mid e \in Min(\mathcal{E}_1)\}$ is a partition of E_2. In order to show that, for each $e \in Min(\mathcal{E}_1)$, $h\lceil\mathcal{E}_2(e)$ is a projection of $\mathcal{E}_2(e)$ onto $\uparrow e$, it is sufficient to prove that $h\lceil\mathcal{E}_2(e)$ is surjective. Assume $\bar{e} \in\uparrow e$, i.e. $e \leq_1 \bar{e}$. As h is bijective, there exists $e' \in E_2$ such that $h(e') = \bar{e}$. It follows, by the definition of $\mathcal{E}_2(e)$, that $e' \in \mathcal{E}_2(e)$. Thus $h\lceil\mathcal{E}_2(e)$ is surjective. All the other properties of projection maps follow from the hypothesis that h is itself a projection.

(ii) Assume, towards a contradiction, that $\bar{e} \in \mathcal{E}_2(e)$, $e' \leq_2 \bar{e}$ and $e' \notin \mathcal{E}_2(e)$. Then, by (i), there exists $\hat{e} \in Min(\mathcal{E}_1)$ such that $e' \in \mathcal{E}_2(\hat{e})$. It must be the case that $\hat{e} \neq e$. Then, by the definition of $\mathcal{E}_2(\cdot)$, $e \leq_1 h(\bar{e})$ and $\hat{e} \leq_1 h(e')$. By the monotonicity of h, $e' \leq_2 \bar{e}$ implies $h(e') \leq_1 h(\bar{e})$. Thus, by transitivity, $\hat{e} \leq_1 h(\bar{e})$. As $e \neq \hat{e}$ and $e, \hat{e} \in Min(\mathcal{E}_1)$, we have that $e \smile_{\mathcal{E}_1} \hat{e}$. This contradicts the hypothesis that $\mathcal{E}_1$ is separable.

This completes the proof of the proposition. □

The main consequence of the above proposition is that if $h : E_2 \longrightarrow E_1$ is a projection of $\mathcal{E}_2 = (E_2, \leq_2, l_2)$ onto $\mathcal{E}_1 = (E_1, \leq_1, l_1)$ and $\mathcal{E}_1$ is separable then h determines a partition $\equiv_h = \{\mathcal{E}_2(e) \mid e \in Min(\mathcal{E}_1)\}$ of E_2 such that, for each $e \in Min(\mathcal{E}_1)$, $\uparrow e \lhd \mathcal{E}_2(e)$. This property of projection functions will be very useful in relating $\leq_C$ and $\lhd$. In fact, we can give another useful characterization of the partition $\equiv_h$ as follows:

- for each $e \in Min(\mathcal{E}_1)$, $Min(\mathcal{E}_2, e) =_{def} \{e' \in Min(\mathcal{E}_2) \mid e \leq_1 h(e')\}$,
- for each $e \in Min(\mathcal{E}_1)$, $[Min(\mathcal{E}_2, e)] =_{def} \{e' \in E_2 \mid \exists \bar{e} \in Min(\mathcal{E}_2, e) : \ \bar{e} \leq_2 e'\}$, the upper-closure of $Min(\mathcal{E}_2, e)$.

Proposition 6.3.2 *If $\mathcal{E}_1$ is separable then, for each $e \in Min(\mathcal{E}_1)$, $\mathcal{E}_2(e) = [Min(\mathcal{E}_2, e)]$.*

Proof: Assume $e' \in \mathcal{E}_2(e)$, i.e. $e \leq_1 h(e')$. As E_2 is finite, there exists $\bar{e} \in Min(\mathcal{E}_2)$ such that $\bar{e} \leq_2 e'$. By proposition 6.3.1, we also have that $\bar{e} \in \mathcal{E}_2(e)$. Hence $e' \in [Min(\mathcal{E}_2, e)]$. The converse inclusion easily follows by the monotonicity of h. □

The import of the above proposition is that the partition $\equiv_h$ can be obtained as the upper-closure of a partition over the minimal elements of $\mathcal{E}_2$: this partition being $\equiv_h^{min} = \{Min(\mathcal{E}_2, e) \mid e \in Min(\mathcal{E}_1)\}$.

6.3.2 Interpretation of Computations as A-Posets

In order to interpret computations as A-posets, we shall have to impose the structure of a Σ^{Comp}-algebra over the set of A-posets, where Σ^{Comp} is obviously given by $\{nil, |\} \cup \{a.\_ \mid a \in A\}$. This is all that is needed to give an interpretation of $Comp$ in terms of A-posets.

Let $\mathcal{E}_i = (E_i, \leq_i, l_i)$, $i = 1, 2$, be two A-posets. Then:

- $a.\mathcal{E}_1 = (E, \leq, l)$, where
  (1) $E = E_1 \cup \{r\}$, with $r \notin E_1$,
  (2) $\forall e, e' \in E \; e \leq e'$ iff $e = r$ or $e \leq_1 e'$,
  (3) for each $e \in E$
  $$l(e) = \begin{cases} a & \text{if } e = r \\ l_1(e) & \text{otherwise.} \end{cases}$$
- $\mathcal{E}_1 | \mathcal{E}_2 = (E, \leq, l)$, where
  (1) $E = E_1 \uplus E_2$, the disjoint union of E_1 and E_2,
  (2) $\leq = \{((i, e), (j, e')) \mid i = j \text{ and } e \leq_i e'\}$ $(\leq = \leq_1 \uplus \leq_2)$,
  (3) $l(i, e) = l_i(e)$ $(l = l_1 \uplus l_2)$.
- $NIL = (\emptyset, \emptyset, \emptyset)$.

The unique homomorphism from $Comp$ to the Σ^{Comp}-algebra of A-posets will be denoted by $[\![\cdot]\!]$. It can be given the usual inductive characterization as follows:

$$\begin{aligned} [\![nil]\!] &= NIL \\ [\![a.u]\!] &= a.[\![u]\!] \\ [\![u|v]\!] &= [\![u]\!] | [\![v]\!]. \end{aligned}$$

The following lemma, which can be easily shown by structural induction over $u \in Comp$, states that the interpretations of computations are always separable A-posets.

Lemma 6.3.4 *For each $u \in Comp$, $[\![u]\!]$ is a separable A-poset.*

Let $\lhd$ denote, with abuse of notation, the preorder over $Comp$ defined by:

$$\forall u, v \in Comp \;\; u \lhd v \text{ iff } [\![u]\!] \lhd [\![v]\!].$$

We shall now proceed to show that $\lhd$ and $\leq_C$ concide over $Comp$, i.e. that $\leq_C$ completely axiomatizes $\lhd$. This result will hopefully provide a good motivation for the choice of $\leq_C$ as our preorder over computations. First of all we show that $\lhd$ is a Σ^{Comp}-precongruence over the algebra of A-posets (and, consequently, over $Comp$).

Proposition 6.3.3 $\lhd$ *is a precongruence over the algebra of A-posets.*

The next lemma states the soundness of the syntactically defined relation $\leq_C$ with respect to $\lhd$.

Lemma 6.3.5 (Soundness) *For each* $u, v \in Comp$, $u \leq_C v$ *implies* $u \lhd v$.

Proof: We just show the validity of axiom (SEQ), i.e. that $a.(u|v) \lhd a.u|v$. We shall show that $a.([\![u]\!]|[\![v]\!]) \lhd a.[\![u]\!]|[\![v]\!]$. Let $\mathcal{E}_i = (E_i, \leq_i, l_i)$, $i = 1, 2$, denote $[\![u]\!]$ and $[\![v]\!]$, respectively. Then, by the definition of the operations over A-posets,

(1) $a.([\![u]\!]|[\![v]\!]) = a.(\mathcal{E}_1|\mathcal{E}_2) = (E, \leq, l)$, where $E = \{r\} \cup (E_1 \uplus E_2)$ and $\leq$, l are defined as expected;

(2) $a.[\![u]\!]|[\![v]\!] = a.\mathcal{E}_1|\mathcal{E}_2 = (E', \leq', l')$, where $E' = (E_1 \cup \{r'\}) \uplus E_2$ and $\leq'$, l' are defined as expected.

Then it is easy to see that $h : E \longrightarrow E'$ defined by

$$h(i,e) = \begin{cases} r & \text{if } (i,e) = (1, r') \\ (i,e) & \text{otherwise} \end{cases}$$

is a projection function. Thus $a.(u|v) \lhd a.u|v$. □

We now concentrate on proving that $\leq_C$ is indeed complete with respect to $\lhd$ over $Comp$, i.e. that $u \lhd v$ implies $u \leq_C v$. The proof of completeness relies as usual on the isolation of normal forms for computations.

Definition 6.3.7 (Normal Forms for Computations) *The set of* normal forms NF *is the least subset of Comp which satisfies:*

- $nil \in \mathtt{NF}$,
- $\prod_{i=0}^{n} a_i.u_i \in \mathtt{NF}$ *if* $n \geq 0$ *and, for each* $0 \leq i \leq n$, $u_i \in \mathtt{NF}$.

The following normalization lemma, whose proof requires axioms (PAR1)-(PAR3) only, states that each computation u can be reduced to a normal form $nf(u)$ using the axioms in C.

Lemma 6.3.6 (Normalization) *For each* $u \in Comp$, *there exists a normal form* $nf(u)$ *such that* $u =_C nf(u)$.

The following lemma, which is based upon Propositions 6.3.1 and 6.3.2, states a fundamental decomposition property which will be used in the proof of the completeness theorem.

Lemma 6.3.7 (Decomposition Lemma) *Assume* $u \equiv \prod_{i=0}^{n} a_i.u_i \lhd v \equiv \prod_{j=0}^{m} b_j.v_j$. *Then there exists a partition* $\{J_i \mid 0 \leq i \leq n\}$ *of* $J = \{0, \ldots, m\}$ *such that, for each* $0 \leq i \leq n$, $a_i.u_i \lhd \prod_{j \in J_i} b_j.v_j$.

Proof: By the definition of $\lhd$, $u \lhd v$ iff $[\![u]\!] \lhd [\![v]\!]$ iff $\prod_{i=0}^{n} a_i.[\![u_i]\!] \lhd \prod_{j=0}^{m} b_j.[\![v_j]\!]$. By the definition of the operations over A-posets, $Min([\![u]\!]) = Min(\prod_{i=0}^{n} a_i.[\![u_i]\!]) = \{e_0, \ldots, e_n\}$, where e_i denotes the least element of $a_i.[\![u_i]\!]$. By lemma 6.3.4, $[\![u]\!]$ is a separable A-poset. By propositions 6.3.1 and 6.3.2, a projection h of $[\![v]\!]$ onto $[\![u]\!]$ establishes a partition of the event set of $[\![v]\!]$. Moreover, this partition is actually determined by a partition $\{J_i \mid 0 \leq i \leq n\}$ over the set of minimal elements $\{f_0, \ldots, f_m\}$ of $[\![v]\!]$, where f_j is the least element of $b_j.[\![v_j]\!]$. By propositions 6.3.1 and 6.3.2, for each $0 \leq i \leq n$,

$$a_i.[\![u_i]\!] = \uparrow e_i \lhd [\![v]\!](e_i) = [Min([\![v]\!], e_i)] = \prod_{j \in J_i} b_j.[\![v_j]\!]. \ \Box$$

We can now prove the promised completeness theorem.

Theorem 6.3.1 (Completeness) *For each* $u, v \in Comp$, $u \lhd v$ *iff* $u \leq_C v$.

Proof: The *if* implication is just lemma 6.3.5. Assume $u \lhd v$. By the normalization lemma we may assume, wlog, that u and v are *nf*'s. The proof proceeds by induction of the combined size of u and v (axioms (PAR2)-(PAR3) are used throughout the proof). If $u \equiv nil$ then it is easy to see that also $v \equiv nil$. The claim is then trivial.

Otherwise assume that $u \equiv \prod_{i=0}^{n} a_i.u_i$ and $v \equiv \prod_{j=0}^{m} b_j.v_j$, with $n \geq 0$. If $u \lhd v$ then, by the decomposition lemma, there exists a partition $\{J_i \mid 0 \leq i \leq n\}$ of $J = \{0, \ldots, m\}$ such that, for each $i \in \{0, \ldots, n\}$, $a_i.u_i \lhd \prod_{j \in J_i} b_j.v_j$. We show that, for each $0 \leq i \leq n$, $a_i.u_i \leq_C \prod_{j \in J_i} b_j.v_j$. The claim will then follow by substitutivity and (PAR2)-(PAR3).

Let $i \in \{0, \ldots, n\}$. As $a_i.u_i \lhd \prod_{j \in J_i} b_j.v_j$, there exists a projection h of $[\![\prod_{j \in J_i} b_j.v_j]\!]$ onto $[\![a_i.u_i]\!]$. Let e denote the least element of $[\![a_i.u_i]\!]$, which is labelled by a_i. By proposition 6.3.1, there exists $e' \in Min([\![\prod_{j \in J_i} b_j.v_j]\!])$ such that $h(e') = e$. Assume, wlog, that e' is the least element of $[\![b_k.v_k]\!]$, $k \in J_i$. As h is label-preserving, $b_k = a_i$. It is easy to see that, by the construction of $[\![\prod_{j \in J_i} b_j.v_j]\!]$, $h - \{(e', e)\}$ is a projection which establishes

$$u_i \lhd v_k | \prod_{j \in J_i - \{k\}} b_j.v_j.$$

We may now apply the inductive hypothesis to obtain $u_i \leq_C v_k | \prod_{j\in J_i - \{k\}} b_j.v_j$. Then

$$\begin{array}{rcll} a_i.u_i & \leq_C & a_i.(v_k | \prod_{j\in J_i-\{k\}} b_j.v_j) & \text{by substitutivity} \\ & \leq_C & a_i.v_k | \prod_{j\in J_i-\{k\}} b_j.v_j & \text{by (SEQ)} \\ & =_C & \prod_{j\in J_i} b_j.v_j & \text{by (PAR2)-(PAR3) and } b_k = a_i. \end{array}$$

This completes the proof of the theorem. □

Similarly, one can show that, for each $u, v \in Comp$, $[\![u]\!] \cong [\![v]\!]$ iff $u =_C v$. An interesting consequence of this observation is that, for A-posets which are interpretations of computations, $[\![u]\!] \lhd [\![v]\!] \lhd [\![u]\!]$ implies $[\![u]\!] \cong [\![v]\!]$, i.e. $\lhd$ is a partial order over $[\![Comp]\!]/\cong$.

We can now rephrase the results in lemma 6.3.2 in model-theoretic terms.

Proposition 6.3.4 *Let $\mathcal{E}_i = (E_i, \leq_i, l_i)$, $i = 1, 2$, be nonempty A-posets. Then:*

(1) *if $\mathcal{E}_2$ is totally ordered and $\mathcal{E}_1 \lhd \mathcal{E}_2$ then $\mathcal{E}_1 \cong \mathcal{E}_2$;*

(2) *$\leq_1 = Id_{E_1}$ and $\mathcal{E}_1 \lhd \mathcal{E}_2$ imply $\mathcal{E}_1 \cong \mathcal{E}_2$.*

We shall end this section with a result which expresses an intuitive property of the preorder $\leq_C$ which will be useful in the following section.

Proposition 6.3.5 (Linearization) *Let $u, v \in Comp$. Assume that $a_0. \ldots .a_n \leq_C u$ and $b_0. \ldots .b_m \leq_C v$. Then $a_0. \ldots .a_n.b_0. \ldots .b_m \leq_C u|v$.*

Proof: Assume $u, v \in Comp$, $a_0. \ldots .a_n \leq_C u$ and $b_0. \ldots .b_m \leq_C v$. We prove, first of all, that

$$a_0 \ldots a_n.b_0 \ldots b_m \quad \leq_C \quad a_0 \ldots a_n | b_0 \ldots b_m. \tag{6.1}$$

The proof of (6.1) is by induction on n.

- Basis, $n = 0$. Then

$$\begin{array}{rcll} a_0.b_0 \ldots b_m & =_C & a_0.(nil | b_0 \ldots b_m) & \text{by (PAR1)} \\ & \leq_C & a_0.nil | b_0 \ldots b_m & \text{by (SEQ).} \end{array}$$

- Inductive step, $n > 0$. By the inductive hypothesis we have that

$$a_1 \ldots a_n.b_0 \ldots b_m \leq_C a_1 \ldots a_n | b_0 \ldots b_m.$$

  Then

$$\begin{array}{rcll} a_0 \ldots a_n.b_0 \ldots b_m & \leq_C & a_0.(a_1 \ldots a_n | b_0 \ldots b_m) & \text{by substitutivity} \\ & \leq_C & a_0 \ldots a_n | b_0 \ldots b_m & \text{by (SEQ).} \end{array}$$

  By induction this implies (6.1). It is now easy to see that the claim follows by (6.1) and substitutivity. □

6.4 Relating Nondeterminism and Concurrency

In §6.3 we have introduced a syntactic preorder $\leq_C$ over the set of computations *Comp* and studied some of its elementary properties. The aim of this section is to show how $\leq_C$ can be used to induce an interesting preorder over the set of processes $\mathcal{P}$. This preorder will allow us to relate sequential nondeterministic processes with concurrent ones, which may be thought of as their implementations. However, this preorder does not reduce concurrency to sequential nondeterminism as it will be clear from what follows. The preorder on $\mathcal{P}$ will be defined by means of the notion of *bisimulation*, [Pa81],[Mil83,88].

Let *Rel* denote the set of binary relations over $\mathcal{P}$. The functional $\mathcal{F} : Rel \longrightarrow Rel$ is defined, for each $\mathcal{R} \in Rel$, as follows:

$(p,q) \in \mathcal{F}(\mathcal{R})$ if, for each $u \in Comp$,

(1) $p \xrightarrow{u}_{\mathcal{P}} p'$ implies, for some $q' \in \mathcal{P}$ and $v \in Comp$ such that $u \leq_C v$, $q \xrightarrow{v}_{\mathcal{P}} q'$ and $(p',q') \in \mathcal{R}$;

(2) $q \xrightarrow{u}_{\mathcal{P}} q'$ implies, for some $p' \in \mathcal{P}$ and $v \in Comp$ such that $v \leq_C u$, $p \xrightarrow{v}_{\mathcal{P}} p'$ and $(p',q') \in \mathcal{R}$.

A relation $\mathcal{R} \in Rel$ will be called an *$\mathcal{F}$-bisimulation* if $\mathcal{R} \subseteq \mathcal{F}(\mathcal{R})$. Let $\sqsubseteq_C$ denote $\bigcup\{\mathcal{R} \mid \mathcal{R} \subseteq \mathcal{F}(\mathcal{R})\}$. The following proposition is then standard.

Proposition 6.4.1 *$\mathcal{F}$ is a monotonic endofunction over the complete lattice $(Rel, \subseteq)$ and $\sqsubseteq_C$ is its largest fixed-point.*

Intuitively, for processes p and q, $p \sqsubseteq_C q$ if, for each computational step u,

- if p may perform u and thereby be transformed into p' then q may perform a computation v, which is a "possibly more parallel version" of u, and enter a state q' such that $p' \sqsubseteq_C q'$; conversely
- if $q \xrightarrow{u}_{\mathcal{P}} q'$ then there exist a computation v, which is a "possibly more sequential version " of u, and a state p' such that $p \xrightarrow{v}_{\mathcal{P}} p'$ and $p' \sqsubseteq_C q'$.

This informal explanation of the nature of $\sqsubseteq_C$ should give a hint of the importance of the preorder on computations $\leq_C$ in this setting. Before giving a few examples of the relationships among processes that can be established using $\sqsubseteq_C$, examples that will highlight the rôle played by $\leq_C$ in relating process behaviours, we show that $\sqsubseteq_C$ is indeed a preorder. However, it is easy to check that $Id_{\mathcal{P}}$ is an $\mathcal{F}$-bisimulation and that the composition of $\mathcal{F}$-bisimulations is again an $\mathcal{F}$-bisimulation. As a corollary of these observations we obtain the following:

Fact 6.4.1 *$\sqsubseteq_C$ is a preorder over $\mathcal{P}$.*

A few examples are now in order.

Example 6.4.1

(1) $p \equiv a.b + b.a \sqsubseteq_C a|b + a.b + b.a \equiv q$. *In fact, it is easy to see that the relation*

$$\mathcal{R} = \{(p,q),(b, nil|b),(a, a|nil),(nil, nil|nil)\} \cup Id_{\mathcal{P}}$$

is an $\mathcal{F}$-bisimulation. The key point is that the move $q \xrightarrow{a|b}_{\mathrm{p}} nil|nil$ can be matched by $p \xrightarrow{a.b}_{\mathrm{p}} nil$ as $a.b \leq_C a|b$ (or, equivalently, by $p \xrightarrow{b.a}_{\mathrm{p}} nil$).

(2) $q \sqsubseteq_C a|b \equiv r$. *In fact, it is easy to check that the relation*

$$\mathcal{S} = \{(q,r),(b, nil|b),(a, a|nil),(nil, nil|nil)\} \cup Id_{\mathcal{P}}$$

is an $\mathcal{F}$-bisimulation.

(3) $q \not\sqsubseteq_C p$. *In fact $q \xrightarrow{a|b}_{\mathrm{p}} nil|nil$, but $p \stackrel{a|b}{\not\longrightarrow}_{\mathrm{p}}$ and $p \stackrel{b|a}{\not\longrightarrow}_{\mathrm{p}}$.*

(4) $r \not\sqsubseteq_C q$. *In fact, $q \xrightarrow{a.b}_{\mathrm{p}} nil$, but $r \stackrel{a.b}{\not\longrightarrow}_{\mathrm{p}}$. By lemma 6.3.2 this is sufficient to establish that, for no $v \leq_C a.b$, $r \xrightarrow{v}_{\mathrm{p}}$.*

(5) $r \not\sqsubseteq_C p$. *In fact, $r \xrightarrow{a|b}_{\mathrm{p}} nil|nil$, but $p \stackrel{a|b}{\not\longrightarrow}_{\mathrm{p}}$ and $p \stackrel{b|a}{\not\longrightarrow}_{\mathrm{p}}$. Again, by lemma 6.3.2 this is sufficient to establish that, for no v such that $a|b \leq_C v$, $p \xrightarrow{v}_{\mathrm{p}}$.*

Proposition 6.4.2 *Let $p, q, r \in \mathcal{P}$. Assume $p \sqsubseteq_C q$. Then $a.p \sqsubseteq_C a.q$, $p + r \sqsubseteq_C q + r$ and $p|r \sqsubseteq_C q|r$.*

The above proposition tells us that $\sqsubseteq_C$ is a well-behaved preorder with respect to all the operators of the algebra apart from recursion. In order to show that $\sqsubseteq_C$ is also preserved by recursive contexts, we first extend it to open terms.

Definition 6.4.1 *A* closed substitution *is a map $\rho : X \longrightarrow \mathcal{P}$. Then, for open terms t_1, t_2, $t_1 \sqsubseteq_C t_2$ iff, for every closed substitution ρ, $t_1\rho \sqsubseteq_C t_2\rho$.*

The following proposition, whose proof follows standard lines [Mil83,89], states that $\sqsubseteq_C$ is closed with respect to recursive contexts.

Proposition 6.4.3 *Let $t_1 \sqsubseteq_C t_2$. Then $rec\,x.\,t_1 \sqsubseteq_C rec\,x.\,t_2$.*

The preorder $\sqsubseteq_C$ that we have just defined enjoys some interesting relationships with Milner's strong bisimulation equivalence $\sim$. As already mentioned in the introduction, $\sim$ will be used to assess the reasonableness of our proposal. Intuitively, it will be required that $\sqsubseteq_C$ relates processes p and q only when $p \sim q$, i.e. only when p and q have the same "functional behaviour". Moreover, given the intuition that we are trying to capture, it might be expected that, for sequential nondeterministic processes, $\sqsubseteq_C$ coincides with $\sim$. We shall now show that both these properties are true of $\sqsubseteq_C$. These results should hopefully reinforce $\sqsubseteq_C$ as a reasonable notion of preorder over $\mathcal{P}$.

Lemma 6.4.1 *For each $p \in \mathcal{P}$, $a \in A$, $p \xrightarrow{a.nil}_p p'$ iff $p \xrightarrow{a} p'$.*

The above lemma is all that is needed to prove that $\sqsubseteq_C$ is contained in $\sim$.

Theorem 6.4.1 *For each $p, q \in \mathcal{P}$, $p \sqsubseteq_C q$ implies $p \sim q$.*

Proof: It is sufficient to show that $\sqsubseteq_C$ is an interleaving bisimulation. Assume $p \sqsubseteq_C q$ and $p \xrightarrow{a} p'$. Then, by lemma 6.4.1, $p \xrightarrow{a.nil}_p p'$. As $p \sqsubseteq_C q$, there exist $v \in Comp$ and q' such that $a.nil \leq_C v$, $q \xrightarrow{v}_p q'$ and $p' \sqsubseteq_C q'$. By lemma 6.3.2, $v \equiv a.nil$ and thus, by lemma 6.4.1, $q \xrightarrow{a} q'$.

Assume $q \xrightarrow{a} q'$. Then, again by lemma 6.4.1, $q \xrightarrow{a.nil}_p q'$. As $p \sqsubseteq_C q$, there exist $v \leq_C a.nil$ and p' such that $p \xrightarrow{v}_p p'$ and $p' \sqsubseteq_C q'$. By lemma 6.3.2, $v \equiv a.nil$ and, by lemma 6.4.1, $p \xrightarrow{a} p'$. Thus $\sqsubseteq_C$ is an interleaving bisimulation and $p \sim q$. □

It follows from the above theorem and examples that the containment is indeed strict.

Intuitively, it may be expected that $\sqsubseteq_C$ and $\sim$ coincide over processes that do not contain occurrences of the parallel composition operator. We shall call these processes *sequential* and the subset of sequential processes will be denoted by $\mathcal{P}_{Seq}$.

Lemma 6.4.2 *Let $p \in \mathcal{P}_{Seq}$. Then, for each $u \in Comp$, $p \xrightarrow{u}_p$ implies u has the form $a_0. \ldots .a_n.nil$.*

In order to prove that $\sqsubseteq_C$ coincides with $\sim$ over $\mathcal{P}_{Seq}$, we shall need to relate derivations of the form $p \xrightarrow{u}_p q$, $u \equiv a_0. \ldots .a_n$, with Milner's multi-step derivations $p \xrightarrow{a_0} \cdots \xrightarrow{a_n} q$. This is done in the following lemma.

Lemma 6.4.3 *Let $p \in \mathcal{P}_{Seq}$. Then:*

(1) $p \xrightarrow{a}_p p' \xrightarrow{u}_p q$ *implies* $p \xrightarrow{a.u}_p q$;

(2) *for* $u \equiv a_0.\ldots.a_n$, $p \xrightarrow{u}_p q$ *iff* $p \xrightarrow{a_0} \cdots \xrightarrow{a_n} q$.

Proof: Statement (1) is proven by induction on the proof of the derivation $p \xrightarrow{a.nil}_p p'$. For what concerns statement (2), the *only if* implication is shown by induction on the proof of the derivation $p \xrightarrow{u}_p q$. The *if* implication is shown by induction on n, using (1) and lemma 6.4.1 for the inductive step. The details are omitted. □

We can now show that $\sqsubseteq_C$ and $\sim$ coincide over $\mathcal{P}_{Seq}$, i.e. that for sequential processes $\sqsubseteq_C$ reduces to Milner's strong bisimulation equivalence.

Theorem 6.4.2 *For each* $p, q \in \mathcal{P}_{Seq}$, $p \sqsubseteq_C q$ *iff* $p \sim q$.

Proof: The *only if* implication follows by Theorem 6.4.1. In order to prove that, for $p, q \in \mathcal{P}_{Seq}$, $p \sim q$ implies $p \sqsubseteq_C q$, it is sufficient to show that the relation $\mathcal{R} = \sim \cap \mathcal{P}^2_{Seq}$ is an $\mathcal{F}$-bisimulation. Assume $(p, q) \in \mathcal{R}$.

Suppose that $p \xrightarrow{u}_p p'$. Then, by lemma 6.4.2, $u \equiv a_0.\ldots.a_n.nil$. By lemma 6.4.3, $p \xrightarrow{u}_p p'$ implies $p \xrightarrow{a_0} \cdots \xrightarrow{a_n} p'$. As $p \sim q$, it is easy to see that there exists q' such that $q \xrightarrow{a_0} \cdots \xrightarrow{a_n} q'$ and $p' \sim q'$. By lemma 6.4.3, $q \xrightarrow{u}_p q'$. Obviously, $u \leq_C u$ and, as $p', q' \in \mathcal{P}_{Seq}$, $(p', q') \in \mathcal{R}$. The other defining clause of $\sqsubseteq_C$ is checked by a similar argument. Thus $\mathcal{R}$ is an $\mathcal{F}$-bisimulation. □

Indeed, one can prove a slightly stronger statement than Theorem 6.4.2. Our aim is to show that, for $p \in \mathcal{P}_{Seq}$ and $q \in \mathcal{P}$, $p \sim q$ implies $p \sqsubseteq_C q$. This statement is useful for verification purposes and gives a pleasing proof technique for $\sqsubseteq_C$ which will be used in §6.6. The following lemma will be useful in the proof of the above mentioned result.

Lemma 6.4.4 *Let* $p \in \mathcal{P}$. *Then the following statements hold:*

(1) $p \xrightarrow{a} p' \xrightarrow{u}_p q$ *implies* $p \xrightarrow{v}_p q$ *for some* v *such that* $a.u \leq_C v$;

(2) $p \xrightarrow{\sigma} q$, $\sigma \equiv a_0 \ldots a_n$, *implies* $p \xrightarrow{u}_p q$ *for some* u *such that* $a_0.\ldots.a_n \leq_C u$;

(3) $p \xrightarrow{u}_p q$ *implies* $p \xrightarrow{\sigma} q$, $\sigma \equiv a_0 \ldots a_n$ *for some* $a_0.\ldots.a_n \leq_C u$.

Proof: We prove each statement in turn.

(1) By induction on the proof of the derivation $p \xrightarrow{a} p'$. The only interesting case is the following:

- $p \equiv p_1|p_2 \xrightarrow{a} p_1'|p_2$ because $p_1 \xrightarrow{a} p_1'$. We proceed by analyzing the move $p_1'|p_2 \xrightarrow{u}_p q$. By the pomset operational semantics, $p_1'|p_2 \xrightarrow{u}_p q$ iff

  (a) $p_1' \xrightarrow{u}_p p_1''$ and $q \equiv p_1''|p_2$, or

  (b) $p_2 \xrightarrow{u}_p p_2'$ and $q \equiv p_1'|p_2'$, or

  (c) $p_1' \xrightarrow{u_1}_p p_1''$, $p_2 \xrightarrow{u_2}_p p_2'$, $u \equiv u_1|u_2$ and $q \equiv p_1''|p_2'$.

  If (a) holds then, by the inductive hypothesis, $p_1 \xrightarrow{v}_p p_1''$ for some v such that $a.u \leq_C v$. Hence, by the pomset operational semantics, $p_1|p_2 \xrightarrow{v}_p p_1''|p_2 \equiv q$.

  If (b) holds then, as by lemma 6.4.1 $p_1 \xrightarrow{a} p_1'$ implies $p_1 \xrightarrow{a.nil}_p p_1'$, by the pomset operational semantics $p_1|p_2 \xrightarrow{a|u}_p p_1'|p_2' \equiv q$. Moreover,

  $$\begin{array}{rcll} a.u & =_C & a.(nil|u) & \text{by (PAR1) and substitutivity} \\ & \leq_C & a.nil|u & \text{by (SEQ).} \end{array}$$

  If (c) holds then, by the inductive hypothesis, $p_1 \xrightarrow{v}_p p_1''$ for some v such that $a.u_1 \leq_C v$. Hence, by the pomset operational semantics, $p_1|p_2 \xrightarrow{v|u_2}_p p_1''|p_2' \equiv q$. Moreover,

  $$\begin{array}{rcll} a.u \equiv a.(u_1|u_2) & \leq_C & a.u_1|u_2 & \text{by (SEQ)} \\ & \leq_C & v|u_2 & \text{by substitutivity.} \end{array}$$

(2) By induction on n. The basis of the induction easily follows by lemma 6.4.1. For the inductive step, assume $p \xrightarrow{\sigma} q$, $\sigma \equiv a_0 \ldots a_n$, $n \geq 1$. Then $p \xrightarrow{a_0} p' \xrightarrow{a_1 \ldots a_n} q$, for some p'. By the inductive hypothesis, $p' \xrightarrow{u}_p q$ for some u such that $a_1 \ldots a_n \leq_C u$. By statement (1) of the lemma, $p \xrightarrow{a_0} p' \xrightarrow{u}_p q$ implies $p \xrightarrow{v}_p q$ for some v such that $a_0.u \leq_C v$. By transitivity and substitutivity, $a_0 \ldots a_n \leq_C a_0.u \leq_C v$.

(3) By induction on the proof of the derivation $p \xrightarrow{u}_p q$. The only interesting case is the following:

- $p \equiv p_1|p_2 \xrightarrow{u}_p q$. By symmetry and the pomset operational semantics, we may restrict ourselves to considering the following two cases.

  (a) $p_1 \xrightarrow{u}_p q_1$ and $q_1|p_2 \equiv q$. By the inductive hypothesis, $p_1 \xrightarrow{a_0.\ldots.a_n} q_1$ for some $a_0.\ldots.a_n \leq_C u$. Then $p_1|p_2 \xrightarrow{a_0.\ldots.a_n} q_1|p_2 \equiv q$.

  (b) $p_1 \xrightarrow{u_1}_p q_1$, $p_2 \xrightarrow{u_2}_p q_2$, $q \equiv q_1|q_2$ and $u \equiv u_1|u_2$. By the inductive hypothesis, there exist $\sigma = a_0.\ldots.a_n \leq_C u_1$ and $\omega = b_0.\ldots.b_m \leq_C u_2$ such that $p_1 \xrightarrow{\sigma} q_1$ and $p_2 \xrightarrow{\omega} q_2$. Obviously, $p_1|p_2 \xrightarrow{\sigma} q_1|p_2 \xrightarrow{\omega} q_1|q_2 \equiv q$. It remains to be shown that $a_0.\ldots.a_n.b_0.\ldots.b_m \leq_C u_1|u_2$, but this follows from proposition 6.3.5. □

Theorem 6.4.3 *Let $p \in \mathcal{P}_{Seq}$ and $q \in \mathcal{P}$. Then $p \sim q$ iff $p \sqsubseteq_C q$.*

Proof: Again, the *if* implication follows from Theorem 6.4.1. In order to prove the *only if* implication, it is sufficient to show that the relation

$$\mathcal{R} =_{def} \{(p,q) \mid p \in \mathcal{P}_{Seq} \text{ and } p \sim q\}$$

is an $\mathcal{F}$-bisimulation. We check that the $\mathcal{F}$-bisimulation clauses are met for $(p,q) \in \mathcal{R}$.

- Assume $p \xrightarrow{u}_p p'$. Then, by lemma 6.4.2, $u \equiv a_0.\ldots.a_n$. By lemma 6.4.3, $p \xrightarrow{u}_p p'$ implies $p \xrightarrow{a_0} \cdots \xrightarrow{a_n} p'$. As $p \sim q$, there exists q' such that $q \xrightarrow{a_0} \cdots \xrightarrow{a_n} q'$ and $p' \sim q'$. By statement (2) of the previous lemma, $q \xrightarrow{a_0} \cdots \xrightarrow{a_n} q'$ implies $q \xrightarrow{v}_p q'$ for some v such that $u \leq_C v$. Moreover, $(p', q') \in \mathcal{R}$ as $p' \in \mathcal{P}_{Seq}$.
- Assume $q \xrightarrow{v}_p q'$. Then, by statement (3) of the above lemma, $q \xrightarrow{a_0} \cdots \xrightarrow{a_n} q'$ for some $u \equiv a_0.\ldots.a_n \leq_C v$. As $p \sim q$, there exists p' such that $p \xrightarrow{a_0} \cdots \xrightarrow{a_n} p'$ and $p' \sim q'$. As $p \in \mathcal{P}_{Seq}$, by lemma 6.4.3, $p \xrightarrow{a_0} \cdots \xrightarrow{a_n} p'$ implies $p \xrightarrow{u}_p p'$. Moreover $(p', q') \in \mathcal{R}$ as $p' \in \mathcal{P}_{Seq}$.

Hence $\mathcal{R}$ is an $\mathcal{F}$-bisimulation. □

A rather pleasing consequence of this result is that, whenever trying to establish that a process q is a correct implementation of a sequential specification p, i.e. that $p \sqsubseteq_C q$, it is sufficient to exhibit a standard strong bisimulation containing them.

Pomset bisimulation equivalence, $\approx_{Pom}$, has been proposed in [BC87,88] as a modification of the standard notion of bisimulation equivalence which clearly distinguishes concurrency from sequential nondeterminism. As the definition of the preorder $\sqsubseteq_C$ presented in this section has much in common with that of $\approx_{Pom}$, it is natural to investigate the relationships between the kernel of $\sqsubseteq_C$, $\simeq$, and $\approx_{Pom}$.

Let $=_P$ denote the least congruence over *Comp* which satisfies axioms (PAR1)-(PAR3). Then $\approx_{Pom}$ is defined as the largest binary, symmetric relation over $\mathcal{P}$ which satisfies the following condition:

$p \approx_{Pom} q$ if, for each $u \in Comp$,

$$p \xrightarrow{u}_p p' \text{ implies, for some } q' \text{ and } v =_P u,\ q \xrightarrow{v}_p q' \text{ and } p' \approx_{Pom} q'.$$

It is easy to see that, because $=_P \subseteq \leq_C$ over *Comp*, $\approx_{Pom}$ is indeed an $\mathcal{F}$-bisimulation. As a consequence of this observation we have the following:

Fact 6.4.2 *For each $p, q \in \mathcal{P}$, $p \approx_{Pom} q$ implies $p \simeq q$.*

However, the converse implication does not hold, i.e. $\simeq$ is strictly weaker than $\approx_{Pom}$. Consider, in fact, the following two processes:

$$p \equiv a.b.c + c.a.b + a|b|c,$$

$$q \equiv p + a.b|c.$$

Obviously, $p \not\approx_{Pom} q$ as $q \xrightarrow{a.b|c}_{\mathcal{P}}$, but $p \xrightarrow{u}_{\mathcal{P}}$ for no $u =_{\mathcal{P}} a.b|c$. However, we do have that $p \simeq q$. This essentially depends on the fact that, because $a.b.c \leq_C a.b|c \leq_C a|b|c$, the transition $q \xrightarrow{a.b|c}_{\mathcal{P}} nil|nil$ can be matched "in a more sequential fashion" by $p \xrightarrow{a.b.c}_{\mathcal{P}} nil$ and "in a more parallel one" by $p \xrightarrow{a|b|c}_{\mathcal{P}} nil|nil|nil$.

6.5 Algebraic Characterization of the Preorder

The purpose of this section is to axiomatize the preorder $\sqsubseteq_C$ defined in the previous section over the set of finite processes $\mathcal{P}_{Fin}$. As noted by G. Boudol and I. Castellani in [BC87,88], the interpretation of processes given by a pomset transition system semantics is just an ordinary labelled transition system over a set of actions. The only difference being that the actions are "structured".

Following M. Hennessy and R. Milner [HM85], there is a standard way of axiomatizing bisimulation-like relations over ordinary, finite, acyclic labelled transition systems. Their method involves the reduction of terms to so-called *sumforms* over the set of actions into consideration. As in our pomset transition system semantics for the language $\mathcal{P}$ processes evolve by performing computations and computations are not in the signature for processes themselves, this method is not directly applicable to the language $\mathcal{P}$.

In order to apply Hennessy and Milner's technique to provide an axiomatization for the preorder $\sqsubseteq_C$ over $\mathcal{P}_{Fin}$, we thus need to extend the language $\mathcal{P}$ to $\mathcal{P}'$, where $\mathcal{P}'$ is built as $\mathcal{P}$ with the additional formation rule:

$$u \in Comp \text{ and } p \in \mathcal{P}' \text{ imply } u : p \in \mathcal{P}'.$$

Thus the signature of the language $\mathcal{P}$ has been extended by allowing prefixing operators of the form $u : \_$, for $u \in Comp$. The language $\mathcal{P}'$ thus allows one to prefix computations to processes and this is what will be needed to define a suitable set of sumforms for $\mathcal{P}'$. The operational semantics for $\mathcal{P}'$ is obtained by extending the rules in definition 2.3 with the axiom

$$\textbf{(PRE)} \quad u : p \xrightarrow{u}_{\mathcal{P}} p.$$

It is easy to see that $\sqsubseteq_C$ can be conservatively extended to the language $\mathcal{P}'$ and that the following proposition holds:

Proposition 6.5.1 *$\sqsubseteq_C$ is a $\mathcal{P}'_{Fin}$-precongruence.*

Definition 6.5.1 (Sumforms) *The set of* sumforms *over Comp,* $\mathtt{SF}(Comp)$*, is the least subset of $\mathcal{P}'_{Fin}$ which satisfies:*

(1) $nil \in \mathtt{SF}(Comp)$,

(2) $u \in Comp$ *and* $p \in \mathtt{SF}(Comp)$ *imply* $u : p \in \mathtt{SF}(Comp)$,

(3) $p, q \in \mathtt{SF}(Comp)$ *imply* $p + q \in \mathtt{SF}(Comp)$.

In order to give a complete axiomatization for $\sqsubseteq_C$ over $\mathcal{P}'_{Fin}$ (and, consequently, over $\mathcal{P}_{Fin}$), it will be sufficient to devise a set of axioms which allow us to reduce terms in $\mathcal{P}'_{Fin}$ to sumforms and which are complete for $\sqsubseteq_C$ over $\mathtt{SF}(Comp)$. Formally, the theory we shall consider is the two-sorted theory consisting of the set of axioms C over $Comp$, the set of axioms $\mathcal{A}$ given in Figure 6.2 together with the following rule

$$\textbf{(SUB)} \quad u \leq_C v,\, p \leq q \quad \text{imply} \quad u : p \leq v : q.$$

Note that the axioms in C are only needed to relate computations; indeed, (PAR1)-(PAR3) are valid for processes as well (although we have no use for them in the equational theory for processes) whilst (SEQ) is *not* sound for processes. In fact, $a.b \leq_C a|b$, but $a.b \not\sqsubseteq_C a|b$. It is worth pointing out that all the axioms in $\mathcal{A}$ have an equational nature, i.e. they essentially express properties of the kernel of $\sqsubseteq_C$. Thus all the essence of the preorder $\sqsubseteq_C$ is captured by the inequational theory of computations and namely by axiom (SEQ). Axiom (R1) expresses the interplay between the two kinds of prefixing operators in the language $\mathcal{P}'$. It will be used in reducing processes to sumforms to eliminate occurrences of the action-prefixing operators $a.\_$ in terms in favour of the new computation-prefixing operators $u : \_$.

From now onwards, sumforms will be written as $\sum_{i \in I} u_i : p_i$, where, for each $i \in I$, $u_i \in Comp$ and p_i is a sumform. This notation is justified by axioms (A1)-(A2). If $I = \emptyset$ then, by convention, $\sum_{i \in I} u_i : p_i \equiv nil$. Let $\leq_{\mathcal{A}}$ be the least precongruence over $\mathcal{P}'_{Fin}$ generated by the set of axioms in Figure 6.2 and rule (SUB). The following proposition, whose proof is omitted, states the soundness of the proof system with respect to $\sqsubseteq_C$.

Proposition 6.5.2 (Soundness) *For each $p, q \in \mathcal{P}'_{Fin}$, $p \leq_{\mathcal{A}} q$ implies $p \sqsubseteq_C q$.*

As usual, the key of the proof of completeness is to show that each term may be reduced to a suitable normal form using the axioms. Following [HM85], our normal forms will be the sumforms of Definition 6.5.1.

The set of axioms C

$$
\begin{array}{llcl}
(PAR1) & x|nil & = & x \\
(PAR2) & x|y & = & y|x \\
(PAR3) & (x|y)|z & = & x|(y|z) \\
(SEQ) & a.(x|y) & \leq & a.x|y
\end{array}
$$

The set of axioms $\mathcal{A}$

$$
\begin{array}{llcl}
\textbf{(A1)} & (x+y)+z & = & x+(y+z) \\
\textbf{(A2)} & x+y & = & y+x \\
\textbf{(A3)} & x+nil & = & x \\
\textbf{(A4)} & x+x & = & x \\
\textbf{(R1)} & a.(\sum_{i\in I} u_i : p_i) & = & (a.nil):(\sum_{i\in I} u_i : p_i) + \sum_{i\in I}(a.u_i) : p_i \\
\textbf{(R2)} & (\sum_{i\in I} u_i : p_i)|(\sum_{j\in J} v_j : q_j) & = & \sum_{i\in I} u_i : (p_i|\sum_{j\in J} v_j : q_j) + \sum_{j\in J} v_j : (\sum_{i\in I} u_i : p_i|q_j) \\
 & & & + \sum_{i\in I, j\in J}(u_i|v_j) : (p_i|q_j)
\end{array}
$$

Figure 6.2. A complete set of axioms for $\sqsubseteq_C$

Lemma 6.5.1 (Normalization) *For each $p \in \mathcal{P}'_{Fin}$, there exists a sumform $sf(p)$ such that $p =_{\mathcal{A}} sf(p)$.*

Proof: By induction on the size of p. The proof proceeds by a further induction on the structure of p.

- $p \equiv nil$. Then nil is already a sumform.

- $p \equiv u : q$. Then, by the inductive hypothesis, $q =_{\mathcal{A}} sf(q)$. Thus, by substitutivity, $p =_{\mathcal{A}} u : sf(q)$ which is a sumform.

- $p \equiv a.q$. By the inductive hypothesis, $q =_{\mathcal{A}} sf(q)$. If $sf(q) \equiv nil$ then

$$
\begin{array}{llll}
p & =_{\mathcal{A}} & a.nil & \text{by substitutivity} \\
 & =_{\mathcal{A}} & (a.nil) : nil + nil & \text{by (R1)} \\
 & =_{\mathcal{A}} & (a.nil) : nil & \text{by (A3).}
\end{array}
$$

  If $sf(q) \equiv \sum_{i\in I} u_i : q_i \not\equiv nil$ then

$$
\begin{array}{llll}
p & =_{\mathcal{A}} & a.(\sum_{i\in I} u_i : q_i) & \text{by substitutivity} \\
 & =_{\mathcal{A}} & (a.nil) : (\sum_{i\in I} u_i : q_i) + \sum_{i\in I}(a.u_i) : q_i & \text{by (R1).}
\end{array}
$$

- $p \equiv q + r$. By the inductive hypothesis, there exist sumforms $sf(q)$ and $sf(r)$ such that $q =_{\mathcal{A}} sf(q)$ and $r =_{\mathcal{A}} sf(r)$. If $sf(q) \equiv nil$ then $p =_{\mathcal{A}} sf(r)$ by (A2)-(A3).

Symmetrically $p =_{\mathcal{A}} sf(q)$ if $sf(r) \equiv nil$. If $sf(q), sf(r) \not\equiv nil$ then $p =_{\mathcal{A}} sf(q) + sf(r)$ which is a sumform.

- $p \equiv q|r$. By the inductive hypothesis, there exist sumforms $sf(q) \equiv \sum_{i \in I} u_i : q_i$ and $sf(r) \equiv \sum_{j \in J} v_j : r_j$ such that $q =_{\mathcal{A}} sf(q)$ and $r =_{\mathcal{A}} sf(r)$. Then

$$\begin{aligned} p &=_{\mathcal{A}} (\textstyle\sum_{i \in I} u_i : q_i)|(\sum_{j \in J} v_j : r_j) && \text{by substitutivity} \\ &=_{\mathcal{A}} \textstyle\sum_{i \in I} u_i : (q_i|sf(r)) + \sum_{j \in J} v_j : (sf(q)|r_j) \\ &\quad + \textstyle\sum_{i \in I, j \in J} (u_i|v_j) : (q_i|r_j) && \text{by (R2).} \end{aligned}$$

Again by the inductive hypothesis, for each $i \in I$, $j \in J$ and $(i,j) \in I \times J$ there exist sumforms π_i, π_j and π_{ij} such that

$$q_i|sf(r) =_{\mathcal{A}} \pi_i, \; sf(q)|r_j =_{\mathcal{A}} \pi_j \text{ and } q_i|r_j =_{\mathcal{A}} \pi_{ij}.$$

Thus $p =_{\mathcal{A}} \sum_{i \in I} u_i : \pi_i + \sum_{j \in J} v_j : \pi_j + \sum_{i \in I, j \in J} (u_i|v_j) : \pi_{ij}$, which is a sumform. □

We can now prove the promised completeness theorem.

Theorem 6.5.1 (Completeness) *For each $p, q \in \mathcal{P}'_{Fin}$, $p \sqsubseteq_C q$ iff $p \leq_{\mathcal{A}} q$.*

Proof: The *if* implication follows by Proposition 6.5.2. Assume then that $p, q \in \mathcal{P}'_{Fin}$ and $p \sqsubseteq_C q$. The proof is by induction on the combined size of the terms. By the normalization lemma we may assume, wlog, that p and q are sumforms. Thus assume that $p \equiv \sum_{i \in I} u_i : p_i$ and $q \equiv \sum_{j \in J} v_j : q_j$.

- First of all we show that $p + q \leq_{\mathcal{A}} q$. As $p \sqsubseteq_C q$, for each $i \in I$ there exists $h(i) \in J$ such that $u_i \leq_C v_{h(i)}$ and $p_i \sqsubseteq_C q_{h(i)}$. As p_i and $q_{h(i)}$ are sumforms and their combined size is less than that of p and q, we may apply the inductive hypothesis to obtain $p_i \leq_{\mathcal{A}} q_{h(i)}$. By rule (SUB), $u_i \leq_C v_{h(i)}$ and $p_i \leq_{\mathcal{A}} q_{h(i)}$ imply $u_i : p_i \leq_{\mathcal{A}} v_{h(i)} : q_{h(i)}$. As this is the case for each $i \in I$, by substitutivity

$$p \equiv \sum_{i \in I} u_i : p_i \leq_{\mathcal{A}} \sum_{i \in I} v_{h(i)} : q_{h(i)}.$$

Thus,

$$\begin{aligned} p + q &\leq_{\mathcal{A}} \textstyle\sum_{i \in I} v_{h(i)} : q_{h(i)} + q && \text{by substitutivity} \\ &=_{\mathcal{A}} q && \text{by (A1)-(A4).} \end{aligned}$$

- We show that $p \leq_{\mathcal{A}} p + q$. As $p \sqsubseteq_C q$, for each $j \in J$ there exists $k(j) \in I$ such that $u_{k(j)} \leq_C v_j$ and $p_{k(j)} \sqsubseteq_C q_j$. By the inductive hypothesis, $p_{k(j)} \leq_{\mathcal{A}} q_j$ and, by rule (SUB), $u_{k(j)} : p_{k(j)} \leq_{\mathcal{A}} v_j : q_j$. As this is the case for each $j \in J$, by substitutivity

$$\sum_{j \in J} u_{k(j)} : p_{k(j)} \leq_{\mathcal{A}} \sum_{j \in J} v_j : q_j \equiv q.$$

Thus, again by substitutivity, $p + \sum_{j\in J} u_{k(j)} : p_{k(j)} \leq_{\mathcal{A}} p + q$ and, by (A1)-(A4),

$$p =_{\mathcal{A}} p + \sum_{j\in J} u_{k(j)} : p_{k(j)} \leq_{\mathcal{A}} p + q.$$

Thus we have proved that $p \leq_{\mathcal{A}} p + q \leq_{\mathcal{A}} q$ and, by transitivity, $p \leq_{\mathcal{A}} q$. □

6.6 A Simple Example

This section is devoted to the discussion of a simple example of the application of the theory developed in the previous sections to the specification of concurrent systems. The example is based upon the definition of an n-ary semaphore, which admits any sequence of get's and put's in which the number of get's minus the number of put's lies in the range 0 to n inclusive. The following description of an n-ary semaphore is taken from [Mil89]:

$$\begin{aligned} Sem_n(0) &\Leftarrow \texttt{get}.Sem_n(1) \\ Sem_n(k) &\Leftarrow \texttt{get}.Sem_n(k+1) + \texttt{put}.Sem_n(k-1) \quad \text{if } 0 < k < n \\ Sem_n(n) &\Leftarrow \texttt{put}.Sem_n(n-1), \end{aligned}$$

where get and put correspond to Dijkstra's P and V operations on a semaphore, respectively. The simplest semaphore is the unary one, whose definition reduces to

$$Sem \Leftarrow \texttt{get.put}.Sem.$$

Let us assume now that we want to give the specification of an $(n+1)$-semaphore, $n \geq 1$, which is to be implemented on a multiprocessor having at least k processors, for some $k > 1$. Moreover, we want to specify the $(n+1)$-semaphore in such a way that an implementation be "forced" to exploit this feature of our system, i.e. we require that the degree of parallelism of any actual implementation be "at least k" in some formal sense. This is *not* possible in the semantic theory of standard bisimulation equivalence because of the reduction of parallelism to sequential nondeterminism it enforces, but, as shown in what follows, it can be done by using the preorder $\sqsubseteq_C$ as a formal "implementation ordering" over $\mathcal{P}$. As already remarked, the semantic statement $p \sqsubseteq_C q$ will be interpreted as meaning that q is an implementation of p and we require implementations to exhibit at least all the parallelism present in the specification.

One way to incorporate the above-given requirement into the specification of our semaphore is to stipulate, for instance, that

$$SPEC \Leftarrow Sem_{n-k+2}(0) | \underbrace{Sem | \cdots | Sem}_{(k-1)\text{-times}},$$

where $Sem_{n-k+2}(0)$ is the sequential nondeterministic description of an $(n-k+2)$-semaphore. It is easy to see that $SPEC$ is indeed functionally equivalent to an $(n+1)$-semaphore in the theory of bisimulation equivalence as

$$SPEC \sim Sem_{n+1}(0).$$

If during the implementation process it is realized, for instance by means of performance tests, that our multiprocessor system is capable of supporting the efficient, concurrent execution of more than k processes, the efficiency of the semaphore might be increased by maximizing the parallelism in its implementation. For instance, $Sem_{n-k+2}(0)$ might itself be implemented by

$$Sem^{n-k+2} \Leftarrow \underbrace{Sem|\cdots|Sem}_{(n-k+2)\text{-times}},$$

the system consisting of $(n-k+2)$ unary semaphores running in parallel. In order to show that

$$IMP \Leftarrow \underbrace{Sem|\cdots|Sem}_{(n+1)\text{-times}}$$

is a correct implementation of $SPEC$ with respect to $\sqsubseteq_C$, i.e. that $SPEC \sqsubseteq_C IMP$, it is sufficient to prove that

$$Sem_{n-k+2}(0) \sqsubseteq_C \underbrace{Sem|\cdots|Sem}_{(n-k+2)\text{-times}}, \tag{6.2}$$

as the claim will then follow by substitutivity. A proof of (6.2) can be given by exhibiting an $\mathcal{F}$-bisimulation $\mathcal{R}$ containing the pair $(Sem_{n-k+2}(0), Sem^{n-k+2})$.

Moreover, as $Sem_{n-k+2}(0)$ is in $\mathcal{P}_{Seq}$, by Theorem 6.4.3 it is indeed sufficient to check that the two systems are bisimilar. This can be done by using the standard techniques supported by the theory of bisimulation equivalence.

This proof technique supported by the preorder $\sqsubseteq_C$ is quite pleasing because refinements of specifications into implementations frequently proceed by substituting parallel processes for sequential nondeterministic ones in a context. Hence, in most practical cases, checking that $SPEC \sqsubseteq_C IMP$ reduces to showing bisimilarity between a sequential nondeterministic process SEQ and a process $IMPSEQ$ or to a combination of such proofs. Moreover, several automated tools for checking bisimilarity between labelled transition systems are now available, e.g. [CPS89].

Note, however, that $Sem_{n+1}(0)$ would *not* be a correct implementation of $SPEC$ with respect to $\sqsubseteq_C$. In fact, $SPEC \not\sqsubseteq_C Sem_{n+1}(0)$ as $SPEC \xrightarrow{\mathtt{get}^k}_p$, where

$$\mathtt{get}^k \equiv \underbrace{\mathtt{get}|\cdots|\mathtt{get}}_{k\text{-times}},$$

whilst $Sem_{n+1}(0) \xrightarrow{u}_p$ for no u such that $\mathtt{get}^k \leq_C u$. Intuitively, the specification permits k concurrent accesses to the semaphore, but $Sem_{n+1}(0)$ permits only one access at the time. This formalizes the intuitive fact that $Sem_{n+1}(0)$ does not capture all the parallelism which is present in the specification $SPEC$.

We hope that this very simple example shows at least that, by using $\sqsubseteq_C$ as a formal implementation ordering, it is possible to require that implementations preserve all the parallelism already present in the specification of a system. As already remarked, this is not possible in the theory of standard bisimulation equivalence. On the other hand, semantic statements of the form $SPEC \sqsubseteq_C IMP$ may be often shown by employing the elegant and compositional proof techniques supported by such an interleaving equivalence.

6.7 Concluding Remarks

In this chapter, we have presented a semantic theory for a simple **CCS**-like language which allows us to relate concurrency and nondeterminism in the behaviour of concurrent processes without semantically reducing the former to the latter. The theory is based on a simple behavioural preorder, $\sqsubseteq_C$, which relates processes p and q, $p \sqsubseteq_C q$, if p and q have the same observable behaviour and q is at least as parallel as p. The preorder has been defined by means of a slight modification of the notion of pomset bisimulation equivalence [BC87,88], obtained by parameterizing it with respect to a natural preorder over computations. This preorder on computations measures their relative degree of parallelism. Moreover, we have shown how this preorder can be axiomatized over the set of recursion-free processes by means of standard techniques used in the literature to give equational characterizations of bisimulation-like relations.

The semantic theory for processes developed in this chapter enjoys some pleasing relationships with that of strong bisimulation equivalence and supports many of the proof techniques which are familiar from the theory of such an interleaving equivalence. On the other hand, it has the advantage of allowing us to take into consideration the relative degree of parallelism between specifications and implementations, for instance by forcing implementations to exhibit at least all the parallelism which is present in the specification. This is *not* possible in the theory of standard strong bisimulation because of the semantic reduction of parallelism to sequential nondeterminism it enforces.

As it is often the case with bisimulation-like relations, the preorder presented in this chapter may be given a pleasing logical characterization by means of a variation of Hennessy-Milner logic, HML. Let HML($Comp$) denote an infinitary Hennessy-Milner logic whose modalities are indexed by computations $u \in Comp$ (infinite conjunctions and disjunctions are needed as the pomset transition system semantics for our language is neither *image-finite* [HM85] nor *sort-finite* [Ab87a,b]). The satisfaction relation $\models \subseteq \mathcal{P} \times \mathtt{HML}(Comp)$ is defined in a standard way following the above given references. The only interesting clauses of its definition are the ones regarding the modalities $<u>$ and $[u]$; these read as follows:

$$p \models <u>\phi \quad \text{iff } \exists v \in Comp, q \in \mathcal{P} \text{ such that } u \leq_C v, p \xrightarrow{v} q \text{ and } q \models \phi$$
$$p \models [u]\phi \quad \text{iff } \forall v \in Comp \;\; v \leq_C u \text{ and } p \xrightarrow{v} q \text{ imply } q \models \phi.$$

It is then possible to prove using standard techniques that, for processes p and q,

$$p \sqsubseteq_C q \text{ iff, for each } \phi \in \mathtt{HML}(Comp),\ p \models \phi \text{ implies } q \models \phi.$$

The work presented in this chapter is undoubtedly just a first attempt at giving a semantic theory for processes, which reconciles both an interleaving and a "truly concurrent" approach to the semantics of concurrency. More work remains to be done in extending the results presented in this chapter to languages incorporating features like communication, silent moves and restriction/hiding operators, as these concepts have been shown to be of great significance in the specification of realistic systems [Hoare85], [Mil89]. The approach followed in this chapter relied on a pomset operational semantics [BC87,88] and a bisimulation-like preorder. It is a challenging open problem to study whether the methods used in this chapter extend to more powerful languages or whether alternative forms of operational semantics, e.g. the *distributed operational semantics* of [CH89] or the *timed operational semantics* of [H88c], are more viable in general.

We end this section with a brief discussion of related work. The preorder presented in this chapter may be seen as an instance of the general scenario developed in [Thom87][1]. There the author studies a bisimulation preorder over Labelled Transition Systems induced by a given, uninterpreted preorder on actions. Our $\sqsubseteq_C$ may be seen as arising from this setting by interpreting all the parameters of Thomsen's proposal and, in particular, by using $\leq_C$ as the preorder on actions.

The preorder on computations $\leq_C$ used in this chapter is related to the notion of *augmentation* presented by several authors in the literature on pomsets and partial words [Gr81], [Gi84], [Pr86]. Essentially, for pomsets P and Q, P is an augmentation of Q if P is obtained from Q by adding more causal dependencies to those of Q. This notion is intuitively related to the model-theoretic counterpart of $\leq_C$, $\lhd$. However, no attempt has been made in the above-given references to build on top of this intuitive notion of measure of parallelism in computations a theory which, like the one presented in this chapter, allows us to compare nondeterminism

[1] I thank Kim Larsen and Arne Skou for pointing out this reference.

and parallelism in the behaviour of concurrent processes. Recently, S. Abramsky has independently defined a notion of "concurrent refinement" based on pomset transitions for a **CCS**-like language, [Ab90]. Although the pomset operational semantics presented in [Ab90] is different from the one we use, the preorder on processes defined by Abramsky is based on essentially the same ideas underlying the work presented in this chapter.

Chapter 7

Conclusions

We present here a brief summary of the main results presented in the thesis and a discussion of some of the problems which were left open.

In Chapter 2, we presented a semantic theory for a process algebra incorporating explicit constants for successful termination, deadlock and divergence. The semantic theory was based on the assumption that processes evolve by performing actions which are atomic. We gave both a behavioural and a denotational semantics for the language and proved that the denotational semantics is fully abstract with respect to the behavioural one. Moreover, in the proof of full-abstraction we established two results of independent interest; namely, the finite approximability of the behavioural preorder and a partial completeness result for the set of inequations E with respect to the behavioural preorder.

In Chapter 3, we developed a simple process algebra which incorporates an operator for the refinement of actions by processes and proposed a suitable semantic theory for it. We proved that the largest congruence over the simple language contained in strong bisimulation equivalence can be characterized in terms of Hennessy's timed equivalence [H88c], $\sim_t$. We presented both an algebraic and a behavioural proof of this characterization theorem. The algebraic proof relies on a complete axiomation of $\sim_t$ over the simple language; in the proof of the completeness theorem for $\sim_t$, following R. Milner and F. Moller, we also established unique factorization results for processes, modulo $\sim_t$, with respect to the operators of sequential and parallel composition. The behavioural proof relied on showing that $\sim_t$ is preserved by action refinement over the simple language and made an essential use of refine equivalence, $\sim_r$. We proved that $\sim_t$ and $\sim_r$ coincide over the simple language and that $\sim_r$ is preserved by action refinement.

In Chapter 4, we studied action refinement for a richer process language including hand-shake communication, an internal τ action and a restriction operator à la **CCS**. We studied suitable notions of action refinements for the language and how they are to be applied to processes. In particular, the application of action refinements to processes was defined by considering restriction as a binding operator. We showed that the natural version of $\sim_r$ which abstracts from the internal evolution of processes, $\approx_r$, is preserved by action refinements over the extended language. Moreover, we discussed an example, due to Frits Vaandrager, which shows that $\approx_r$ and the weak version of $\sim_t$ are different equivalences over the richer language.

In Chapter 5, we applied the theory of action refinement developed in Chapter 3 to provide a behavioural characterization of the class of series-parallel pomsets. More precisely, we proved that series-parallel pomsets are fully abstract with respect to strong bisimulation equivalence over a simple language with action refinement. Moreover, we showed that series-parallel pomsets can be given a trace-theoretic characterization in terms of ST-traces; namely, we proved that two series-parallel pomsets are equal iff they have the same set of ST-traces.

In Chapter 6, we defined and axiomatized a behavioural preorder for a subset of **CCS** based on actions which have a pomset structure. This preorder allows us to relate concurrency to nondeterminism in the behaviour of processes without reducing the former to the latter. We also investigated the relationships of the preorder with strong bisimulation equivalence and pomset bisimulation.

* * *

In the remainder of this conclusion we shall briefly discuss some of the problems that were left open in the thesis.

The fully abstract model for a weak version of prebisimulation presented in Chapter 2 gave us a c.p.o. model for a weak version of bisimulation. However, the model we proposed is a so-called *term model* and, although it gives us a complete proof system for reasoning about the behavioral preorder, does not shed much light on the nature of the preorder. For this purpose, it would be helpful to find a "natural" model which is fully abstract with respect to the preorder. Such a model might be obtained by generalizing the ones proposed by S. Abramsky in [Ab87b,90].

In Chapter 4, we studied action refinement in the rather rich setting of a finite process algebra, essentially obtained by extending finite **CCS** with a general sequential composition operator. We proved that a suitable notion of semantic equivalence for this language can be obtained by means of the largest congruence over the language contained in weak refine equivalence,

$\approx_r^c$. However, we were unable to prove (or disprove) that $\approx_r^c$ is the largest congruence over the richer language contained in weak bisimulation equivalence, $\approx$. This would give us a characterization result for $\approx_r^c$ similar to the one we showed for $\sim_t$ in Chapter 3. We have been able to prove that, for all $p, q \in \mathbf{P}^\Gamma$ which only use action symbols in Λ,

Fact 7.0.1 $p \approx_r^c q$ iff for all $\mathbf{P}^\Gamma$-contexts $C[\cdot]$, $C[p] \approx C[q]$.

The extension of the above result to the whole of $\mathbf{P}^\Gamma$ is made difficult by the extra complexity introduced in the behaviour of refined processes of the form $p\rho$ by synchronization. In particular, single step transitions of the form $p \xrightarrow{\tau} q$ in the behaviour of p correspond, in general, to to multi-step transitions of the form $p\rho \overset{\tau}{\Longrightarrow} \sim q\rho$ in the behaviour of $p\rho$.

Another interesting open problem related to $\approx_r^c$ is the development of proof systems for this congruence relation. We believe that, following the proof of the equational characterization for $\sim_t$ given in Chapter 3 and adapting equations from [H88c], it should be possible to provide a complete equational characterization for $\approx_r^c$ over the restriction-free sublanguage of $\mathbf{P}^\Gamma$. Indeed, we have been able to obtain a finite, complete axiomatization for $\approx_r^c$ over the set of finite, restriction-free CCS processes. This axiomatization, like several others presented in the literature on non-interleaving equivalences, see e.g. [CH89], makes use of Bergstra and Klop's auxiliary operators, the *left-merge*, denoted by $\|$, and the *communication-merge*, denoted by $|_c$. Formally, let $\Sigma_{\mathbf{CCS}} = \{\Sigma_{\mathbf{CCS}}^n\}_{n \in \omega}$ denote the signature given by

$$\Sigma_{\mathbf{CCS}}^0 = \{nil\}$$
$$\Sigma_{\mathbf{CCS}}^1 = \{\mu.\_ \mid \mu \in Act(\Lambda)\}$$
$$\Sigma_{\mathbf{CCS}}^2 = \{+, |, \|, |_c\}$$
$$\Sigma_{\mathbf{CCS}}^n = \emptyset \text{ for all } n > 2.$$

We let $\mathbf{P}_{\mathbf{CCS}}$ denote the word algebra over $\Sigma_{\mathbf{CCS}}$. The operational semantics for all the operators in $\Sigma_{\mathbf{CCS}} - \{\|, |_c\}$ is given as in [Mil89]. The rules defining the operational semantics for the auxiliary operators are the following:

(LM) $\quad p \overset{\mu}{\Longrightarrow} p' \qquad\qquad$ implies $\quad p\|q \overset{\mu}{\Longrightarrow} p'|q$

(CM) $\quad p \overset{a}{\Longrightarrow} p'$ and $q \overset{\bar{a}}{\Longrightarrow} q'$ imply $\quad p|_c q \overset{\tau}{\Longrightarrow} p'|q'$,

where $\overset{\mu}{\Longrightarrow}$ denotes the "weak" transition relation defined in, e.g. Chapter 2. The congruence relation $\approx_r^c$ may now be defined over $\mathbf{P}_{\mathbf{CCS}}$ along the lines followed in Chapter 4. We then have the following theorem, which gives a complete axiomatization of $\approx_r^c$ over $\mathbf{P}_{\mathbf{CCS}}$:

Theorem 7.0.1 $\approx_r^c$ is the least $\Sigma_{\mathbf{CCS}}$-congruence which satisfies the equations in Figure 7.1.

$$
\begin{array}{rlcl}
\textbf{(A1)} & (x+y)+z & = & x+(y+z) \\
\textbf{(A2)} & x+y & = & y+x \\
\textbf{(A3)} & x+x & = & x \\
\textbf{(A4)} & x+nil & = & nil \\
\\
\textbf{(LM1)} & (x+y)\lfloor z & = & x\lfloor z + y\lfloor z \\
\textbf{(LM2)} & (x\lfloor y)\lfloor z & = & x\lfloor(y|z) \\
\textbf{(LM3)} & x\lfloor nil & = & x \\
\textbf{(LM4)} & nil\lfloor x & = & nil \\
\\
\textbf{(T1)} & x+\tau.x & = & \tau.x \\
\textbf{(T2)} & \mu.\tau.x & = & \mu.x \\
\textbf{(T3)} & \tau.x\lfloor y & = & \tau.(x|y) \\
\textbf{(T4)} & x\lfloor\tau.y & = & x\lfloor y \\
\textbf{(T5)} & x\lfloor(y+\tau.z) & = & x\lfloor(y+\tau.z)+x\lfloor z \\
\\
\textbf{(CM1)} & (x+y)|_c z & = & x|_c z + y|_c z \\
\textbf{(CM2)} & x|_c y & = & y|_c x \\
\textbf{(CM3)} & x|_c nil & = & nil \\
\textbf{(CM4)} & (a.x\lfloor x')|_c(b.y\lfloor y') & = & \begin{cases} \tau.(x|x'|y|y') & \text{if } a=\bar{b} \\ nil & \text{otherwise} \end{cases} \\
\textbf{(CM5)} & \tau.x|_c y & = & x|_c y \\
\\
\textbf{(PAR)} & x|y & = & x\lfloor y + y\lfloor x + x|_c y
\end{array}
$$

Figure 7.1. Equations for $\approx_r^c$ over $\mathbf{P_{CCS}}$

It is moreover possible to prove that the equations in Figure 7.1 also completely axiomatize $\approx_t$ over $\mathbf{P}_{\mathbf{CCS}}$. Hence, contrary to what happens over $\mathbf{P}^{\Gamma}$, $\approx_t$ and $\approx_r^c$ coincide over $\mathbf{P}_{\mathbf{CCS}}$. Unfortunately, an adequate algebraic treatment of restriction in the setting of non-interleaving equivalences has so far proved elusive.

In Chapter 5, we proved that series-parallel pomsets are characterized by their set of ST-traces. We conjecture that the same result holds by considering split-traces in place of ST-traces. This result would have some interesting consequences. For instance, it would imply that, contrary to what happens for standard bisimulation and trace equivalence, $\sim_t$ and split-trace equivalence coincide over the language $\mathcal{SP}$. Moreover, pomset equality would be the largest $\mathcal{SP}$-congruence contained in trace equivalence which is preserved by action refinement.

Finally, the preorder for relating concurrency to nondeterminism presented in Chapter 6 should be extended to full **CCS** and applied to the verification of some complex concurrent systems.

Chapter 8

Bibliography

[Ab87a] S. Abramsky, Observation Equivalence as a Testing Equivalence, TCS 53, pp. 225-241, 1987

[Ab87b] S. Abramsky, A Domain Equation for Bisimulation, Imperial College Technical Report, 1987

[Ab90] S. Abramsky, Causal Semantics in Process Algebra, Unpublished Manuscript, 1990

[AB84] D. Austry and G. Boudol, Algebre de Processus et synchronisations, TCS 30(1), pp. 91-131, 1984

[Ace89] L. Aceto, On Relating Concurrency and Nondeterminism, Report 6/89, Dept. of Computer Science, University of Sussex, October 1989. To appear in the Proceedings of the 7th Conference on Mathematical Foundations of Programming Semantics, 1991.

[Ace90] L. Aceto, Full Abstraction for Series-Parallel Pomsets, in the Proceedings TAPSOFT '91, Volume 1: Colloquium on Trees in Algebra and Programming (CAAP '91), LNCS 493, pp. 1-25, Springer-Verlag 1991. A full version of the paper appeared as Report 2/90, Dept. of Computer Science, University of Sussex, March 1990.

[AH88] L. Aceto and M. Hennessy, Termination, Deadlock and Divergence, Report 6/88, Dept. of Computer Science, University of Sussex, December 1988. To appear in the Journal of the ACM. An extended abstract appeared in Proceedings of the 5th Worshop on Mathematical Foundations of Programming Semantics 1989, LNCS 442, pp. 301-318, Springer-Verlag, 1990.

[AH89] L. Aceto and M. Hennessy, Towards Action-refinement in Process Algebras, Proc. 4th LICS, pp. 138-145, IEEE Computer Society Press, 1989. A full version of the paper appeared as Computer Science Report 3/88, University of Sussex, April 1988. To appear in Information and Computation.

[AH90] L. Aceto and M. Hennessy, Adding Action Refinement to a Finite Process Algebra, Computer Science Report 6/90, University of Sussex, November 1990. An extended abstract appeared in the Proceedings of ICALP '91, LNCS 510, Springer-Verlag, 1991.

[BBK87] J. C. M. Baeten, J. A. Bergstra and J. W. Klop, On the Consistency of Koomen's Fair Abstraction Rule, TCS 51, pp. 129-176, 1987

[BC87] G. Boudol and I. Castellani, On the Semantics of Concurrency: Partial Orders and Transition Systems, in Proc. TAPSOFT 87, LNCS 249, pp. 122-137, Springer Verlag, 1987

[BC88] G. Boudol and I. Castellani, Concurrency and Atomicity, TCS 59, pp. 25-84, 1988

[BG87a] J. C. M. Baeten and R. J. van Glabbeek, Abstraction and Empty Process in Process Algebra, Report CS-R8721, CWI Amsterdam, 1987 (to appear in Fundamenta Informaticae)

[BG87b] J. C. M. Baeten and R. J. van Glabbeek, Merge and Termination in Process Algebra, Proceedings 7th Conference on Foundations of Software Technology and T.C.S. (K.V. Nori ed.), LNCS 287, pp. 153-172, Springer Verlag, 1987

[BHR84] S. D. Brookes, C. A. R. Hoare and A. W. Roscoe, A Theory of Communicating Sequential Processes, J. ACM 31,3, pp. 560-599, 1984

[BK82] J. A. Bergstra and J. W. Klop, Fixed Point Semantics in Process Algebra, Report IW 206/82, Centre for Mathematics and Computer Science, Amsterdam, 1982

[BK84] J. A. Bergstra and J. W. Klop, Process Algebra for Synchronous Communication, Information and Control, 60, pp. 109-137, 1984

[BK85] J. A. Bergstra and J. W. Klop, Algebra of Communicating Processes with Abstraction, TCS 37,1, pp. 77-121, 1985

[BK88] J. A. Bergstra and J. W. Klop, Process Theory Based on Bisimulation Semantics, Proceedings REX School '88, LNCS 354, pp. 50-122, Springer-Verlag, 1988

[BKO87] J. A. Bergstra, J. W. Klop and E.-R. Olderog, Failures Without Chaos: a New Process Semantics for Fair Abstraction, Formal Description of Programming Concepts-III (M. Wirsing ed.), North-Holland, 1987

[Bou89] G. Boudol, Atomic Actions (note), Bulletin of the EATCS 38, pp. 136-144, 1989

[BV89] J. C. M. Baeten and F. Vaandrager, An Algebra for Process Creation , Report CS-R8907, CWI, Amsterdam, 1989

[Ca88] I. Castellani, *Bisimulations for Concurrency*, Ph. D. Thesis CST-51-88, University of Edinburgh, April 1988

[CDP87] L. Castellano, G. de Michelis and L. Pomello, Concurrency vs Interleaving: an Instructive Example, Bulletin of the EATCS 31, pp. 12-15, 1987

[CH89] I. Castellani and M. Hennessy, Distributed Bisimulations, JACM. October 1989

[CN76] B. Courcelle and M. Nivat, Algebraic Families of Interpretations, Proceedings 17th IEEE Symposium on Foundations of Computer Science, 1976

[CPS89] R. Cleaveland, J. Parrow and B. Steffen, The Concurrency Workbench: A Semantics-Based Verification Tool for Finite-State Systems, Report ECS-LFCS-89-83, University of Edinburgh, 1989

[DD89a] P. Darondeau and P. Degano, Causal Trees, Proc. ICALP 89, LNCS 372, pp. 234-248, Springer-Verlag, 1989

[DD89b] P. Darondeau and P. Degano, About Semantic Action Refinement, to appear in Fundamenta Informaticae

[DD90] P. Darondeau and P. Degano, Event Structures, Causal Trees and Refinements, Unpublished Manuscript, 1990

[DDM86] P. Degano, R. de Nicola and U. Montanari, Observational Equivalences for Concurrency Models, in Proc. IFIP TC2 Workshop on Formal Description of Programming Concepts IV (M. Wirsing ed.), pp. 105-137, North-Holland, 1987

[DDM88] P. Degano, R. De Nicola and U. Montanari, Partial Ordering Semantics for CCS, Technical Report TR-3/88, Università di Pisa, 1988

[DeN85] R. De Nicola, Two Complete Sets of Axioms for a Theory of Communicating Sequential Processes, Information and Control 64(1-3), pp. 136-176, 1985

[DH84] R. de Nicola and M. Hennessy, Testing Equivalences for Processes, TCS 34,1, pp. 83-134, 1984

[Eng90] U. Engberg, *Partial Orders and Fully Abstract Models for Concurrency*, Ph. D. Thesis, DAIMI PB-307, University of Aarhus, March 1990

[Gi84] J. L. Gischer, *Partial Orders and the Axiomatic Theory of Shuffle*, Ph. D. Thesis, Stanford University, 1984

[GG88] R. van Glabbeek and U. Goltz, Equivalence Notions for Concurrent Systems and Refinement of Actions, Proc. 14th MFCS, LNCS 379, pp. 237-248, Springer-Verlag, 1988

[Gl90] R. van Glabbeek, The Refinement Theorem for ST-bisimulation, to appear in Proc. IFIP Working Group, Sea of Galilee, 1990

[Gl90a] R. van Glabbeek, *Comparative Concurrency Semantics and Refinement of Actions*, Ph. D. Thesis, Free University of Amsterdam, 1990

[GMM88] R. Gorrieri, S. Marchetti and U. Montanari, A^2CCS: a Simple Extension of CCS for Handling Atomic Actions, Proceedings CAAP '88, LNCS 299, pp. 258-270, Springer-Verlag, 1988

[Go88] U. Goltz, On Representing CCS Programs by Finite Petri Nets, Arbeitspapiere der GMD 290, February 1988

[Gr81] J. Grabowski, On Partial Languages, Fundamenta Informaticae IV.2, pp. 427-498, 1981

[Gro90] J. F. Groote, A New Strategy for Proving ω-Completeness applied to Process Algebra, in Proceedings CONCUR '90, LNCS 458, pp. 314-331, Springer-Verlag, 1990

[GrV89] J. F. Groote and F. Vaandrager, Structured Operational Semantics and Bisimulation as a Congruence, in *Proceedings ICALP 89* (G. Ausiello , M. Dezani-Ciancaglini and S. Ronchi della Rocca eds.), LNCS 372, pp. 423-438, Springer-Verlag, 1989

[GR83] G. Golsen and W. Rounds, Connections between Two Theories of Concurrency: Metric Spaces and Synchronization Trees, Information and Control 57, pp. 102-124, 1983

[GTWW77] J. A. Goguen, J. W. Thatcher, E. G. Wagner and J. B. Wright, Initial Algebra Semantics and Continuous Algebras, JACM 24,1, pp. 68-95, 1977

[Gue81] I. Guessarian, *Algebraic Semantics*, Lecture Notes in Computer Science vol. 99, Springer-Verlag, Berlin, 1981

[GV87] R. van Glabbeek and F. Vaandrager, Petri Net Models for Algebraic Theories of Concurrency, in Proc. PARLE Conference 1987 (J. de Bakker et al. eds.), LNCS 259, Springer Verlag, 1987

[GW89a] R. van Glabbeek and P. Weijland, Branching Time and Abstraction in Bisimulation Semantics, Information Processing '89, Elsevier Science Publishers B.V., pp. 613-618, 1989

[GW89b] R. van Glabbeek and P. Weijland, Refinement in Branching Time Semantics, Proceedings AMAST Conference, May '89, Iowa, U.S.A., pp. 197-201

[H81] M. Hennessy, A Term Model for Synchronous Processes, Information and Control 51,1, pp. 58-75, 1981

[H85] M. Hennessy, Acceptance Trees, JACM 32,4, pp. 896-928, 1985

[H88a] M. Hennessy, *Algebraic Theory of Processes*, MIT Press, 1988

[H88b] M. Hennessy, Observing Processes, Proc. REX Workshop 88, LNCS 3 54, pp. 173-200, Springer-Verlag, 1988

[H88c] M. Hennessy, Axiomatising Finite Concurrent Processes, SIAM Journal on Computing, October 1988

[HM85] M. Hennessy and R. Milner, Algebraic Laws for Nondeterminism and Concurrency, J. ACM 32,1, pp. 137-161, 1985

[Hoare85] C. A. R. Hoare, *Communicating Sequential Processes*, Prentice-Hall, 1985

[HP79] M. Hennessy and G. Plotkin, Full Abstraction for a Simple Parallel Programming Language, Proc. MFCS, Lecture Notes in Computer Science vol. 74, Springer-Verlag, 1979

[HP80] M. Hennessy and G. Plotkin, A Term Model for CCS, Proceedings 9th MFCS, Lecture Notes in Computer Science vol. 88, Springer-Verlag, 1980

[K88] A. Kiehn, Petri Net System and their Closure Properties, TU Munchen, Institut fur Informatik, 1988 (to appear in Advances in Petri Nets 1989)

[Kel76] R. Keller, Formal Verification of Parallel Programs, C. ACM 19,7, pp. 561-572, 1976

[LS87] J. Loeckx and K. Sieber, *The Foundations of Program Verification (2nd edition)*, Wiley-Teubner Series in Computer Science, 1987

[Main88] M. Main, Trace, Failure and Testing Equivalences for Communicating Systems, International Journal of Parallel Programming 16, pp. 383-401, 19 88

[Mil77] R. Milner, Fully Abstract Models of Typed Lambda-Calculi, TCS 4, pp. 1-22, 1977

[Mil80] R. Milner, *A Calculus of Communicating Systems*, LNCS 92, Springer Verlag, 1980

[Mil81] R. Milner, A Modal Characterization of Observable Machine-Behaviour, Proc. 6th CAAP, Lecture Notes in Computer Science vol. 112, pp. 23-34, Springer-Verlag, 1981

[**Mil83**] R. Milner, Calculi for Synchrony and Asynchrony, TCS 25, pp. 267-310, 1983

[**Mil88**] R. Milner, Operational and Algebraic Semantics of Concurrent Processes, Report ECS-LFCS-88-46, University of Edinburgh, February 1988 (to appear as a chapter of the *Handbook of Theoretical Computer Science*)

[**Mil89**] R. Milner, *Communication and Concurrency*, Prentice-Hall, 1989

[**MM90**] R. Milner and F. Moller, Unique Decomposition of Processes, Bulletin of the EATCS 41, pp. 226-232, 1990

[**Mol89**] F. Moller, *Axioms for Concurrency*, Ph. D. Thesis, University of Edinburgh, 1989

[**NEL88**] M. Nielsen, U. Engberg and K. S. Larsen, Fully Abstract Models for a Process Language with Refinement, Proc. REX Workshop 88, LNCS 354, pp. 52 3-548, Springer-Verlag, 1988

[**OC84**] INMOS Limited, *OCCAM Programming Manual*, Prentice-Hall, London, 1984

[**Pa81**] D. Park, Concurrency and Automata on Infinite Sequences, LNCS 104, Springer Verlag, 1981

[**Pl77**] G. Plotkin, LCF Considered as a Programming Language, TCS 5, pp . 223-255, 1977

[**Pl81**] G. Plotkin, A Structural Approach to Operational Semantics, Report DAIMI FN-19, Computer Science Dept. , Aarhus University, 1981

[**Pn85**] A. Pnueli, Linear and Branching Structures in the Semantics and Logics of Reactive Systems, LNCS 194, Springer Verlag, pp. 14-32, 1985

[**Pr86**] V. Pratt, Modelling Concurrency with Partial Orders, International Journal of Parallel Programming 15, pp. 37-71, 1986

[**Rei85**] W. Reisig, *Petri Nets*, EATCS Monographs on Theoretical Computer Science, Springer Verlag, 1985

[**Rou85**] W. Rounds, On the Relationships between Scott Domains, Synchronization Trees and Metric Spaces, Information and Control, 1985

[**Sco82**] D. Scott, Domains for Denotational Semantics, Proceedings ICALP 82 (M. Nielsen and E. M. Schmidt eds.), LNCS vol. 140, pp. 577-613, 1982

[**St87**] C. Stirling, Modal Logics for Communicating Systems, TCS 49, pp. 311-347, 1987

[**Sto88**] A. Stoughton, *Fully Abstract Models of Programming Languages*, Research Notes in TCS, Pitman-Wiley, 1988

[Sto88a] A. Stoughton, Substitution Revisited, TCS 59, pp. 317-325, 1988

[Tau89] D. Taubner, *The Finite Representation of Abstract Programs by Automata and Petri Nets*, LNCS 369, Springer-Verlag, 1989

[Thom87] B. Thomsen, An Extended Bisimulation Induced by a Preorder on Actions, M.Sc. Thesis in Computer Science, Aalborg University Centre, 1987

[Ts88] S. Tschantz, Languages Under Concatenation and Shuffling, Vanderbilt University Report, 1988

[Va88] F. Vaandrager, Determinism $\longrightarrow$ (Event Structure Isomorphism = Step Sequence Equivalence), Report CS-R8839, CWI Amsterdam, October 1988

[Va90] F. Vaandrager, Personal Communication, 1990

[Va90a] F. Vaandrager, *Algebraic Techniques for Concurrency and their Application*, Ph. D. Thesis, University of Amsterdam, 1990

[Vra86] J. L. M. Vrancken, The Algebra of Communicating Processes with Empty Process, Report FVI 86-01, Dept. of Computer Science, University of Amsterdam, 1986

[Vo90] W. Vogler, Failures Semantics Based on Interval Semiwords is a Congruence for Refinement, Proceedings STACS '90, LNCS, Springer-Verlag, 1990

[Wal87] D. Walker, Bisimulation Equivalence and Divergence in CCS, LFCS Report Series, ECS-LFCS-87-29, June 1987 (extended abstract in Proc. LICS 1988)

[Win80] G. Winskel, *Events in Computation*, Ph. D. Thesis, University of Edinburgh, 1980

[Win82] G. Winskel, Event Structure Semantics for CCS and Related Languages, Proc. ICALP 82, LNCS, Springer-Verlag, 1982

[Win87] G. Winskel, Event Structures, in Advances in Petri Nets 1986, LNCS 255, pp. 325-392, Springer Verlag, 1987

Index